T0180206

Nanoparticles and their Biomedical Applications

Ashutosh Kumar Shukla
Editor

Nanoparticles and their Biomedical Applications

Editor
Ashutosh Kumar Shukla
Physics Department
Ewing Christian College
Prayagraj, Uttar Pradesh, India

ISBN 978-981-15-0393-1   ISBN 978-981-15-0391-7   (eBook)
https://doi.org/10.1007/978-981-15-0391-7

This Springer imprint is published by the registered company Springer Nature Singapore Pte Ltd.
The registered company address is: 152 Beach Road, #21-01/04 Gateway East, Singapore 189721, Singapore

To my parents

Preface

This collection intends to highlight the potential applications of nanoparticles in the field of medicine. Rare-earth based nanoparticles, metal-oxide nanoparticles, metal nanoparticles, graphene oxide and lignin based nanoparticles have been covered in different chapters. Ecofriendly greener synthesis approaches have been given due consideration in view of increasing environmental concerns. Biomedical and bioengineering applications of nanoparticles with impact on health issues and related therapeutic measures have been described.

I am thankful to the expert contributors to manage their time out of their busy schedule. It is only their sincere effort which enabled me to present this volume before the audience. My sincere thanks are due to reviewers from different countries who have contributed a lot to improve the content quality through their constructive comments.

I sincerely thank Dr. Naren Aggarwal and Dr. Gaurav Singh from Springer Nature for giving me the opportunity to present this book. I also thank Ms. Vaishnavi Venkatesh for her support during the production process.

I could learn many things while going through individual chapters and hope that it will be a good experience for the audience too.

Prayagraj, India
October 2019

Ashutosh Kumar Shukla

Contents

About the Editor

Ashutosh Kumar Shukla obtained his B.Sc., M.Sc., and D.Phil. degrees from the University of Allahabad. He has been a University Educator and Researcher for more than 17 years and is currently an Associate Professor of Physics at Ewing Christian College, Prayagraj, a constituent institution of the University of Allahabad. Dr. Shukla has successfully completed numerous research projects and published several edited volumes in collaboration with prominent experts.

Rare Earth-Based Nanoparticles: Biomedical Applications, Pharmacological and Toxicological Significance

Susheel Kumar Nethi, Vishnu Sravan Bollu,
Neeraja Aparna Anand P., and Chitta Ranjan Patra

Abstract

It is well established that nanomaterials play an important role in addressing various unresolved problems in biomedical research. Among these, the rare earth or lanthanide-based metal nanoparticles are increasingly explored for their potential biomedical applications. Several researchers across the globe including us reported the biological applications (magnetic resonance imaging, anticancer, antimicrobial, fluorescence, antioxidant properties) of rare earth-based nanomaterials. Furthermore, in spite of their excellent photoluminescence, magnetic resonance properties, photostability and therapeutic efficacy, there is still an issue of growing concern for the clinical applications of these rare earth nanomaterials. The desirable properties of these elements might also be equally associated with unexpected and biohazardous toxicities. Therefore, many groups have reported the comprehensive toxicity assessment (*in vitro*—cell lines and *in vivo*—rodent models) of rare earth nanomaterials. This chapter focuses on the detailed discussion of the biomedical applications of these rare earth-based nanomaterials along with their toxicological evaluation and pharmacokinetics significances. The future opportunities and challenges are also incorporated to highlight the importance for their clinical translation.

S. K. Nethi · V. S. Bollu · C. R. Patra (✉)
Department of Applied Biology, CSIR-Indian Institute of Chemical Technology,
Hyderabad, Telangana State, India

Academy of Scientific & Innovative Research (AcSIR), Ghaziabad, Uttar Pradesh, India
e-mail: crpatra@iict.res.in

Neeraja Aparna Anand P.
Department of Applied Biology, CSIR-Indian Institute of Chemical Technology,
Hyderabad, Telangana State, India

© Springer Nature Singapore Pte Ltd. 2020
A. K. Shukla (ed.), *Nanoparticles and their Biomedical Applications*,
https://doi.org/10.1007/978-981-15-0391-7_1

Keywords

Biomedical applications · Clearance · Pharmacokinetics · Pharmacology · Rare earth-based nanoparticles · Toxicology

Abbreviations

786-O	Human renal adenocarcinoma
ALT	Alanine Aminotransferase
AST	Aspartate Aminotransferase
ATP	Adenosine triphosphate
BDNF	Brain-derived neurotropic factor
BEAS-2B	Transformed human bronchial epithelial cells
bFGF	basic fibroblast growth factor
CCl_4	Carbon tetrachloride
CeO_2	Cerium oxide
CF	ciprofloxacin
CHO	Chinese hamster ovary
CI	Cerebral ischaemia
CNP	Cerium nanoparticle
CPNPs	Coordination polymer nanoparticles
CT	Computed tomography
DNA	Deoxy ribonucleic acid
DOX	Doxorubicin
DU145	Human prostate cancer cell line
Dy_2O_3	Dysprosium oxide
EC50	Median effective concentration
EDS	Energy dispersive spectroscopy
EHNs	Europium hydroxide nanorods
Eu	Europium
Eu_2O_3	Europium oxide
$Eu^{III}(OH)_3$	Europium hydroxide
Gd_2O_3	Gadolinium oxide
GOx	Glucose oxidase enzyme
H_2O_2	Hydrogen peroxidase
HCAEC	Primary human coronary artery endothelial cells
HUVEC	Human umbilical vein endothelial cells
I.P.	Intraperitoneal
I.V	Intravenous
ICPMS	Inductively coupled plasma mass spectroscopy
ICPOES	Inductively coupled plasma optical emission spectroscopy
IHN	Inhalation
IR	Infrared

IT	Intratracheal
$KBrO_3$	Potassium bromate
La_2CO_3	Lanthanum carbonate
La_2O_3	Lanthanide oxide
$LaPO_4$	Lanthanum phosphate
LD50	Median lethal dose
LDH	Lactate Dehydrogenase
LIBS	Laser-induced breakdown elimination
LNMs	Lanthanide nanomaterials
LRET	Luminescence resonance energy transfer
MCF-7	Human breast adenocarcinoma cell line
MDA-MB 231	Human breast adenocarcinoma cell line
MRI	Magnetic resonance imaging
MSCs	Mesenchymal stem cells
NCTC1469	Murine liver cell line
NO	Nitric oxide
NPs	Nanoparticles
NSCLC	Non-small-cell lung carcinoma
P.O.	Peroral
PCD	Polycystic kidney disease
PDGF	Platelet-derived growth factor
PET	Positron emission tomography
PMN	Polymorphonuclear neutrophils
PPARβ	Peroxisome-proliferator-activated receptor β
REBNPs	Rare earth-based nanoparticles
ROS	Reactive oxygen species
RT-PCR	Reverse transcription polymerase chain reaction
TEM	Transmission electron microscopy
TrkB	Tropomyosin receptor kinase B
VEGF	Vascular endothelial growth factor

1.1 Background: Lanthanide Elements and Properties

The rare-earth group of elements usually referrers to the lanthanide series in the periodic table. The lanthanide/rare-earth elements are mostly stable in the trivalent ion form with a stable oxidation state of +3 (Ln^{+3}). The characteristic properties of the lanthanides such as magnetic moments and magnetic susceptibilities are derived from the unpaired electrons in the 4f orbitals (Gschneidner et al. 2002; Dong et al. 2015). In the past decade, several research groups are actively working in the area of design and development of rare earth-based nanoparticles (REBNPs) with characteristic optical and biological properties for biomedical applications (Shen et al. 2008). The exceptional properties of fluorescence, high photostability, pharmacological potency, low toxicity and biocompatibility make rare earth-based nanoparticles (REBNPs) as excellent candidates for applications in the fields of biology and medicine (Bollu et al. 2015; Dong et al. 2015; Li et al. 2016b; Nethi et al. 2015; Patra et al. 2008, 2011). In spite of the numerous reports on the biomedical applications of rare-earth nanoparticles, there are few comprehensive reports highlighting the biomedical applications and toxicological significance of the lanthanide nanoparticles (Bouzigues et al. 2011; Dong et al. 2015). In this chapter the synthesis, various biomedical applications and toxicological aspects of rare-earth nanoparticles along with future opportunities and challenges are discussed in detail.

1.2   Synthesis Aspects

Rare earth-based nanoparticles (REBNPs) demonstrate vital advantages over other nanomaterials owing to their photostability, inherent fluorescence, excellent optical transparency, low toxicity, sharp emission spectra, high thermal stability (Wang et al. 2011). It is well-known that size, shape and morphology parameters affect the characteristic properties of nanomaterials and play vital role in modulating the biological responses (Albanese et al. 2012). Henceforth, design and synthesis of lanthanide-based nanomaterials is another important area of research. Various approaches such as hydrothermal synthesis (Wei et al. 2014), microwave irradiation method (Nethi et al. 2015), thermal decomposition (Boyer et al. 2006), cation exchange technique (Dong and van Veggel 2009) and co-precipitation (Yi et al. 2004) are some of the well-known reported procedures for the synthesis of REBNPs. Recently, the green chemistry has evolved as an excellent approach for the preparation of stable, pharmacologically potent and eco-friendly nanoparticles and as an alternate to the conventional synthesis methods. Various research groups have evaluated the green synthesis of REBNPs using natural sources (Bae et al. 2010; Iram et al. 2016; Mendoza-Mendoza et al. 2012). However, the synthesis of REBNPs using the above methods is beyond the scope of this chapter and is not discussed in detail in this chapter.

1.2.1   Functionalization of REBNPs

The surface modification the nanoparticles is very essential to improve the interaction of the nanoparticles with the cells and clearance from the body (Qie et al. 2016). The surface decoration of the nanoparticles helps in (1) loading of drug molecules/targeting agents/fluorescent moieties (depending on the nature of surface charge and functional groups), (2) delivering therapeutic drug molecules/gene, (3) enhancing the drug loading and efficiency, (4) modifying the physiological and biological properties, (5) enhancing in vivo circulation time, etc. (Siafaka et al. 2016). The surface functionalization also maintains homogeneous dispersion of the REBN in biological fluids, water and physiological buffers. Secondly, these nanoparticles are decorated with organic moieties which aid in targeting specific cellular receptors

(Zhang et al. 2015). Several research groups also demonstrated the therapeutic applications of bare and non-functionalized REBNPs (Das et al. 2012; Patra et al. 2008, 2011).

1.2.2 Rare Earth-Doped Nanoparticles

The rare earth metal doped nanomaterials in recent years are gaining rapid attention due to their unique applications in electronics, sensing, therapeutics and diagnostics. When the rare earth elements are doped into a host structure, they exhibit excellent light emitting properties with sharp absorption and emission characteristic peaks without any photobleaching (Perera et al. 2015; Zhao et al. 2016a). These characteristic properties highlight the rare earth-doped nanoparticles as excellent candidates over the conventional fluorescent dyes, aiding their applications in biomedical imaging and therapy (Deng et al. 2011; Xiong et al. 2010). The light absorbing and emitting rare earth ions are well protected from the external environment in the doped nanostructure. Hence, there will be little or no fluorescence quenching of these rare earth ions *in vivo*.

1.3 Biomedical Applications

The application of novel nanoparticles for biological and therapeutic applications is an emerging area of research. Among such various nanoparticles, the lanthanide nanoparticles are widely being investigated by several scientists globally, for their wide biomedical applications. The unique physico-chemical properties of REBNPs aid in significantly improving the conventional approaches for disease diagnosis, therapy and development novel approaches for human heath and welfare. The various applications of the lanthanide nanomaterials are discussed below.

1.3.1 Bioimaging

The magnetic resonance imaging (MRI) became a very powerful imaging tool for the bioimaging as it provides high spatial and temporal resolution of anatomical and physiological systems of organisms (Geschwind and Konopka 2009). MRI does not involve the use of high-energy electromagnetic waves or radioactive agents like other imaging techniques (PET: positron emission tomography, CT: computed tomography etc.). It is a non-invasive technique, with high penetration capability. Recently the application of lanthanide-based systems for MRI imaging and diagnostics has gained enormous importance. Several researchers across the world have investigated in detail the application of gadolinium as MRI contrast agents. For the first time, Donald and his co-workers have evaluated the physico-chemical and magnetic resonance properties of gadolinium oxide nanoparticles and established the area of gadolinium nanoparticle-based contrast imaging (McDonald and Watkin 2003). The presence of seven unpaired electrons in gadolinium, yielding high paramagnetic stability along with slow electron relaxation strongly supports its application for MRI (Datta and Raymond 2009). Several gadolinium-based small chelates are commercially available for MRI imaging as positive contrast agents such as DOTAREM, MAGNEVIST, etc. (Geraldes and Laurent 2009). Mignot et al. developed novel nanoparticles named as AGuIX, which consists of a polysiloxane core covered by 1,4,7,10-tetraazacyclododecane-1,4,7,10-tetraacetic [DOTA(Gd)] derivatives, which are bound covalently to inorganic matrix (Mignot et al. 2013). Further they have intravenously injected these gadolinium nanoparticles in rodents and monitored the excretion mode by MRI, which revealed that elimination of these nanoparticles from the body depends exclusively on renal route. The signal of gadolinium was detected at 5 min post-intravenous injection, reached maximum at 4 h and gradually reduced up to 1 week demonstrating effective renal clearance with time. The laser-induced breakdown elimination (LIBS) and ICP analysis confirmed the above observation of the elimination behaviour of the nanoparticles by the kidneys (Fig. 1.1) (Sancey et al. 2014).

The main problem associated with the treatment of lung cancer is the late diagnosis of its pathology. Bianchi et al. developed ultrasmall gadolinium-based contrast agents to detect the presence of non-small-cell lung cancer (NSCLC), by non-invasive approach (Bianchi et al. 2014). The MRI acquisitions were performed

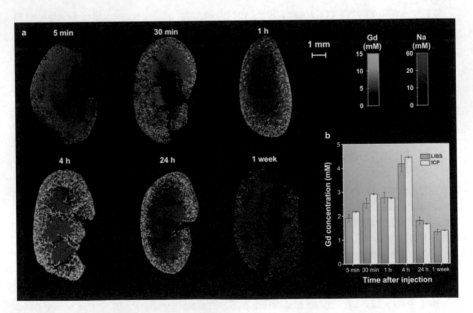

Fig. 1.1 Distribution of gadolinium-based nanoparticles in the kidney as a function of the time elapsed since administration. (**a**), Quantitative imaging of Gd and Na in kidney coronal sections. The images were recorded at a 40-mm resolution and represent 30,000 pixels. (**b**), Agreement between the Gd concentrations measured via LIBS (green) and ICP (yellow). Figure reproduced with permission from Sancey et al. (2014). Copyright © 2014, Nature Publishing Group

before and after administration of nanoparticles to determine the tumour. Colocalization of the tumour tissue with bioluminescence, fluorescence imaging and MRI aided in the detection and monitoring the development of NSCLCs. Taken together the non-invasiveness and absence of ionizing radiation widely encourage the clinical translation of this technique. Li et al. demonstrated a theranostic approach that combines an anticancer drug gemcitabine- and gadolinium-based MRI contrast agent following a supramolecular self-assembly synthesis (Li et al. 2016b). The authors demonstrated that this theranostic formulation exhibited enhanced retention and a strong T1 MRI contrast signal in a mouse model along with potent growth inhibition of *in vivo* MDA-MB-231 tumours. Additionally, researchers have explored the MRI properties of other lanthanide nanomaterials. For example, Kattel et al. have demonstrated the one-pot design and synthesis of ultrasmall lanthanide oxide Ln_2O_3 nanoparticles (consisting of Eu, Gd, Dy, Ho and Er) coated with D-glucuronic acid (Kattel et al. 2012). Further to determine their MRI ability the authors assessed their water proton relaxivities. Among these series of nanoparticles, the gadolinium oxide (Gd_2O_3) nanoparticles only showed highest contrasting ability compared to other Ln_2O_3 nanoparticles. The ultrasmall Dy_2O_3 nanoparticles coated with D-glucuronic acid showed the maximum transverse (r_2) water proton relaxivities and applied as T2 MRI contrast agent for 3 T T2 MRI of mouse. Taken together, this study establishes these ultrasmall lanthanide nanoparticles as potential T2 contrast agents for MRI.

1.3.2   Biosensing

Biosensing is usually the detection of the biologically active molecules or parameters which have critical importance for biomedical and environmental applications. In the recent decades, various researchers thoroughly investigated the scope of REBNPs for biosensing applications. For instance, Stipic et al. demonstrated a sensitive and versatile technique called luminescence resonance energy transfer (LRET) (Stipic et al. 2015). In this technique, functionalized lanthanide-based nanoparticles are used for the detection of anti-okadaic acid rabbit polyclonal IgG, arising due to environmental toxins (okadaic acid) exposure. These studies might aid in the development of novel and robust biosensors for identifying biotoxin-triggered immune responses. Tan and his co-workers recently prepared the terbium (Tb^{3+})-based coordination polymer nanoparticles (CPNPs) and demonstrated the detection of the drug ciprofloxacin (CF) in urine samples at lower concentrations (Tan et al. 2013). This strategy is cheap, simple and involves no sample-pretreatment and could widely expand the environmental and biological application of lanthanide-based polymer nanoparticles. Zeng et al. illustrated the simple and rapid synthesis of novel cerium coordination polymer nanoparticles (ATP-Ce-Tris CPNs), which helps in the selective detection of hydrogen peroxidase (H_2O_2) at very low concentrations (Zeng et al. 2016). Further, the authors extended the application of this nanoparticle-based system to detect glucose which forms H_2O_2 by glucose oxidation. Peng and his colleagues further demonstrated the design and development of luminescent hybrid nanoparticles of size 20–30 nm, using visible-light-sensitized $Eu^{(III)}$ chelates (Peng et al. 2010). The authors demonstrated that these nanoparticles aid in detecting physiological temperatures by imaging because they exhibit temperature dependence and fluorescence intensity over a physiological range of temperature.

REBNPs have been used for the detection of nucleic acids such as DNA using various approaches. Using the wet chemical method, Wang et al. successfully prepared the luminescent LaF_3-Ce^3/Tb^{3+} nanocrystals which exhibited size-independent emissions along with stable photocycles. These rare earth-doped nanomaterials were applied as fluorescence probes for rapid quantitative analysis of DNA to overcome the fluorescence quenching exhibited by DNA. The results exhibited proportionality between the fluorescence intensity of the nanospheres and the DNA concentration from various sources at microgram level (Wang et al. 2009). Using a microarray platform, Son et al. demonstrated application of rare earth nanoparticles for DNA quantification as an alternate to conventional organic fluorescence probes. The REBNPs system was used for detection of bacteria which degrades methyl tertiary-butyl ether (MTBE), a groundwater contaminant. The fluorescence spot intensities showed a direct linear relationship with the bacterial 16S rDNA over a varied target DNA. Additionally, the fluorescence intensity was much stronger compared to commercially available fluorescent dyes. Henceforth, these Eu:Gd_2O_3 nanoparticles could be developed as simple, cheap, non-toxic and rapid alternative approaches to the DNA microarrays (Son et al. 2008). Van De Rijke et al. further reported the advantage of REBN in DNA detection over conventional fluorophores. They showed that very low concentration of target DNA was detected by labelling the biotinylated target DNA with employing the fluorescence intensities, which is 5 times more sensitive compared to the commercial dyes (van De Rijke et al. 2001). The detection of DNA sequence variations such as single nucleotide polymorphism is essential to diagnose several diseases such as polycystic kidney disease (PCD). Techniques like reverse transcription polymerase chain reaction (RT-PCR) are generally employed for this purpose which are highly expensive and time-taking. Son et al. demonstrated the rare earth-based nanoparticles consisting of iron oxide/Eu: Gd_2O_3 core-shell nanoparticles which are used to successfully identify the SNPs (Son et al. 2007). The authors demonstrated the successful detection of SNPs related to PCD by direct hybridization of genomic DNA isolated from blood and tissue samples of patients, with a very sensitive detection range. Henceforth, the above studies firmly emphasize the lanthanide-based nanoparticles for sensing applications in biology and medicine.

1.3.3 Therapeutic Applications

The characteristic properties of rare earth lanthanide nanomaterials firmly encourage their role in several biomedicinal applications (Abdukayum et al. 2013; Chen et al. 2015a, b; Estevez and Erlichman 2014). Their principal therapeutic properties include antioxidant (Kwon et al. 2016), anticancer (Gao et al. 2014; Luchette et al.

2014), angiogenic (Patra et al. 2011; Zhao et al. 2016b), radioprotectant (Colon et al. 2009), etc. along with diagnostic and imaging applications (Chen et al. 2015a, b; Hagan and Zuchner 2011). These features explain their role in curing certain cardiovascular, neurodegenerative, diabetic, vascular, inflammatory, immunological disorders etc. and the following studies highlight their significant theranostic applications.

1.3.3.1 Drug Delivery

Recently, REBNPs are reported to hold promising significance for acting as drug delivery systems as compared to conventional therapeutics. This is because therapeutic moieties at bulk scale possess certain limitations like minimal selectivity, low specificity and poor solubility. Hence, lanthanide nanoparticles have successfully proved to overcome the major challenges of therapeutic bulk materials by showing therapeutic efficacy at lower dosages with minimal toxicities and higher precision. The target specificity onto the ligands has actually brought the huge impact on the choice of the researchers in considering them for drug delivery (Dong et al. 2015). The following researchers explained their importance in serving as agents for dreadful diseases and disorders. Rajendiran et al. synthesized lanthanum fluoride doped terbium nanoparticles, functionalized with chitosan (precipitation method) and used as a carrier for methotrexate. The binding of methotrexate on to the nanoparticles is facilitated by surface active functional groups of the carrier. The Van der Waals attraction between lanthanide carrier and drug supports the quick dislodging of drug at the target when compared to other nanoparticles. This facilitated targeted drug delivery to cancer cells displaying a greater extent of cytotoxicity (MCF-7) than that of free drug (Mangaiyarkarasi et al. 2015). Yang et al. synthesized $NaREF_4$ nanoparticles (oil-water two phase methods) with rare earth (RE) elements including neodymium, lutetium and ytterbium. The group confirmed the biocompatibility of the complex with certain *in vitro* assays and mentioned its application for the delivery of anticancer drug doxorubicin (DOX) to HeLa cells (Yang et al. 2013a). *Singh* et al. synthesized europium-based nanoparticles (YVO_4:Eu3+) encapsulated inside the matrix of mesoporous silica nanoparticles (sol gel method). These particles served as carriers for DOX which confirmed efficient cytotoxicity (90%) towards HeLa and MCF-7 cell lines (Shanta Singh et al. 2013). Li et al. synthesized gadolinium-based nanoparticles loaded with gemcitabine monophosphate, an anticancer drug (with 55% loading). They observed significant *in vivo* tumour inhibition (MDA MB-231 injected immunodeficient nu/nu mice model) with higher retention time and better diagnosis with improved MRI contrasting (Li et al. 2016b). Wu et al. synthesized β-$NaYF_4$:$Ce^{3+}Tb^{3+}$ complex and encapsulated inside mesoporous silica nanocomposites. They proposed an application as a drug carrier for controlled release of DOX and confirmed significant cytotoxicity (towards A549 cells) through *in vitro* assays. In addition, the nanocomposite serves as a bio-probe to study the patterns of drug action (Fig. 1.2) (Wu et al. 2013). Kang et al. synthesized $GdVO_4$:Dy^{3+} nanospheres for their use as therapeutically applicable drug (DOX) carrier. The group proved that the hollow nature of the spheres facilitate efficient drug loading capacity and porous nature conferring controlled release (Kang et al. 2013).

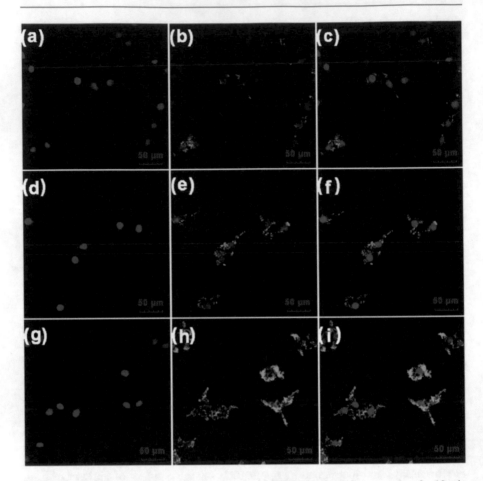

Fig. 1.2 CLSM images of A549 cells incubated with FITC-NPs@mSiO$_2$ composites for 10 min (**a–c**), 1 h (**d–f**), and 6 h (**g–i**) at 37 °C. Each series can be classified to the nuclei of cells (being dyed in blue by Hoechst 33342 for visualization), FITC-NPs@mSiO$_2$, and a merge of the two above channels, respectively. Figure reproduced with permission from Wu et al. (2013). Reproduced by permission of The Royal Society of Chemistry

1.3.3.2 Fluorescence (Image-Guided Therapy)

In theranostics, the luminescent property of lanthanide nanoparticles exhibited as emission signals during their upconversion helps in identifying the exact site of the therapeutic need, called the 'Image guided therapy'. The luminescent or fluorescent properties of lanthanide nanoparticles are brought about by 4f-4f transitions thereby contributing diverse applications in therapeutic techniques (Li et al. 2008). Lanthanide nanoparticles act as multimodal platforms in serving as systems for imaging and also for curing (Dong et al. 2015).

Hagan et al. reported that the usage of lanthanum nanoparticles in the form of lanthanum chelate labels in immunoassays increases the sensitivity and overcomes the problem of auto-luminescence. The persistent luminescence of lanthanum labels with highly minimal interferences becomes the chief factor in considering lanthanide nanoparticles in therapeutic techniques (Hagan and Zuchner 2011). Victor et al. synthesized a complex of hydroxyl apatite conjugated alginic acid as a carrier of the drug, 4-acetyl salicylic acid in the treatment of colon cancer. The complex when doped with neodymium attained the near IR fluorescence ability, a diagnostic application for early detection and targeted treatment of tumours (Victor et al. 2016). Chen et al. reported that gadolinium oxide nanoparticles, when doped with terbium ions (1% doping concentration), act as MRI contrast agent in addition to fluorescence imaging with no significant cytotoxicity (Chen et al. 2015a). Abdukayum et al. stated that praseodymium ions when doped with zinc gallogermanate (citrate sol-gel technique) shown luminescence (near IR) persistently for over 360 h. With improved biocompatibility upon PEGylation, these can be used for *in vivo* imaging of targeted site (Abdukayum et al. 2013). Rocha et al. reported that neodymium ions, when doped with lanthanum fluoride nanoparticles, possess promising application as probes for bioimaging of the targeted site because of high penetration ability (up to 1 cm), high image contrast and less auto-fluorescence (Rocha et al. 2014). Foucault-Collet et al. synthesized a samarium-based dendrimer, bio-compatible (upto a size of 2.5 μm) for imaging live cells, since it exhibits luminescence in visible and near IR regions (Foucault-Collet et al. 2014). Kattel et al. reported dysprosium (Dy) oxide nanoparticles coated with D-glucuronic acid, as effective MRI agents as compared to that of nanorod morphology since the former exhibits high sensitivity and renal excretion, with a good *in vitro* biocompatibility (Kattel et al. 2012). Singh et al. synthesized a nanohybrid with SnO_2 nanoparticles doped with Tb^{3+}ions, incorporated in polyvinyl alcohol matrix. The group suggested that nanohybrid at higher concentrations exhibits biocompatibility towards HeLa cells in addition of possessing luminescent properties, thereby aiding in optical imaging of targeted area, both *in vitro* and *in vivo* (Singh et al. 2015). Our group illustrated the design and fabrication of inorganic lanthanide phosphate $LnPO_4$. H_2O (Eu, Tb) nanorods using microwave technology (Patra et al. 2007). By employing transmission electron microscopy (TEM) and confocal microscopy techniques, we demonstrated the internalization of these nanorods in the cytoplasm of 786-O (human renal adenocarcinoma) and HUVEC (human umbilical vein endothelial cells) cells, without any significant cytotoxicity. Taken together, we believe that these $LnPO_4$ nanorods could be developed as potential candidates for visualizing the live cell components and facilitating detection of cancerous cells. Bridot and his co-workers prepared luminescent hybrid nanoparticles containing gadolinium oxide core inside polysiloxane shell for dual functionality purpose of magnetic resonance and fluorescence imaging (Bridot et al. 2007). The gadolinium oxide core acts as a positive contrast agent for MRI, whereas the polysiloxane shell functionalization inside with organic dyes and outer side by PEG: poly-ethylene glycol aids in fluorescence imaging. The set of images captured at various time points, post-intravenous (i.v.) injection of these nanoparticles in mice and rats demonstrated their accumulation majorly in the kidneys and urinary bladder with no undesirable accumulation in liver and lungs (Fig. 1.3). The above observations were found to be in accordance with the MRI obtained by tracking the body circulation of these nanoparticles.

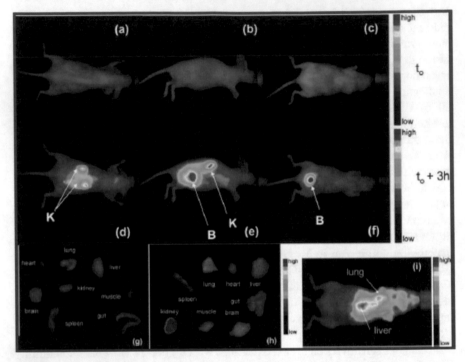

Fig. 1.3 Fluorescence reflectance imaging of a nude mouse (**a**, **b**, **c**) before and (**d**, **e**, **f**) 3 h after the injection of GadoSiPEG2C (K, kidneys; B, bladder). Fluorescence reflectance imaging of some organs after dissection (**g**) of a control mouse (no particles injection) and (**h**) of the nude mouse visualized on pictures (**a–f**). (**i**) Fluorescence reflectance imaging of a nude mouse after the injection of GadoSi2C (particles without PEG). Each image is acquired with an exposure time of 200 ms. Figure reproduced with permission from Bridot etal. (2007). Copyright © 2007 American Chemical Society

1.3.3.3 Antioxidant Property

Reactive oxygen species (ROS) are reported to be the chief causative agents in triggering several dreadful disorders, by altering the cellular metabolism (Estevez and Erlichman 2014). It has therefore become the necessity to develop new antioxidant moieties for meeting the therapeutic demand. Lanthanide nanoparticles possess the capability of scavenging the ROS since they exhibit switching over between dual valency states (Niu et al. 2011) and thereby presenting significant outcomes over the therapeutic challenges in competition with conventional therapeutics (Kwon et al. 2016). Antioxidant behaviour mainly depends on surface valency states of nanoparticles which ultimately determine its effect on the biological activity (Pulido-Reyes et al. 2015). Wang et al. reported that a hybrid of lanthanum oxide nanorods with silver nanoparticles deposited on them possess significant antioxidant property. Additionally, the group has found that the hybrid presents characteristic antibacterial property against Gram-positive and Gram-negative bacteria (Wang et al. 2014). The antioxidant property of cerium oxide nanoparticles has brought tremendous therapeutic significance for the treatment of various neurological (Das et al. 2007; Estevez and Erlichman 2014; Kwon et al. 2016), cardiovascular (Niu et al. 2011; Niu et al. 2007; Patra et al. 2008), ocular (Kong et al. 2011) and hepatic disorders (Oro et al. 2016). Estevez et al. incubated nanoceria in hippocampal brain slice models and observed a reduction in peroxynitrite levels, and proposed their application for the treatment of cerebral ischaemia (Estevez et al. 2011). *Das* et al. proved the neuroprotective mechanism of nanoceria by inducing *in vitro* oxidative damage in rat spinal cord neurons followed by incubation with nanoceria, resulting in protection of neuronal cells from injury (Das et al. 2007). Similarly, the ocular disorders like retinal degeneration were observed to be mitigated by nanoceria treatment as demonstrated by Kong et al. The group injected C57BL/6 J mice (retinal degeneration model) with nanoceria and found interesting results as these lanthanide nanoparticles protected retinal cells from ROS-induced damage by acting as an efficient antioxidant (Kong et al. 2011). An added interesting antioxidant application of nanoceria was demonstrated by Denise et al. The group injected nanoceria to rats treated with carbon tetrachloride, CCl_4 (portal hypertension model), and the observed response at histopathological analysis was reduction in liver steatosis indicating the hepato-protective application of lanthanide nanoparticles (Oro et al. 2016). Kwon et al. demonstrated the design and fabrication of positive-charged triphenylphosphonium (TPP)-conjugated nanoceria with specific ability to localize in the mitochondria of neuronal cells. The authors demonstrated that these biocompatible nanoparticles exhibited potent antioxidant property by scavenging the Aβ-induced intracellular mitochondrial ROS both *in vitro* and *in vivo* (transgenic mouse model of Alzheimer's disease). Further, the TPP-nanoceria were able to mitigate the mitochondrial damage and the reactive gliosis in mouse model, projecting these nanoparticles as potential candidates for neurodegenerative disease therapy (Fig. 1.4) (Kwon et al. 2016). The experimental results act as evidence to state that lanthanide nanoparticles can be used as potential antioxidant moieties for treating various disorders induced by oxidative stress.

1.3.3.4 Angiogenesis/Anti-Angiogenesis Activity

Angiogenesis is the process of growth and development of new blood vessels from pre-existing vasculature, which is tightly regulated in several pathophysiological processes. Under physiological conditions, it is triggered by growth factors such as VEGF (vascular endothelial growth factor), bFGF (basic fibroblast growth factor), PDGF (platelet-derived growth factor), etc. Under certain diseased conditions, when endogenous growth factor production or activity is altered, exogenous delivery of growth factors/molecules is needed to restore the angiogenesis process. This approach of exogenous administration of growth factors initially proved to be beneficial but later reported to cause several adverse effects of thrombosis, fibrosis, tumorigenesis, oedema, etc. The application of nanoparticles to restore the functional angiogenesis as a substitute for growth factors, i.e., the nanomedicine approach has been studied by many researchers, globally. Recently, several research groups including ours have widely investigated the design, development and therapeutic application of several pro-angiogenic rare earth lanthanide nanoparticles.

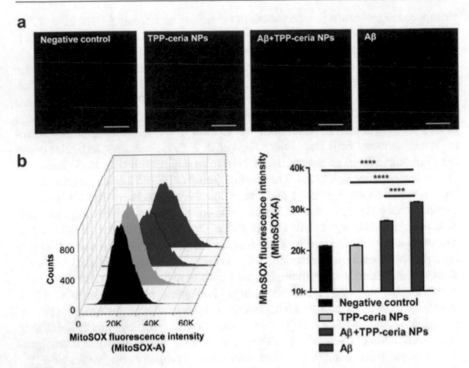

Fig. 1.4 TPP-ceria NPs significantly inhibited Aβ-induced mitochondrial ROS *in vitro*. (**a**) Confocal fluorescence images of mitochondrial-ROS accumulation in SH-SY5Y cells obtained by MitoSOXR. SH-SY5Y cells were stained using 5 μM MitoSOXR without any treatment (negative control), and after exposed to 0.1 mM TPP-ceria NPs (TPP-ceria NPs), to 5 μM Aβ and 0.1 mM TPP-ceria NPs (Aβ + TPP-ceria NPs), and to 5 μM Aβ (Aβ) for 12 h. Scale bar = 50 μm. (**b**) MitoSOXR fluorescence intensity measured in SH-SY5Y cells by flow cytometry. Statistical analysis was performed using an ANOVA test with ∗∗∗∗ marking $p < 0.0001$. Error bars represent 95% confidence intervals. Figure reproduced with permission from Kwon et al. (2016). Copyright © 2016 American Chemical Society

Our group has thoroughly established the pro-angiogenic properties of europium hydroxide nanorods (EHNs) using several *in vitro* (endothelial cells) and *in vivo* assay (chick embryo and zebra fish) systems (Augustine et al. 2017; Bollu et al. 2015; Kim et al. 2011; Nethi et al. 2015, 2017; Patra et al. 2008, 2011). EHNs were demonstrated to promote the endothelial cell proliferation, cell migration and tube formation which are the hallmarks of angiogenesis process. EHNs were reported to induce the *in vivo* blood vessel formation in chick embryo and transgenic zebra fish models (Fig. 1.5) (Patra et al. 2008, 2011). Further in-depth mechanistic studies revealed the formation of intracellular ROS: H_2O_2 and reactive nitrogen species (RNS): nitric oxide (NO) were found to be the key signalling molecules underlying the EHNs-induced pro-angiogenesis (Fig. 1.6) (Nethi et al. 2015; Patra et al. 2011). The *in vivo* toxicity studies also validated that EHNs are mild to non-toxic in nature towards rodent models (Bollu et al. 2015; Patra et al. 2009), which will be discussed in detail in the toxicology part of this chapter. Very recently, Augustine et al.

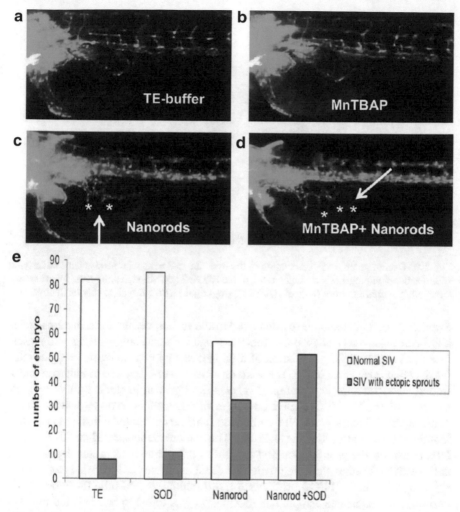

Fig. 1.5 (**a–d**): *In vivo* angiogenesis study in a transgenic FLI-1: EGFP zebrafish model. Nanorods in combination with MnTBAP induce ectopic sprouting from the SIV: lateral view of embryos at 72 hpf. The vehicle control was Tris-EDTA (TE), to which was added 4.5 ng of MnTBAP and/or 50 ng of nanorods. (**e**) The number of embryos showing normal SIVs, and ectopic sprouting from SIVs is summarized. Figure reproduced with permission from Patra et al. (2011). Copyright © 2011 American Chemical Society

demonstrated the successful embedding of EHNs into poly-caprolactone scaffolds and demonstrated their pro-angiogenic properties as potential candidates for tissue engineering and revascularization (Augustine et al. 2017). Similarly, Zhao et al. have demonstrated the shape-dependent pro-angiogenic properties of lanthanide nanoparticles such as europium hydroxide/terbium hydroxide nanoparticles in transgenic zebra fish model (Zhao et al. 2016b). Additionally, the authors have demonstrated that these nanoparticles trigger angiogenesis through redox signalling.

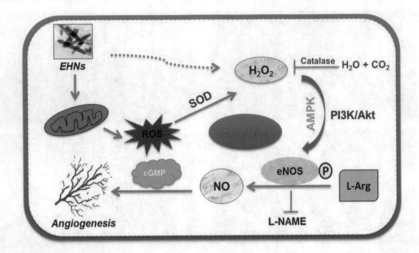

Fig. 1.6 Overall graphical representation of the hypothesized molecular mechanisms underlying EHNs-induced angiogenesis mediated through the ROS-NO-cGMP signalling axis. Figure reproduced with permission from Nethi et al. (2015). Copyright © 2015 Royal Society of Chemistry

Similarly, another lanthanide oxide nanoparticle such as cerium oxide nanoparticles has been extensively studied for their angiogenesis/anti-angiogenesis properties. Das et al. reported the stimulation of angiogenesis by cerium oxide nanoparticles (CeO$_2$ NPs) (Das et al. 2012). These nanoparticles were reported to enhance endothelial tube formation and promoted blood vessel growth in chick embryo by activating HIF-1α, VEGF through modulation of intracellular oxygen levels. Xiang et al. reported that the CeO$_2$ NPs embedded scaffolds exhibited enhanced vascularization of bone grafts (Xiang et al. 2016). The authors demonstrated that these CeO$_2$ NPs enhanced the proliferation and inhibition of apoptosis of mesenchymal stem cells (MSCs). Further, the upregulation of VEGF by these CeO$_2$ NPs led to endothelial progenitor cells (EPCs) proliferation, differentiation and tube formation. This cerium oxide enhanced ectopic bone formation was found to be mediated through calcium channel activation of MSCs. Lord et al. designed cerium oxide nanoparticles functionalized with heparin and reported their anti-angiogenic properties. This heparin functionalization enhanced the cytoplasmic and lysosomal localization, promoted scavenging of cellular ROS and subsided the proliferation of HCAEC cells (primary human coronary artery endothelial cells). The extent of functionalization of heparin on CeO$_2$ NPs was the key factor determining their antioxidant potential, which in turns affects the cell proliferation (Lord et al. 2013). Further, Giri et al. demonstrated the wet chemical synthesis of anti-angiogenic CeO$_2$ NPs and their thorough physicochemical characterization (Giri et al. 2013). The authors demonstrated that CeO$_2$ NPs exhibited significant inhibition of ROS and reduced the growth factor mediated cell migration/invasion of SKOV3 cells. Further, these CeO$_2$ NPs also inhibited the VEGF-induced cell proliferation, tube formation and VEGFR2/MMP2 activation in HUVECs. Altogether these CeO$_2$ NPs exhibited potent anti-angiogenic properties following the above mechanistic events and could be used as potent candidates for the treatment of ovarian cancer.

1.3.3.5 Anticancer Activity

The disorder resulting from unrestricted proliferation and survival of cells due to impaired machinery of cell cycle is termed cancer (Hanahan and Weinberg 2011). The treatment of cancer ever necessitates the scope to new therapeutic strategies. This paved the way for exploring the potential of lanthanide nanoparticles as able anticancer agents. The following reports present the various therapeutic approaches driven by the subjected lanthanide nanoparticles. Lai et al. synthesized lanthanum hexaboride nanoparticles with a coating of carbon doped silica (LaB_6@C-SiO_2) and treated on Hela cell lines, shown significant cytotoxicity when exposed to near infra-red radiation. They stated that LaB_6@C-SiO_2 provides an alternative source for the usage of gold nanoparticles since they act as an economical photothermal treatment in cancer therapy (Lai and Chen 2013). In another study, as mentioned in the above section, Lord et al. functionalized the nanoceria with heparin and demonstrated their ROS scavenging ability along with improved cytoplasmic and lysosomal localization. These nanoparticles were able to profoundly inhibit the endothelial cell proliferation which shows their anti-angiogenic property and therefore can find application in the treatment of cancer (Lord et al. 2013). Bakht et al. experimented on non-small cell type lung cancer insensitive to chemotherapy and observed dual therapeutic significances of nano Pr_2O_3, nano Nd_2O_3 particles in cancer treatment since nano Pr_2O_3 holds radiotherapeutic property and this when decays to nano Nd_2O_3 attains autophagy-inducing property to cancer cells (Bakht et al. 2013). Miladi et al. shown the localized accumulation of injected (i.v) gadolinium-based nanoparticles inside tumour (gliosarcoma) until 24 h i.v. injection and reported its usage as radiosensitizer of tumour with its ability to cross blood-brain barrier (Miladi et al. 2013). Pasqua et al. demonstrated the application of holmium nanoparticles for the treatment of ovarian cancer. They synthesized radiotherapeutic ^{166}Ho nanoparticles (neutron flux irradiation from ^{165}Ho) with PLGA microspheres and folate bound on their surface (DSPE-PEG5000-Folate) to have targeted delivery (I.P) to ovarian tumours. The bio-distribution of these nanoparticles was confirmed by SPECT imaging (Di Pasqua et al. 2012).

1.3.3.6 Neurodegenerative Disease Therapy

The neurodegenerative disorders account for the progressive deterioration or impairment in neuronal metabolism. The condition holds one of the crucial health ailments necessitating the development of new treatment strategies. Researchers upon continuous exploration identified the importance of lanthanide nanoparticles as capable therapeutic agents for treating neurological ailments like Alzheimer and Parkinson's diseases.

Among the lanthanide nanoparticles, nanoceria has been extensively explored for its notable neurological applications. For example, Estevez et al. proposed the application of CeO_2 NPs as potential therapeutic agents for cerebral ischaemia (CI).

The group experimented using hippocampal brain part of CI mouse model, induced ROS generation (peroxynitrite radicals), and followed by treatment with CeO_2 NPs. The results explained that nanoceria has significantly decreased the cell death by about 50%, thereby presenting its neuroprotective mechanism (Estevez et al. 2011). D'Angelo et al. reported that CeO_2 NPs treatment improved the cell viability, protected the SH-SY5Y (human neuroblastoma) cells from neuronal atrophy and promoted the neuronal survival pathways by expression of PPARβ (peroxisome-proliferator-activated receptor β), BDNF (brain-derived neurotrophic factor) and TrkB (Tropomyosin receptor kinase B) in an *in vitro* model of Alzheimer's disease (D'Angelo et al. 2009). It is known that Parkinson's disease is one among the neurodegenerative disorders caused due to impaired dopamine release metabolism. Ciofani et al. experimented on PC-12 (rat pheochromocytoma) neuronal cells and proved that nanoceria incubation increased dopamine secretion, indicating a potential therapy for Parkinsonism (Ciofani et al. 2013). Similarly Das et al. incubated the nanoceria with isolated cultured spinal cord cells from adult rats and quantified the neuronal as well as glial cells by immune-staining method. An increased survival of neurons at the phases of oxidative stress with the nanoceria treatment was observed, highlighting the importance of REBNPs as alternative therapeutic approach to spinal cord damage using nanomedicine approach (Das et al. 2007). As explained in the above section, Kwon and his co-workers have demonstrated synthesis of positive charged TPP-conjugated cerium oxide nanoparticles along with evaluation of their effect on *in vitro* and *in vivo* models of Alzheimer's disease. These nanoparticles recovered the neuronal cells from Aβ-induced mitochondrial damage, improved the cell viability and scavenging of ROS in an *in vivo* mouse model of Alzheimer's disease. Therefore, these TPP-conjugated nanoceria could be developed as alternative strategies for the treatment of neurodegenerative disorders (Kwon et al. 2016). Upon continuous exploration of the biomedical significances, lanthanide nanoparticles have been proved with promising results in treating various neurological disorders and gained a status of competing with commercial therapeutics, thereby increasing the scope of their further research. Several neurological disorders such as Parkinson's disease and Huntington's disease arise due to the aggregation of misfolded proteins inside the cellular environments (Metcalf et al. 2012).

Autophagy is a physiological process which maintains homeostasis or normal functioning by triggering the protein degradation and removal of the destroyed cell organelles to contribute for new cell formation (Glick et al. 2010). Several chemical inducers of autophagy have been investigated by researchers across the globe to promote the clearance of misfolded proteins through autophagy (Renna et al. 2010). However, these approaches end up with several adverse effects and other limitations. Our group in collaboration with USTC, China, reported that EHNs synthesized by domestic microwave method induced authentic autophagic flux and reduced the aggregation of huntingtin protein in various neuronal cell lines (Neuro 2a, PC12 and HeLa cells), by overexpressing characteristic autophagic marker LC3-II (Wei et al. 2014). We also confirmed the role of autophagy induced by EHNs using various chemical autophagy inhibitors and found the depression of protein

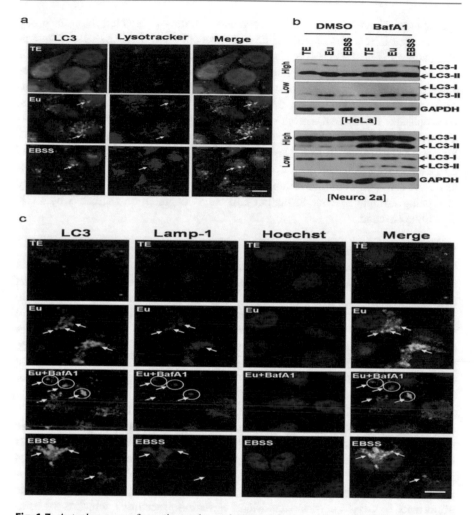

Fig. 1.7 Autophagosome formation and autophagosome lysosome fusion induced by EHNs. (**a**) The increased GFP-LC3 puncta in presence of EHNs (Eu; 50 mg/mL; 24 h) compared to vehicle control experiments (TE: Tris-EDTA buffer) could co-localize with LysoTracker Red (75 nM; 15 min) marked lysosomes in GFP-LC3/HeLa cells suggesting that the nanorods could increase autophagy without inhibiting autophagosome lysosome fusion. Starvation of cells by EBSS (3 h) has been used as a positive control experiment. Scale bar is of 10 mm. (**b**) Western blot analysis reveals that EHNs (Eu; 50 mg/mL; 24 h) could induce more LC3-II accumulation compared to vehicle control experiments (TE: Tris-EDTA buffer) in absence or presence of autophagosome lysosome fusion inhibitor Bafilomycin A1 (BafA1; 400 nM) suggesting the authentic autophagy flux mediated by the nanorods. Starvation of cells by EBSS (3 h) has been used as the positive control experiment. (**c**) Immunofluorescence analysis shows that EHNs (Eu; 50 mg/mL; 24 h) induced colocalization of autophagosome (indicated by green fluorescence of FITC) and lysosome (marked by Lamp-1-RFP) in Lamp-1-RFP stably-expressing HeLa cells could be reduced by autophagosome lysosome fusion inhibitor Bafilomycin A1 (BafA1; 400 nM). Starvation of cells by EBSS (3 h) has been used as a positive control experiment to show autophagosome lysosome fusion. Scale bar is of 10 mm. (For interpretation of the references to colour in this figure legend, the reader is referred to the web version of this article.) Figure reproduced with permission from Wei et al. (2015). Copyright © 2015 Elsevier Ltd

aggregation clearance. Detailed mechanistic studies conducted by our group revealed that EHNs stimulate the autophagy process through MEK/ERK1/2 signalling pathway (Fig. 1.7) (Wei et al. 2015). Additionally, a combination treatment of trehalose (mTOR-independent autophagy inducer) along with EHNs lead to the enhanced clearance or degradation of mutant huntingtin protein aggregates in comparison with their individual treatment effects. Considering the above observations, we emphasize that these nanorods could be developed as potent candidates for treating neurodegenerative disorders using nanomedicine approach in near future. Further, Xu et al. developed two spindle-shaped lanthanide-doped mesocrystals YF$_3$:Ce,Eu,Gd(YEG) and YF$_3$:Ce,Tb,Gd (YTG) (Xu et al. 2016). These mesocrystals were reported to induce autophagy in a time- and dose-dependent manner by promoting autophagosome formation and cargo degradation mediated by PI3K signalling cascade.

1.3.3.7 Antidiabetic Activity

Diabetes has become much prevalent in the developing countries throughout the world. Among the various types, the commonly observed diabetes mellitus patients are usually treated by the daily multiple injections. In this context, researchers are trying to develop carrier systems for efficient delivery of insulin, to avoid the painful process of multiple injections and improve the patient's life expectancy (Sharma et al. 2015). Zhai et al. demonstrated that cerium nanoparticles (CeNPs) inhibit the oxidative stress induced by Cu^{+2}/hydrogen peroxide (H$_2$O$_2$) in β-cells. Based on these suggestions the authors suggest that these antioxidant CeNPs can be useful for the prevention of diabetes (Zhai et al. 2016).

Pourkhalili et al. examined the effect of sodium selenite (Na$_2$SeO$_3$), CeNPs (approximately 100 nm in size) and a combination of both on isolated pancreatic islets in a time-dependent manner (1–6 days) (Pourkhalili et al. 2012). The authors examined an increase in cell viability, insulin secretion, reduction in ROS and increase in mitochondrial energy (ATP/ADP ratio) by a combination of sodium selenite and CeNPs incubation, compared to other treatments. This study lays the importance of CeNPs in pancreatic islets translation procedures, which is the commonly followed approach in major insulin-dependent diabetes mellitus cases. In another study, the same group has studied the effect of combination of sodium selenite and CeNPs by intraperitoneal injection in streptozotocin-induced diabetic rats (Pourkhalili et al. 2011). They have performed various assays in blood and liver tissue collected from rats post-treatment and demonstrated that the combination therapy has ameliorated the antioxidant enzymes, high density lipoprotein levels reduced by diabetes induction. The combination formulation also reduced the cholesterol, triglycerides, low density lipoprotein levels and oxidative stress, which are augmented in response to diabetes.

1.3.3.8 Antimicrobial Activity

The microbial infections are well-known as global threat for mortality and chronic life-threatening infections. Since few decades, the use of antibiotics has been the preferred treatment to combat the microbial infections. However, it has been revealed that extensive use of antibiotics result in several adverse effects (Branch-Elliman et al. 2017). In the recent years metal nanoparticles have been developed as emerging candidates in treating microbial infections in regard to their potent antimicrobial activity (Hajipour et al. 2012; Wang et al. 2017). In this context, different research groups initiated the exploration of antimicrobial activity of these lanthanide nanomaterials and proposed their use as potent agents to treat dreadful diseases caused by microorganisms. For example, Patil et al. reported the antibacterial effect of CeO_2 NPs on Gram-positive (*B. subtilis*) and Gram-negative (*E. coli*) bacteria and compared to cerium oxide microparticles and bulk salts (Patil et al. 2016). Bokare et al. reported that titanium oxide nanoparticles when doped with neodymium shown better antibacterial effect against the tested *S. aureus* and *E. coli* as compared to individual TiO_2 nanoparticles (Bokare et al. 2013). Hameeed et al. showed the neodymium doped with ZnO nanoparticles exhibits increased antibacterial activity against *E. coli* and *K. pneumoniae* bacteria compared to bare ZnO nanoparticles (Hameed et al. 2016). The cell shrinkage, loss of cell membrane integrity and viability of bacterial cells treated with Nd-doped ZnO NPs, were reported to be the mechanism behind the antimicrobial activity. The lanthanide nanomaterials have been widely reported for their antimicrobial activity. Chatterjee et al. demonstrated an eco-friendly method for the synthesis of lanthanide nanomaterials using extract of *Vigna radiata* and showed its antimicrobial as well as anticancer activity observed by various *in vitro* assays (Chatterjee et al. 2016). Balusamy et al. conducted a comparative study of antibacterial effect of lanthanum at bulk and nanoscale and reported the effective antimicrobial activity of lanthanum nanoparticles against Gram-positive bacteria (*S. aureus*) (Balusamy et al. 2012). Hence, lanthanide nanoparticles can act as therapeutic agents for treating specifically and selectively on Gram-positive and Gram-negative bacteria.

1.4   Toxicological and Pharmacokinetic Aspects

The outstanding physicochemical characteristics and properties make NPs more unique than their bulk counterparts (Zheng et al. 2015). With increased use of NPs in various biomedical and industrial applications, it is crucial to draw their safety and health concerns which can be understood by analysing the toxicological aspects (Rim et al. 2013a, b). In general humans and animals are exposed to NPs through dermal, ingestion, inhalation, injection, etc. (Stern and McNeil 2008). Irrespective of route of entry of NPs, the toxicity and subsequent health risk can be inevitable. So, in this perspective thorough knowledge on exposure of NPs and associated toxicities can be useful in their classification, to reduce adverse effects and improve the therapeutic benefits.

1.4.1   General Exposure Routes of Nanoparticles

The major routes of nanoparticles (NPs) exposure are through respiratory tract, skin, gastrointestinal tract (Stern and McNeil 2008). Inhalation route is one of the main entry points for many airborne NPs (Charron 2003) followed by dermal exposure of many metal and metal oxide NPs present in most cosmetics products (Crosera et al. 2009; Schulz et al. 2002). On the other hand, NPs intended for therapeutic and diagnostics purposes need to achieve effective systemic distribution are administered through ingestion (P.O) and/or injection (intravenous (i.v.), intraperitoneal (i.p.) or in the form of implants (Bonner 2010). NPs which enter the body through different routes, their efficient translocation and non-specific deposition in various organs define major challenges in nanotoxicological perspective (Oberdorster 2010). NPs interaction with target organs initiates a series of mechanisms, ensuing membrane damage, cytotoxicity, DNA damage and necrosis (Ho et al. 2011). The interaction and toxicological outcome varies for different types of NPs based on their chemical properties (De Jong and Borm 2008; Lynch et al. 2014). The major toxicological pathways that may be initiated due to NP interaction would ultimately lead to cytotoxicity, membrane damage, loss of protein function, DNA damage (mutation), mitochondrial damage, lysosomal damage, inflammation, fibrinogenesis, platelet abnormalities, oxidative stress (Singh 2016).

1.4.1.1 Pulmonary Exposure of Nanoparticles

The foremost entry point of NPs for effective entry to blood circulation is the respiratory pathway (Donaldson et al. 2002; Oberdorster et al. 2005). Deposition of NPs across the respiratory tract varies significantly according to the size. NPs with a size range of 1 nm will be majorly (90%) deposited in the nasopharyngeal region where as with higher size 20 nm retain the ability to efficiently translocate (~50%) to the alveolar region (Sahu and Casciano 2009). Endocytosis mechanism by alveolar epithelial cells plays a crucial role in effective absorption of NPs into the bloodstream (Yacobi et al. 2010), from where NPs gain access for effective distribution to the liver, spleen, heart and possibly other organs (Choi et al. 2010).

1.4.1.2 Dermal Exposure of Nanoparticles

The skin is the largest organ in body possessing three major layers epidermis, dermis and hypodermis (Singh 2016). It is considered as one of the potential routes for NP entry into body system (Crosera et al. 2009). Dermal entry of NPs is majorly governed by their physicochemical characteristics such as size, surface area, aggregation state and charge (Smijs and Bouwstra 2010). Majorly, toxicity studies till date on insoluble NPs have focused on the size of NPs and their noxious behaviour post-penetration through skin (Smijs and Bouwstra 2010). However, there are very few prime research activities over the toxicity of NPs that support the principle that lesser the particle size, greater is the effect on skin and other tissues (Nohynek et al. 2007).

1.4.2 Lanthanide Nanomaterials (LNMs) Actions at Cellular Level

In order to understand the toxicological behaviour of lanthanide-based NPs, it is important to understand how these materials transform and interact with the body physiological system. Influence of lanthanide elements at the physiological level was studied in the early 1990s. The role and effect of lanthanides in the physiological system might be due to ionization and consequent smaller size ionic radii of lanthanide ions (Lansman 1990; Palasz and Czekaj 2000). Early reports demonstrated that some of the lanthanide elements compete for one of the major ion channel receptors for elucidating their effect on tissues and organs. For example, calcium (Ca^{2+}) ions' physiological function (muscle contractility) at the cellular level was reported to be obstructed by lanthanum (La^{3+}) and gadolinium (Gd^{3+}) ions (Wadkins et al. 1998). Further, Coirault, C et al. (Coirault et al. 1999) reported the possible role of gadolinium (Gd^{3+}) as a potent inhibitor of stretch-sensitive ionic channels (SAC) blocker by patch clamp experiments conducted on diaphragm muscles isolated from adult hamsters. They state that in diaphragm the channels sensitive to gadolinium get activated and refrain the physiological activity. The influence of lanthanides affecting the physiological processes was also stated by Pałasz A and co-worker in their mini review which gives a brief idea of how lanthanides (lanthanum, cerium, neodymium, gadolinium, holmium, erbium and ytterbium) interact with various ion channels thereby altering the physiological mechanisms (Palasz and Czekaj 2000). However, the reports generated were solely corresponding to ionic forms of lanthanides. The nature and effect of the element at the nanoscale range vary from ionic forms that could result in varied biological consequences.

The possible biological activity of cerium and CeO_2 NPs can be explained by the defective lattice which enacts them to behave as free radical scavengers (Das et al. 2013; Hirst et al. 2013). The antibacterial activity of CeO_2 NPs was reported by Thill and co-workers in the year 2006 where Gram-negative bacteria *E. coli* adsorbed the positively charged CeO_2 NPs. This lead to the induction of oxidative stress ultimately causing death of bacteria (Thill et al. 2006). Further, Lin et al. conducted a study to understand the mechanism of cerium oxide NPs toxicity in human lung cancer cells (A549). The results displayed decreased cell viability upon treatment with cerium oxide NPs with 20 nm size range which correlates increased ROS generation and consequent interaction with cellular microenvironment (Lin et al. 2006).

1.4.3 Toxicological Aspects of Lanthanide-Based Nanoparticles

The role of lanthanide elements has not been fully understood in the normal physiological process of plants and animals (Rim et al. 2013a, b). However, the consequential effects of lanthanides to induce toxicity depend on the nature of their interactions with various receptors and biochemical reactions (Nesmerak 2013; Oliveira et al. 2014). The use of lanthanide nanoparticles towards various biomedical applications (as described in earlier sections) triggers a concern over their toxicity to safeguard humans and environment. In this perspective, the present section of this chapter mainly focuses on understanding toxicity studies (*in vitro* and *in vivo*) conducted by various researchers on lanthanide series nanoparticles.

1.4.3.1 Lanthanum Nanoparticles

Lanthanum nanoparticles usually exist in different forms viz. lanthanum oxide (La_2O_3), lanthanum carbonate (La_2CO_3) and lanthanum phosphate ($LaPO_4$). Of these, oxide and phosphate forms are used for various applications and hence their toxicological aspects are discussed. Balusamy and co-workers reported that lanthanum oxide nanoparticles (La_2O_3) show significant toxicity against *S. aureus* bacterium displaying the antibacterial activity. The probable mechanism for this activity of La_2O_3 NPs might be due to the isomorphic capabilities of lanthanide ions to replace the Ca^{2+} ions at the binding sites of staphylococcal nucleases (Balusamy et al. 2015). A further mechanistic approach to demonstrate the antimicrobial activity of La_2O_3 NPs was demonstrated by Gerber and team. In the absence of sufficient amounts of phosphates, NPs showed severe antimicrobial activity. Contrastingly, excess phosphate presence showed no toxic response. This shows that presence of optimum amounts of phosphates in the growth medium is crucial for displaying the toxicity of La_2O_3 NPs (Gerber et al. 2012).

Apart from the toxicities studies over unicellular organisms which elucidated their antibacterial activity, studies pertaining to aquatic organisms were also performed. The acute toxic effects of La_2O_3 NPs over two different aquatic species were reported by Balusamy and co-workers. The results of these studies suggest that La_2O_3 NPs caused no toxic response even at a highest concentration of 1000 mg/L in Chlorella species exposed for 72 h. Correspondingly, in small planktonic crustacean Daphnia magna no to less toxic symptoms were observed with higher EC50 (500 mg/L) and LD50 (1000 mg/L) values (Balusamy et al. 2015).

The acute toxic effects of La_2O_3 NPs in mice model were demonstrated by Balusamy et al. Even at highest dose of exposure (2000 mg/kg) no mortality was evident in mice. However, mice that received 5 or 50 mg/kg showed moderate illness with lethargic states. The biochemical estimations for illustrating the liver functioning capacity post-treatment with La_2O_3 NPs clearly show elevated levels of ALT and AST which also corroborated with damaged areas observed in liver histology sections depicting necrotic areas and infiltration of inflammatory cells (Fig. 1.8) (Brabu et al. 2015). On the other hand, the *in vivo* toxicity analysis of La_2O_3 NPs performed through inhalation route of exposure at two different doses of 10 and 30 mg/m³ in C57BL/6 J mice displayed chronic inflammatory changes and minimal fibrosis in lung tissues. Increased levels of LDH release observed after 1 and 7 days post-exposure were found to decrease to base levels till 56 days. However, induction of chronic inflammatory response by La_2O_3 NPs was observed with elevated levels of polymorphonuclear cells, eosinophils, lymphocytes and alveolar macrophages at two exposed doses. Histopathological examination of the lung tissues post-1-day exposure revealed acute inflammatory changes (Sisler et al. 2016). In another study La_2CO_3 exposure in rats with chronic renal failure showed decreased liver

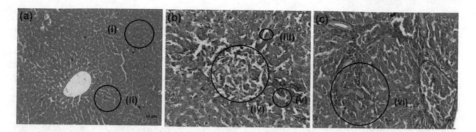

Fig. 1.8 Optical image of the liver cells (**a**) control animal – (i) normal hepatic cells, (ii) normal sinusoidal space; (**b** & **c**) La$_2$O$_3$ nanoparticles treated animal – (iii) Kupffer cell, (iv) necrosis, (v) sinusoidal distension, (vi) mononuclear cell infiltration. Figure reproduced with permission from Brabu et al. (2015). Reproduced by permission of The Royal Society of Chemistry

functioning along with prominent reduction in weight and significant accumulation of lanthanum in liver tissues (Nikolov 2010).

1.4.3.2 Cerium Nanoparticles

One of the few physicochemical characteristics that affect the internalization of NPs is their surface charge. Three different polymer-coated CeO$_2$ NPs with negative [PNC(−)], positive [ANC(+)] and neutral [DNC(0)] charge were studied for internalization into both cancerous and non-cancerous cells. Positive- and neutral-charged particles were mostly internalized into normal cell lines, while particles with negative charge were taken up by cancer cells (Asati et al. 2010). Acute and chronic toxicity studies of CeO$_2$ NPs in three ciliated protists (*Loxocephalus sp.*, *Paramecium aurelia* and *Tetrahymena pyriformis*) showed considerable toxic response which was evident by retardation in growth and carrying capacities. The highest exposure of CeO$_2$ NPs also caused extinction of two species *Loxocephalus* and *Paramecium* microcosms, which were survived in absence of CeO$_2$ NPs (Peng et al. 2017). Another report pertaining to aquatic toxicity of CeO$_2$ NPs studied by García et al. demonstrates growth inhibition (>80%) of Daphnia magna even at very low concentration with LC50 = 0.012 mg/mL (García et al. 2011). Rats exposed for a 28-day period with two nanoscale range cerium oxide particles NM-211 and NM-212 through inhalation route displayed a dose-dependent pulmonary inflammation and lung cell damage. However, no systemic inflammatory response and haematological changes were observed post-exposure (Gosens et al. 2014). An inhalation toxicity study performed in male CD1 mice for different time points of exposure (0–28 days) displayed a severe chronic inflammatory response suggesting exposure to CeO$_2$ NPs could lead to induction of pulmonary and extrapulmonary toxicity (Aalapati et al. 2014). Yet another report by Demokritou P et al. suggests that CeO$_2$ NPs exposed to rats through inhalation route caused significant lung damage and induced inflammation which was evidenced by increased PMN and LDH levels in the bronchoalveolar lavage fluid. In the same study CeO$_2$ NPs coated with SiO$_2$ have not induced any major changes in the lung tissues indicating the surface of nanoparticles could play a crucial role in inhibiting the toxic response (Demokritou et al. 2013). A study in male Sprague-Dawley rats which received 0.5 or 1.0 mg/kg of CeO$_2$ NPs through intratracheal instillation also showed lung damage after 24 h of exposure which sustained for 7 days and subsided after 84 days of exposure period. Contrastingly, no effect on inflammatory signalling and lipid peroxidation was evidenced post-NP exposure (Dunnick et al. 2016). Further, an *in vivo* study employing Drosophila melanogaster showed no toxic and genotoxic effects post-exposure with CeO$_2$ NPs. Instead a decrease in genotoxic effect induced by potassium dichromate was evidenced in presence of CeO$_2$ NPs suggesting its anti-genotoxic effect and their protective nature. Internalization of CeO$_2$ NPs into lumen of intestine of larval body was also determined which displayed uptake of NPs (Alaraby et al. 2015). However, a study performed by Benameur et al. 2015 depicts that generation of free radicals (H$_2$O$_2$) is the lead mechanism for clastogenic effect of CeO$_2$ NPs. The human dermal fibroblast cells exposed to varied concentrations of CeO$_2$ NPs caused significant genotoxic effects through generation of cellular H$_2$O$_2$ (Benameur et al. 2015).

Sub-lethal oxidative damage to tissues of Corophium volutator, an amphipod grown in marine sediments was induced by CeO$_2$ NPs through redox cycling between Ce(III) and Ce(IV). Significant induction of DNA single strand breaks was observed in test organism which was exposed to CeO$_2$ NPs for a 10-day period at 12.5 mg/l concentration. This was further supported by increased lipid peroxidation and super oxide dismutase activity (Dogra et al. 2016). A repeated dose oral toxicity study of CeO$_2$ NPs at doses 30, 300 and 600 mg/kg body weight for a period of 28 days performed by Kumari M et al. demonstrated significant DNA damage at higher dose groups (Kumari et al. 2014a). Female albino Wistar rats demonstrated significant DNA damage in bone marrow and liver cells post-oral exposure of CeO$_2$NPs (1000 mg/kg b.w.). The serum biochemical analysis revealed negative effects of nanoparticles on liver and kidney functioning capacity (Kumari et al. 2014b). An interesting report over the effect of CeO$_2$ NPs on the reproductive system was analysed by Preaubert L. The internalization (endocytosis) of NPs at 100 mg/l dose and 2 h post-exposure into the cumulus cells (cluster of cells that surround the oocyte both in the ovarian follicle) was confirmed by TEM analysis. The results showed that CeO$_2$ NPs caused significant decrease in the fertilization rate even at very low concentrations (0.01 mg/L). Further, significant DNA damage was also observed in mouse spermatozoa and oocytes suggesting the severe toxic effects of CeO$_2$ NPs which might cause impact over reproductive functioning (Preaubert et al. 2016). Antioxidant and anti-genotoxic potential of CeO$_2$ NPs in human lung epithelial cell line (BEAS-2B) was studied by Rubio L and co-workers. Significant decrease in induction of DNA damage and ROS was observed in cultures pre-treated with CeO$_2$ NPs followed by KBrO$_3$ (oxidative stress inducer). These results clearly suggest the CeO$_2$ NPs role in cell protection (Rubio et al. 2016). A molecular approach to access the toxicity of CeO$_2$ NPs in unicellular green alga *Chlamydomonas* reinhardtii was demonstrated by Taylor NS et al. The results suggest that the CeO$_2$ NPs were internalized into intracellular organelles (vesicles), analysed by using energy dispersive spectroscopy (EDS). However, molecular studies clearly indicated the role of CeO$_2$ NPs in reducing the photosynthesis and carbon fixation of alga (Taylor et al. 2016). A molecular mechanistic approach to investigate the toxic potential of three different types of CeO_2 NPs (CNP1 and CNP2, HMT-CNP1) was accessed by Dowding and co-workers. HMT-CNP1 type of particles was readily internalized into endothelial cells which were evidenced by light microscopy. Further, this reduced the cell viability and ATP levels by substantial increase of ATPase (phosphatase) activity. This clearly suggests that increase in uptake and ATPase activity might be the underlying mechanism of toxic potential of CeO_2 NPs (Dowding et al. 2013). The importance of length and width of nanoparticles along with the aspect ratio in determining the toxic potential of CeO_2 NPs was investigated by Ji Z and co-workers. These results clearly indicate that nanoparticles with length up to 200 nm and aspect ratio of 22 induce significant pro-inflammatory effects and cytotoxicity (Ji et al. 2012).

1.4.3.3 Dysprosium Nanoparticles

In vitro cytotoxicity (DU145 and NCTC1469) of surface modified dysprosium oxide and dysprosium hydroxide nanoparticles (Dy_2O_3 NPs) with D-glucuronic acid showed no change in cell viability compared to control untreated cells suggesting their non-toxic nature even at higher concentration of exposure (100 mM) (Kattel et al. 2012). In another *in vitro* cytotoxic estimation of Dy_2O_3 NPs, Tb-doped dysprosium nanorods showed concentration-dependent decrease in cell viability up to 2000 µg/mL (not more than 30% cell death) in both L929 and BEAS-2B cell lines suggesting biocompatible nature of dysprosium nanoparticles (Heng et al. 2010). Dy_2O_3 NPs exposed at 2 mg/L to *E. coli* under NaCl (85 mg/L) and glucose (140 mg/L) showed a decrease in undisturbed cell membrane (UCM) and remaining respiration percentage (RRP) to 88% and 43%, respectively, signifying their toxic potential (Anaya et al. 2016). The genotoxicity of Dy_2O_3 NPs was estimated by the Ames test in five bacterial strains (Salmonella typhimurium TA98, TA100, TA1535, TA1537 and Escherichia coli WP-2 uvrA strains) with (rat liver S9 fraction) and without metabolic activation. The results demonstrated a strong mutagenesis induced by Dy_2O_3 NPs with increase in the number of revertant colonies dose-dependently (20, 40 and 80 mg/mL) in all strains tested with and without metabolic activation (Hasegawa et al. 2012).

1.4.3.4 Europium Hydroxide Nanoparticles

Along with cerium oxide nanoparticles, europium-related nanomaterials are employed in various biomedical applications. Few *in vivo* toxicity studies were also performed to access the safety of these materials. Chronic toxicological analysis of europium hydroxide ($Eu^{III}(OH)_3$) nanorods (EHNs) was analysed by Patra et al. (2009) where the nanoparticles were administered through I.P. route through consecutive dosing (1.25, 12.5 and 125 mg/kg body weight) over a period of 7 days and observed for gross pathological changes for 8 and 60 days. The results revealed that EHNs didn't induce any pathological changes in various organs (Fig. 1.9) with no significant variation in haematological and biochemical analysis. However, mild toxicity of nanoparticles was evident at highest dose of exposure. The results altogether demonstrated the non-toxic nature of EHNs in mouse model. In continuation, to understand whether EHNs at acute exposure could cause DNA damage, we have

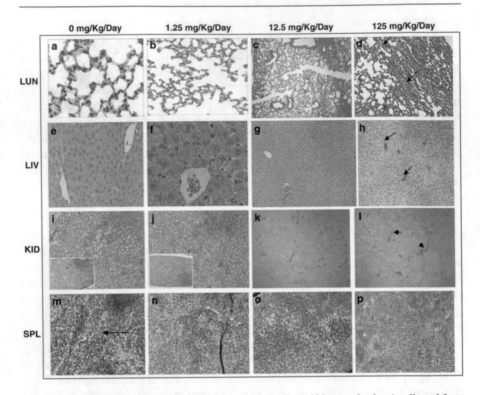

Fig. 1.9 Histologic specimens of mice tissues (the lung, liver, kidney and spleen) collected from mice euthanized on day eight, stained with haematoxylin and eosin (H and E). Histological examination of the liver, kidney, spleen and lungs from nanorod-treated mice showed none or only mild histological changes that indicate mild toxicity at higher doses of nanorods.100% survival of mice was observed even at highest dose of Eu^III(OH)₃ nanorods (125 mg kg − 1 day−1) over more than 60 days of study. (**a**) Control animal lung section showing normal alveolar geometry and normal appearing alveolar septum. (**b**) Normal alveolar geometry and normal appearing alveolar septum with a dose of 1.25 mg kg − 1 day−1. (**c**) Mild thickening of the alveolar membrane is shown (arrow) with a dose of 12.5 mg kg − 1 day−1. (**d**) Parabronchiolarlipophagocytic change (arrow) is detected with a 125 mg kg − 1 day−1 dose. (**e**) Control animal liver section showing normal hepatic architecture, hepatocytes, portal triad and central vein. (**f**) Normal hepatic architecture, hepatocytes, portal triad and central vein are seen with 1.25 mgkg−1 day−1. (**g**) Mild hepatocytes cloudy swelling (arrow) is observed after a dose of 12.5 mg kg − 1 day−1. (**h**) Hepatic sinusoidal congestion (arrow) and mild lobular inflammation were also observed (left bottom) with 125 mg kg − 1 day−1. (**i**) Kidney sections from the control animals are showing normal renal cortex with normal appearing glomerular tufts and tubules and normal renal papilla (left bottom). (**j**) Kidney sections are showing normal renal cortex with normal appearing glomerular tufts and tubules and normal renal papilla (left bottom) after a dose of 1.25 mg kg − 1 day−1. (**k**) Cloudy swelling in renal cortical tubular epithelium(arrow) is seen at 12.5 mg kg − 1 day−1 dose. (**l**) Mild glomerular mesangial cells proliferation (thin arrow) and arteriolar congestion (thick arrow) are detected at 125 dose mg kg − 1 day−1 dose. (**m**) Splenic sections from the control animals are showing normal splenic architecture with normal lymphoid follicles and sinuses (arrow). (**n**) Normal histopathologic findings are still seen after a dose of 1.25 mg kg − 1 day−1. (**o**) No pathological changes are seen with a dose of 12.5 mg kg − 1 day−1. (**p**) At a dose of 125 mg − 1 kg − 1 day,−1 mild follicular hyperplasia is seen. The histological pictures were taken at following magnifications: a → ×40, b → ×100, c → ×10, d → ×40, e → ×40, f → ×100, g → ×10, h → ×10, i → ×40, j → ×40, k → ×10, l → ×10, m → ×100, n → x 40, o → ×40, p → ×10. Figure reproduced with permission from Patra et al. (2009). Copyright © 2009 Elsevier Inc

examined the genotoxic potential of these nanorods in both *in vitro* and *in vivo* models. The results demonstrated that EHNs were not able to induce considerable DNA damage in CHO cells. However, the Swiss mice which received the highest dose (250 mg/kg b.w.) have demonstrated DNA damage evidenced in the form of chromosomal aberrations post-24 h exposure. Altogether, at therapeutic dose level EHNs were found to be safe and non-toxic through exposed route and for specified time period of study (Bollu et al. 2015). Earlier, a similar report on the integrity of buffalo spermatozoa DNA post-exposure with europium oxide (Eu_2O_3) nanoparticles was generated by Pawar and Kaul (2013). The results of their study also demonstrated no significant changes in viability, membrane integrity and DNA damages even at highest concentration (100 µg/mL) of Eu_2O_3 NPs (Pawar and Kaul 2013).

1.4.4 Pharmacokinetics of Lanthanide Nanomaterials (LNMs)

The term pharmacokinetics was derived from Greek words *pharmakon* (drug) and *kinetikos* (movement) which describes the ADME (absorption, distribution, metabolism and excretion) of a compound (Turfus et al. 2017). Basically, the pharmacokinetics can be deduced by employing mathematical equations for estimating the entry (absorption and distribution) and exit (metabolism and excretion) of drugs/nanoparticles (Zou et al. 2012). The pharmacokinetic aspects of drugs and other chemical substances can be evaluated in simple way owing to their profound solubility and distribution patterns. In the case of nanoparticles, understanding the pharmacokinetic aspects would be relatively difficult owing to their complex physiochemical properties (size, surface charge, shape etc.) (Desai 2012). However, for clinical implications of nanoparticles it is important to understand the dose effect relationship which would help in understanding their biological transformation in body physiological system (Eifler and Thaxton 2011; Havel et al. 2016). Depending on the route and mode of application, the entry of nanoparticles into the blood circulation system varies. Intravenously (I.V) injected NPs directly reach the central circulation system making it 100% bioavailable. However, those NPs administered through other routes like oral (P.O), intraperitoneal (I.P), inhalation (IHN) and topical would reach the blood circulation after crossing various biological barriers. Post-absorption phase the NPs distribution usually occur through diffusion into different organs after which they are subjected to metabolism and elimination predominantly by the liver and kidneys, respectively. Figure 1.10 provides an overview of pharmacokinetic process of NPs administered through various routes by illustrating various factors governing their ADME.

The biological, biomedical and theranostic applications of lanthanide-based nanoparticles are gaining at a rapid pace. So, in this regard it is crucial to understand the pharmacokinetics of these materials to estimate their safety and efficacy. In the current section of the chapter, we would like to elucidate the research work performed till date over various lanthanide series nanoparticles.

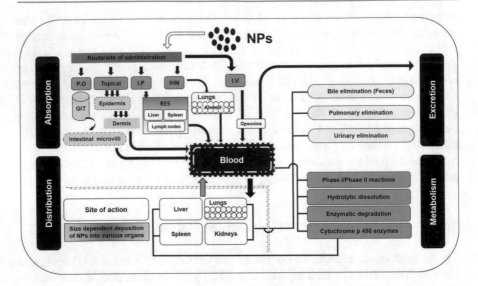

Fig. 1.10 Schematic overview of pharmacokinetics aspects of nanoparticles. The absorption, distribution, metabolism and excretion (ADME) process of NPs administered through various routes (P.O., IHN, I.P., topical and I.V) and factors that govern these processes are illustrated

1.4.4.1 Bio-distribution and Clearance Studies

The distribution of nanoparticles through systemic circulation would reach a state of equilibrium at a faster rate in highly perfused organs compared to organs with low perfusion. The liver is one of the organs with maximum blood flow rate, where nanoparticles entry and deposition would happen at a faster rate as macrophage (kuffer) cells would engulf and convert them into non-toxic or toxic/reactive metabolites (Longmire et al. 2008). Post-distribution phase for effective clearance from body system, NPs can exit from different routes viz. urine, faeces, lungs, sweat and milk. However, the route of elimination of NPs would highly depend on their physicochemical characteristics (Gatoo et al. 2014). Further, greater the rate of elimination, lesser is the retention time in blood which could induce low to no therapeutic response. Contrastingly, more residence time would ultimately lead to adverse effects due to accumulation and deposition in various organs. Consequently, to display efficient therapeutic activity, NPs circulation time period in bloodstream is to be at optimum level followed by elimination stress (Singh 2016).

Various studies were performed on lanthanide-based nanoparticles mostly to understand their effective bio-distribution and elimination profiles. A bio-distribution analysis of orally administered La_2CO_3 NPs in rat model performed by Lacour et al. (2005) demonstrated significant accumulation of lanthanum carbonate in the main metabolic organs viz. the liver, lungs and kidneys (Lacour et al. 2005). In another study, alterations in liver functioning of rats with chronic renal failure were observed upon exposure to La_2CO_3 for a long-time period suggesting its possible toxic response (Nikolov et al. 2010). Further, over dosing of La_2CO_3 at 2000 mg/kg/day for 12 weeks in rats with chronic renal failure showed deposition of La over the

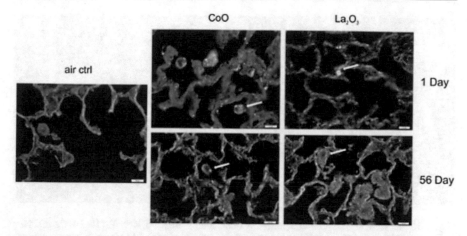

Fig. 1.11 Enhanced-darkfield images-lung deposition. Representative images of lungs from mice 1 and 56 days after exposure to 30 mg/m³ illustrate the lung burden and clearance of nanoparticles from the lung. $n = 6$ male mice per group. Nanoparticles are identified by white arrows. Figure reproduced from Sisler et al. (2016). Copyright © BioMed Central. Open Access Journal

outer surface of bone. This localization of La remained unchanged even after 2–4 weeks of un-exposed period (Behets et al. 2005). In another chronic study, lanthanum oxide nanoparticles (La_2O_3) NPs (25 ± 5 nm) were found to be effectively distributed in the lungs which show the signs of pulmonary elimination visualized by enhanced dark field microscopic images (Fig. 1.11).

The size and surface characteristics of nanoparticles can influence their bio-distribution pattern. CeO_2 NPs exposed through inhalation route to male CD1 mice over a sub-chronic time period displayed significant accumulation of Ce in the lungs and other organs indicating their possible toxic outcome (Aalapati et al. 2014). A repeated dose sub-chronic (28 days) toxicity study of CeO_2 NPs (24.2 ± 1.63 nm) at 30, 300 and 600 mg/kg b.w. doses showed a dose-dependent deposition of Ce in various organs viz., the brain, heart, liver, spleen, kidneys and blood. Urine and faeces collected till 28 days of experimental period also showed the presence of Ce, while 600 mg/kg dose displayed higher elimination than remaining dose groups (Kumari et al. 2014a). A similar result was evident even at high doses (1000 mg/kg b.w.) of CeO_2 NPs (23.2 nm) in rat model. A dose-dependent increase of Ce was observed in all major organs with a profound elimination rate from urine and faeces (Kumari et al. 2014b).

Surface coating and the size of nanoparticle have an influence on their distribution pattern. Variation in the bio-distribution pattern of two radiolabelled CeO_2 NPs; DT10 rCONP (6 nm, −5 mv) and PAA rCONP (2 nm, −40.6 mv) was observed in mouse model. DT10 rCONP showed more accumulation in the liver and spleen, whereas PAA rCONP displayed distribution pattern in the liver and kidneys. Interestingly, the deposition of PAA rCONP was observed to be higher in the skin, intestine, lungs and heart than DT10[141] Ce-rCONPs. Further, *in vivo* SPECT/CT imaging studies showed that DT10 141Ce-rCONPs displayed a gradual decrease in

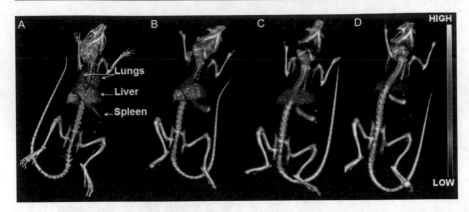

Fig. 1.12 In vivo SPECT/CT imaging of a nude mouse injected with DT10 141Ce-rCONPs (180 mCi, 3.6 nmol, 250 mL), at (**a**) 2 h, (**b**) 24 h, (**c**) 72 h and (**d**) 144 h post-injection. Images shown here were obtained from volume renderings that were adjusted to a uniform scale. Figure reproduced with permission from Yang et al. (2013b). Reproduced by permission of The Royal Society of Chemistry

Ce content with time indicating clearance of NPs till 144 h (Fig. 1.12). These results altogether suggest that surface modifications of nanoparticles will affect their distribution pattern (Yang et al. 2013b).

The route of administration plays a vital role in the distribution profile of nanoparticles. Intrathecally instilled $^{141}CeO_2$ NPs (1 mg/kg) showed a slow pulmonary clearance rate with elimination half-life ($t_{1/2}$) of nearly 140 days. However, CeO_2 NPs administered through oral route (5 mg/kg) showed almost 100% elimination through faeces. Further, intravenously injected (0.1 mg/kg) NPs were predominantly distributed to vital organs like the liver, spleen and bone (Molina et al. 2014). Similarly, another study performed by Konduru et al. demonstrated the effect of surface coating on the pharmacokinetics parameters of nanoparticles. The uncoated and silica (SiO_2) coated zinc oxide (ZnO) nanoparticles were prepared by flame spray pyrolysis method and administered by intra-tracheal (I.T.) and gavage routes to Wistar Han rats. The authors observed that the SiO_2 coating does not affect the pulmonary clearance, whereas it significantly reduces the bio-distribution of nanoparticles to the skeletal muscle, heart and skin tissues and enhances localization to thoracic lymph nodes (Konduru et al. 2014).

Reticuloendothelial system (RES) has a significant impact in understanding the distribution profiles of NPs depending on their size. NPs with size more than 100 nm size will be distributed more readily to the liver and spleen through RES (Petros and DeSimone 2010). A study performed employing various sized CeO_2 NPs displayed different distribution patterns. NPs with size less than 5 nm were not recognized by RES and larger sized NPs with different morphology displayed opsonization and predominantly present in clot areas (Dan et al. 2012). A similar distribution pattern was observed in mice that received 10 nm nanoceria at 1 or 20 mg/kg dose through I.V. route. The results from ICP-MS, TEM and confocal microscopy suggested that the radiolabelled nanoceria found to be effectively distributed to the liver, spleen, lungs, kidneys and also brain with no gross pathological changes. Further, the confocal analysis of brain tissues revealed electron dense nanoceria in the cerebral cortex (Portioli et al. 2013).

On the other hand, silica coated CeO_2 NPs were eliminated through the lungs more readily than bare nanoparticles. The study performed for 28-day period showed that about 35% of silica coated NPs were cleared at a faster rate than bare NPs which were observed to be 19%. Contrarily, I.V. injected silica coated NPs were retained mostly in the liver and spleen may due to opsonization of nanoparticles. Thus, surface modification of NPs caused significant variation in the distribution and elimination pattern of NPs accompanied by its enhanced clearance (Konduru et al. 2016). A major bio-distribution of CeO_2 NPs (25 and 90 nm) was observed post-4 h single inhalation exposure in Sprague-Dawley rats (Li et al. 2016a). On the other hand, the entry of NPs into systemic circulation and subsequent phagocytic metabolism and mucociliary clearance was estimated using a physiologically based pharmacokinetic model. In another study radiotracer technique was employed to study the bio-distribution of CeO_2 NPs. Post-28 days exposure it was observed that 8.2% of instilled NPs were seen in the lungs with an elimination half-life of 103 days (He et al. 2010). Similarly, the bio-kinetics of Eu in various organs was evaluated by our group in a 60-day long-term exposure studies at doses 1.25, 12.5, 125 mg/kg of EHNs exposed for 7 consecutive days (Patra et al. 2009). The results displayed accumulation of Eu in the liver, kidney, spleen and lungs. Similarly, another study was performed in a time-dependent manner for a very short period, where at 250 mg/kg dose of EHNs in Swiss albino mice displayed considerable amount of Eu presence at 6 and 24 h in the spleen, liver, kidney which might be due to activation of RES. Further, the effective circulation and elimination of these nanorods can be understood by detection of Eu in the blood, bone marrow and in faeces, respectively, at 24 h time point (Bollu et al. 2015).

1.5 Conclusions and Future Prospects

Nanotechnology and nanomedicine have been widely explored by several researchers globally to optimize their unique characteristic properties for biological and medical applications. In recent times, several investigators have put tremendous efforts to design and develop various rare earth metal-based nanoformulations towards the diagnosis and therapy of several diseases. However, the evaluation of bio-safety and environmental toxicity of REBNPs is still at a preliminary stage. The toxicological analysis of lanthanide-related nanoparticles was solely concentrated on the basic manifestations and not focused in detail over mechanistic aspects. It is well established that various physico-chemical properties of size, shape, morphology, surface functionalization etc. alter the biological and toxic properties of the nanomaterials. Hence, optimization of the synthesis, surface modification and doping of REBNPs is also an important prerequisite to enhance their cellular interactions for optimal biological activity. Furthermore, the investigation of physiological effects of REBNPs at the sub-cellular, cellular, organ and tissue level is of high need for comprehending their toxicological factors. Most importantly, the route of administration, dose, dosage regimen, pharmacokinetic parameters and immunotoxicity studies need to be optimized for promoting clinical translation of REBNPs. Moreover, to understand the long-term fate of REBNPs in the human system, their bioavailability and clearance studies are very essential. Although few lanthanide (esp. gadolinium) based products are approved by FDA for diagnostic purpose, many therapeutic products are needed to be developed with respect to the potential applications of REBNPs. In summary, we conclude that the researchers should largely focus on the translational research to develop many more marketable theranostic products based on REBNPs for human and environmental applications.

Acknowledgement CRP is thankful to DST-Nanomission, New Delhi (SR/NM/NS-1252/2013; GAP 570), for financial support. SKN is thankful to DST, New Delhi, for supporting with INSPIRE Senior Research Fellowship. The Authors are thankful to the Director, CSIR-IICT for his support and encouragement and for his keen interest in this work. IICT Communication Number: IICT/Pubs./2019/022 is duly acknowledged.

References

Aalapati S, Ganapathy S, Manapuram S, Anumolu G, Prakya BM (2014) Toxicity and bioaccumulation of inhaled cerium oxide nanoparticles in CD1 mice. Nanotoxicology 8:786–798. https://doi.org/10.3109/17435390.2013.829877

Abdukayum A, Chen JT, Zhao Q, Yan XP (2013) Functional near infrared-emitting Cr3+/Pr3+ co-doped zinc gallogermanate persistent luminescent nanoparticles with superlong afterglow for in vivo targeted bioimaging. J Am Chem Soc 135:14125–14133. https://doi.org/10.1021/ja404243v

Alaraby M et al (2015) Antioxidant and antigenotoxic properties of CeO$_2$ NPs and cerium sulphate: studies with *Drosophila melanogaster* as a promising in vivo model. Nanotoxicology 9:749–759. https://doi.org/10.3109/17435390.2014.976284

Albanese A, Tang PS, Chan WC (2012) The effect of nanoparticle size, shape, and surface chemistry on biological systems. Annu Rev Biomed Eng 14:1–16. https://doi.org/10.1146/annurev-bioeng-071811-150124

Anaya NM, Solomon F, Oyanedel-Craver V (2016) Effects of dysprosium oxide nanoparticles on Escherichia coli. Environ Sci Nano 3:67–73

Asati A, Santra S, Kaittanis C, Perez JM (2010) Surface-charge-dependent cell localization and cytotoxicity of cerium oxide nanoparticles. ACS Nano 4:5321–5331. https://doi.org/10.1021/nn100816s

Augustine R, Nethi SK, Kalarikkal N, Thomas S, CR P (2017) Electrospun polycaprolactone (PCL) scaffolds embedded with europium hydroxide nanorods (EHNs) with enhanced vascularization and cell proliferation for tissue engineering applications. J Mater Chem B 5:4660–4672

Bae KH, Kim YB, Lee Y, Hwang J, Park H, Park TG (2010) Bioinspired synthesis and characterization of gadolinium-labeled magnetite nanoparticles for dual contrast T-1- and T-2-weighted magnetic resonance imaging. Bioconjug Chem 21:505–512. https://doi.org/10.1021/bc900424u

Bakht MK, Sadeghi M, Ahmadi SJ, Sadjadi SS, Tenreiro C (2013) Preparation of radioactive praseodymium oxide as a multifunctional agent in nuclear medicine: expanding the horizons of cancer therapy using nanosized neodymium oxide. Nucl Med Commun 34:5–12. https://doi.org/10.1097/MNM.0b013e32835aa7bd

Balusamy B, Kandhasamy YG, Senthamizhan A, Chandrasekaran G, Subramanian MS, Kumaravel TS (2012) Characterization and bacterial toxicity of lanthanum oxide bulk and nanoparticles. J Rare Earths 30:1298–1302

Balusamy B, Tastan BE, Ergen SF, Uyar T, Tekinay T (2015) Toxicity of lanthanum oxide (La2O3) nanoparticles in aquatic environments. Environ Sci Process Impacts 17:1265–1270. https://doi.org/10.1039/c5em00035a

Behets GJ et al (2005) Localization of lanthanum in bone of chronic renal failure rats after oral dosing with lanthanum carbonate. Kidney Int 67:1830–1836. https://doi.org/10.1111/j.1523-1755.2005.00281.x

Benameur L et al (2015) DNA damage and oxidative stress induced by CeO$_2$ nanoparticles in human dermal fibroblasts: evidence of a clastogenic effect as a mechanism of genotoxicity. Nanotoxicology 9:696–705. https://doi.org/10.3109/17435390.2014.968889

Bianchi A et al (2014) Targeting and in vivo imaging of non-small-cell lung cancer using nebulized multimodal contrast agents. Proc Natl Acad Sci U S A 111:9247–9252. https://doi.org/10.1073/pnas.1402196111

Bokare A, Sanap A, Pai M, Sabharwal S, Athawale AA (2013) Antibacterial activities of Nd doped and Ag coated TiO$_2$ nanoparticles under solar light irradiation. Colloids Surf B Biointerfaces 102:273–280. https://doi.org/10.1016/j.colsurfb.2012.08.030

Bollu VS, Nethi SK, Dasari RK, Rao SS, Misra S, Patra CR (2015) Evaluation of in vivo cytogenetic toxicity of europium hydroxide nanorods (EHNs) in male and female Swiss albino mice. Nanotoxicology 10:1–13. https://doi.org/10.3109/17435390.2015.1073398

Bonner JC (2010) Nanoparticles as a potential cause of pleural and interstitial lung disease. Proc Am Thorac Soc 7:138–141

Bouzigues C, Gacoin T, Alexandrou A (2011) Biological applications of rare-earth based nanoparticles. ACS Nano 5:8488–8505

Boyer JC, Vetrone F, Cuccia LA, Capobianco JA (2006) Synthesis of colloidal upconverting NaYF4 nanocrystals doped with Er3+, Yb3+ and Tm3+, Yb3+ via thermal decomposition of lanthanide trifluoroacetate precursors. J Am Chem Soc 128:7444–7445. https://doi.org/10.1021/ja061848b

Brabu B, Haribabu S, Revathy M, Anitha S, Thangapandiyan M, Navaneethakrishnan KR, Gopalakrishnan C, Murugan SS, Kumaravel TS (2015) Biocompatibility studies on lanthanum oxide nanoparticles. Toxicol Res 4:1037–1044

Branch-Elliman W et al (2017) Risk of surgical site infection, acute kidney injury, and Clostridium difficile infection following antibiotic prophylaxis with vancomycin plus a beta-lactam versus either drug alone: a national propensity-score-adjusted retrospective cohort study. PLoS Med 14:e1002340. https://doi.org/10.1371/journal.pmed.1002340

Bridot JL et al (2007) Hybrid gadolinium oxide nanoparticles: multimodal contrast agents for in vivo imaging. J Am Chem Soc 129:5076–5084. https://doi.org/10.1021/ja068356j

Charron A, Harrison RM (2003) Primary particle formation from vehicle emissions during exhaust dilution in the roadside atmosphere. Atmos Environ 37:4109–4119

Chatterjee A, Archana L, Niroshinee V, Abraham J (2016) Biosynthesis of lanthanum nanoparticles using green gram seeds and their effect on microorganisms. Res J Pharm Biol Chem Sci 7:1462–1470

Chen F et al (2015a) Terbium-doped gadolinium oxide nanoparticles prepared by laser ablation in liquid for use as a fluorescence and magnetic resonance imaging dual-modal contrast agent. Phys Chem Chem Phys 17:1189–1196. https://doi.org/10.1039/c4cp04380d

Chen Y, Liu B, Deng X, Huang S, Hou Z, Li C, Lin J (2015b) Multifunctional Nd 3+−sensitized upconversion nanomaterials for synchronous tumor diagnosis and treatment. Nanoscale 7:8574–8583

Choi HS et al (2010) Rapid translocation of nanoparticles from the lung airspaces to the body. Nat Biotechnol 28:1300–1303. https://doi.org/10.1038/nbt.1696

Ciofani G et al (2013) Effects of cerium oxide nanoparticles on PC12 neuronal-like cells: proliferation, differentiation, and dopamine secretion. Pharm Res 30:2133–2145. https://doi.org/10.1007/s11095-013-1071-y

Coirault C, Sauviat MP, Chemla D, Pourny JC, Lecarpentier Y (1999) The effects of gadolinium, a stretch-sensitive channel blocker, on diaphragm muscle. Eur Respir J 14:1297–1303

Colon J et al (2009) Protection from radiation-induced pneumonitis using cerium oxide nanoparticles. Nanomedicine 5:225–231. https://doi.org/10.1016/j.nano.2008.10.003

Crosera M, Bovenzi M, Maina G, Adami G, Zanette C, Florio C, Filon Larese F (2009) Nanoparticle dermal absorption and toxicity: a review of the literature. Int Arch Occup Environ Health 82:1043–1055. https://doi.org/10.1007/s00420-009-0458-x

D'Angelo B et al (2009) Cerium oxide nanoparticles trigger neuronal survival in a human Alzheimer disease model by modulating BDNF pathway. Curr Nanosci 5:167–176

Dan M, Wu P, Grulke EA, Graham UM, Unrine JM, Yokel RA (2012) Ceria-engineered nanomaterial distribution in, and clearance from, blood: size matters. Nanomedicine 7:95–110. https://doi.org/10.2217/Nnm.11.103

Das M, Patil S, Bhargava N, Kang JF, Riedel LM, Seal S, Hickman JJ (2007) Auto-catalytic ceria nanoparticles offer neuroprotection to adult rat spinal cord neurons. Biomaterials 28:1918–1925. https://doi.org/10.1016/j.biomaterials.2006.11.036

Das S et al (2012) The induction of angiogenesis by cerium oxide nanoparticles through the modulation of oxygen in intracellular environments. Biomaterials 33:7746–7755. https://doi.org/10.1016/j.biomaterials.2012.07.019

Das S, Dowding JM, Klump KE, McGinnis JF, Self W, Seal S (2013) Cerium oxide nanoparticles: applications and prospects in nanomedicine. Nanomedicine (London) 8:1483–1508. https://doi.org/10.2217/nnm.13.133

Datta A, Raymond KN (2009) Gd-hydroxypyridinone (HOPO)-based high-relaxivity magnetic resonance imaging (MRI) contrast agents. Acc Chem Res 42:938–947. https://doi.org/10.1021/ar800250h

De Jong WH, Borm PJ (2008) Drug delivery and nanoparticles:applications and hazards. Int J NanomedicineInt J Nanomedicine 3:133–149

Demokritou P et al (2013) An in vivo and in vitro toxicological characterisation of realistic nanoscale CeO(2) inhalation exposures. Nanotoxicology 7:1338–1350. https://doi.org/10.3109/17435390.2012.739665

Deng M, Ma Y, Huang S, Hu G, Wang L (2011) Monodisperse upconversion NaYF 4 nanocrystals: syntheses and bioapplications. Nano Res 4:685–694

Desai N (2012) Challenges in development of nanoparticle-based therapeutics. AAPS J 14:282–295. https://doi.org/10.1208/s12248-012-9339-4

Di Pasqua AJ, Huckle JE, Kim JK, Chung Y, Wang AZ, Jay M, Lu X (2012) Preparation of neutron-activatable holmium nanoparticles for the treatment of ovarian cancer metastases. Small 8:997–1000. https://doi.org/10.1002/smll.201102488

Dogra Y et al (2016) Cerium oxide nanoparticles induce oxidative stress in the sediment-dwelling amphipod Corophium volutator. Nanotoxicology 10:480–487. https://doi.org/10.3109/17435390.2015.1088587

Donaldson K et al (2002) The pulmonary toxicology of ultrafine particles. J Aerosol Med 15:213–220. https://doi.org/10.1089/089426802320282338

Dong C, van Veggel FC (2009) Cation exchange in lanthanide fluoride nanoparticles. ACS Nano 3:123–130. https://doi.org/10.1021/nn8004747

Dong H et al (2015) Lanthanide nanoparticles: from design toward bioimaging and therapy. Chem Rev 115:10725–10815. https://doi.org/10.1021/acs.chemrev.5b00091

Dowding JM et al (2013) Cellular interaction and toxicity depend on physicochemical properties and surface modification of redox-active nanomaterials. ACS Nano 7:4855–4868. https://doi.org/10.1021/nn305872d

Dunnick KM, Morris AM, Badding MA, Barger M, Stefaniak AB, Sabolsky EM, Leonard SS (2016) Evaluation of the effect of valence state on cerium oxide nanoparticle toxicity following intratracheal instillation in rats. Nanotoxicology 10:992–1000. https://doi.org/10.3109/17435390.2016.1157220

Eifler AC, Thaxton CS (2011) Nanoparticle therapeutics: FDA approval, clinical trials, regulatory pathways, and case study. Methods Mol Biol 726:325–338. https://doi.org/10.1007/978-1-61779-052-2_21

Estevez AY, Erlichman JS (2014) The potential of cerium oxide nanoparticles (nanoceria) for neurodegenerative disease therapy. Nanomedicine (London) 9:1437–1440. https://doi.org/10.2217/nnm.14.87

Estevez AY et al (2011) Neuroprotective mechanisms of cerium oxide nanoparticles in a mouse hippocampal brain slice model of ischemia. Free Radic Biol Med 51:1155–1163. https://doi.org/10.1016/j.freeradbiomed.2011.06.006

Foucault-Collet A, Shade CM, Nazarenko I, Petoud S, Eliseeva SV (2014) Polynuclear SmIII polyamidoamine-based dendrimer: a single probe for combined visible and near-infrared live-cell imaging. Angew Chem Int Ed 53:2927–2930

Gao Y, Chen K, Ma JL, Gao F (2014) Cerium oxide nanoparticles in cancer. Onco Targets Ther 7:835–840. https://doi.org/10.2147/OTT.S62057

García A, Espinosa R, Delgado L, Casals E, González E, Puntes V, Barata C, Font X, Sánchez A (2011) Acute toxicity of cerium oxide, titanium oxide and iron oxide nanoparticles using standardized tests. Desalination 269:136–141

Gatoo MA, Naseem S, Arfat MY, Dar AM, Qasim K, Zubair S (2014) Physicochemical properties of nanomaterials: implication in associated toxic manifestations. Biomed Res Int 2014:498420. https://doi.org/10.1155/2014/498420

Geraldes CF, Laurent S (2009) Classification and basic properties of contrast agents for magnetic resonance imaging. Contrast Media Mol Imaging 4:1–23. https://doi.org/10.1002/cmmi.265

Gerber LC, Moser N, Luechinger NA, Stark WJ, Grass RN (2012) Phosphate starvation as an antimicrobial strategy: the controllable toxicity of lanthanum oxide nanoparticles. Chem Commun 48:3869–3871

Geschwind DH, Konopka G (2009) Neuroscience in the era of functional genomics and systems biology. Nature 461:908–915. https://doi.org/10.1038/nature08537

Giri S et al (2013) Nanoceria: a rare-earth nanoparticle as a novel anti-angiogenic therapeutic agent in ovarian cancer. PLoS One 8:e54578. https://doi.org/10.1371/journal.pone.0054578

Glick D, Barth S, Macleod KF (2010) Autophagy: cellular and molecular mechanisms. J Pathol 221:3–12. https://doi.org/10.1002/path.2697

Gosens I, Mathijssen LE, Bokkers BG, Muijser H, Cassee FR (2014) Comparative hazard identification of nano- and micro-sized cerium oxide particles based on 28-day inhalation studies in rats. Nanotoxicology 8:643–653. https://doi.org/10.3109/17435390.2013.815814

Gschneidner KA, Eyring L, Lander GH (eds) (2002) Handbook on the physics and chemistry of rare earths, vol 32. Elsevier, Amsterdam

Hagan AK, Zuchner T (2011) Lanthanide-based time-resolved luminescence immunoassays. Anal Bioanal Chem 400:2847–2864. https://doi.org/10.1007/s00216-011-5047-7

Hajipour MJ et al (2012) Antibacterial properties of nanoparticles. Trends Biotechnol 30:499–511. https://doi.org/10.1016/j.tibtech.2012.06.004

Hameed AS, Karthikeyan C, Ahamed AP, Thajuddin N, Alharbi NS, Alharbi SA, Ravi G (2016) In vitro antibacterial activity of ZnO and Nd doped ZnO nanoparticles against ESBL producing Escherichia coli and Klebsiella pneumoniae. Sci Rep 6:24312. https://doi.org/10.1038/srep24312

Hanahan D, Weinberg RA (2011) Hallmarks of cancer: the next generation. Cell 144:646–674. https://doi.org/10.1016/j.cell.2011.02.013

Hasegawa G, Shimonaka M, Ishihara Y (2012) Differential genotoxicity of chemical properties and particle size of rare metal and metal oxide nanoparticles. J Appl Toxicol 32:72–80. https://doi.org/10.1002/jat.1719

Havel H et al (2016) Nanomedicines: from bench to bedside and beyond. AAPS J 18:1373–1378. https://doi.org/10.1208/s12248-016-9961-7

He XA et al (2010) Lung deposition and extrapulmonary translocation of nanoceria after intratracheal instillation. Nanotechnology 21:285103. https://doi.org/10.1088/0957-4484/21/28/285103

Heng BC, Das GK, Zhao X, Ma LL, Tan TT, Ng KW, Loo JS (2010) Comparative cytotoxicity evaluation of lanthanide nanomaterials on mouse and human cell lines with metabolic and DNA-quantification assays. Biointerphases 5:FA88–FA97. https://doi.org/10.1116/1.3494617

Hirst SM, Karakoti A, Singh S, Self W, Tyler R, Seal S, Reilly CM (2013) Bio-distribution and in vivo antioxidant effects of cerium oxide nanoparticles in mice. Environ Toxicol 28:107–118. https://doi.org/10.1002/tox.20704

Ho M, Wu KY, Chein HM, Chen LC, Cheng TJ (2011) Pulmonary toxicity of inhaled nanoscale and fine zinc oxide particles: mass and surface area as an exposure metric. Inhal Toxicol 23:947–956. https://doi.org/10.3109/08958378.2011.629235

Iram S, Khan S, Ansary AA, Arshad M, Siddiqui S, Ahmad E, Khan RH, Khan MS (2016) Biogenic terbium oxide nanoparticles as the vanguard against osteosarcoma. Spectrochim Acta A Mol Biomol Spectrosc 168:123–131

Ji Z et al (2012) Designed synthesis of CeO_2 nanorods and nanowires for studying toxicological effects of high aspect ratio nanomaterials. ACS Nano 6:5366–5380. https://doi.org/10.1021/nn3012114

Kang X et al (2013) Fabrication of hollow and porous structured GdVO4:Dy3+ nanospheres as anticancer drug carrier and MRI contrast agent. Langmuir 29:1286–1294. https://doi.org/10.1021/la304551y

Kattel K et al (2012) Paramagnetic dysprosium oxide nanoparticles and dysprosium hydroxide nanorods as T(2) MRI contrast agents. Biomaterials 33:3254–3261. https://doi.org/10.1016/j.biomaterials.2012.01.008

Kim J-H et al (2011) Single molecule detection of H_2O_2 mediating angiogenic redox signaling on fluorescent single-walled carbon nanotube array. ACS Nano 5:7848–7857

Konduru NV, Murdaugh KM, Sotiriou GA, Donaghey TC, Demokritou P, Brain JD, Molina RM (2014) Bioavailability, distribution and clearance of tracheally-instilled and gavaged uncoated or silica-coated zinc oxide nanoparticles. Part Fibre Toxicol 11:44. https://doi.org/10.1186/s12989-014-0044-6

Konduru NV et al (2016) Silica coating influences the corona and biokinetics of cerium oxide nanoparticles (vol 12, 31, 2015). Part Fibre Toxicol 13:35. https://doi.org/10.1186/s12989-016-0146-4

Kong L, Cai X, Zhou X, Wong LL, Karakoti AS, Seal S, McGinnis JF (2011) Nanoceria extend photoreceptor cell lifespan in tubby mice by modulation of apoptosis/survival signaling pathways. Neurobiol Dis 42:514–523. https://doi.org/10.1016/j.nbd.2011.03.004

Kumari M, Kumari SI, Grover P (2014a) Genotoxicity analysis of cerium oxide micro and nanoparticles in Wistar rats after 28 days of repeated oral administration. Mutagenesis 29:467–479. https://doi.org/10.1093/mutage/geu038

Kumari M, Kumari SI, Kamal SS, Grover P (2014b) Genotoxicity assessment of cerium oxide nanoparticles in female Wistar rats after acute oral exposure. Mutat Res Genet Toxicol Environ Mutagen 775–776:7–19. https://doi.org/10.1016/j.mrgentox.2014.09.009

Kwon HJ et al (2016) Mitochondria-targeting ceria nanoparticles as antioxidants for Alzheimer's disease. ACS Nano 10:2860–2870. https://doi.org/10.1021/acsnano.5b08045

Lacour B, Lucas A, Auchere D, Ruellan N, Patey NMD, Drueke TB (2005) Chronic renal failure is associated with increased tissue deposition of lanthanum after 28-day oral administration (vol 67, pg 1062, 2005). Kidney Int 68:427–427

Lai BH, Chen DH (2013) LaB6 nanoparticles with carbon-doped silica coating for fluorescence imaging and near-IR photothermal therapy of cancer cells. Acta Biomater 9:7556–7563. https://doi.org/10.1016/j.actbio.2013.03.034

Lansman JB (1990) Blockade of current through single calcium channels by trivalent lanthanide cations. Effect of ionic radius on the rates of ion entry and exit. J Gen Physiol 95:679–696

Li WX, Guo L, Chen LJ, Shi XY (2008) Synthesis and fluorescence properties of lanthanide (III) perchlorate complexes with bis(benzoylmethyl) sulfoxide. J Fluoresc 18:1043–1049. https://doi.org/10.1007/s10895-008-0331-4

Li DS et al (2016a) In vivo biodistribution and physiologically based pharmacokinetic modeling of inhaled fresh and aged cerium oxide nanoparticles in rats. Part Fibre Toxicol 13:45. https://doi.org/10.1186/s12989-016-0156-2

Li L, Tong R, Li M, Kohane DS (2016b) Self-assembled gemcitabine-gadolinium nanoparticles for magnetic resonance imaging and cancer therapy. Acta Biomater 33:34–39. https://doi.org/10.1016/j.actbio.2016.01.039

Lin W, Huang YW, Zhou XD, Ma Y (2006) Toxicity of cerium oxide nanoparticles in human lung cancer cells. Int J Toxicol 25:451–457. https://doi.org/10.1080/10915810600959543

Longmire M, Choyke PL, Kobayashi H (2008) Clearance properties of nano-sized particles and molecules as imaging agents: considerations and caveats. Nanomedicine 3:703–717. https://doi.org/10.2217/17435889.3.5.703

Lord MS, Tsoi B, Gunawan C, Teoh WY, Amal R, Whitelock JM (2013) Anti-angiogenic activity of heparin functionalised cerium oxide nanoparticles. Biomaterials 34:8808–8818. https://doi.org/10.1016/j.biomaterials.2013.07.083

Luchette M, Korideck H, Makrigiorgos M, Tillement O, Berbeco R (2014) Radiation dose enhancement of gadolinium-based AGuIX nanoparticles on HeLa cells. Nanomedicine 10:1751–1755. https://doi.org/10.1016/j.nano.2014.06.004

Lynch I, Weiss C, Jones EV (2014) A strategy for grouping of nanomaterials based on key physicochemical descriptors as a basis for safer-by-design NMs. Nano Today 9:266–270

Mangaiyarkarasi R, Chinnathambi S, Aruna P, Ganesan S (2015) Synthesis and formulation of methotrexate (MTX) conjugated LaF3:Tb(3+)/chitosan nanoparticles for targeted drug delivery applications. Biomed Pharmacother 69:170–178. https://doi.org/10.1016/j.biopha.2014.11.023

McDonald MA, Watkin KL (2003) Small particulate gadolinium oxide and gadolinium oxide albumin microspheres as multimodal contrast and therapeutic agents. Investig Radiol 38:305–310

Mendoza-Mendoza E, Montemayor SM, Escalante-Garcia JI, Fuentes AF (2012) A "green chemistry" approach to the synthesis of rare-earth aluminates: perovskite-type LaAlO3 nanoparticles in molten nitrates. J Am Ceram Soc 95:1276–1283. https://doi.org/10.1111/j.1551-2916.2011.05043.x

Metcalf DJ, Garcia-Arencibia M, Hochfeld WE, Rubinsztein DC (2012) Autophagy and misfolded proteins in neurodegeneration. Exp Neurol 238:22–28. https://doi.org/10.1016/j.expneurol.2010.11.003

Mignot A et al (2013) A top-down synthesis route to ultrasmall multifunctional Gd-based silica nanoparticles for theranostic applications. Chemistry 19:6122–6136. https://doi.org/10.1002/chem.201203003

Miladi I et al (2013) Biodistribution of ultra small gadolinium-based nanoparticles as theranostic agent: application to brain tumors. J Biomater Appl 28:385–394. https://doi.org/10.1177/0885328212454315

Molina RM, Konduru NV, Jimenz RJ, Pyrgiotakis G, Demokritou P, Whollebenb W, Brain JD (2014) Bioavailability, distribution and clearance of tracheally instilled, gavaged or injected cerium dioxide nanoparticles and ionic cerium. Environ Sci Nano 1:561–573. https://doi.org/10.1039/C4EN00034J

Nesmerak K (2013) Lanthanide/actinide toxicity. In: Encyclopedia of metalloproteins. Springer, New York, pp 1098–1103. https://doi.org/10.1007/978-1-4614-1533-6_151

Nethi SK et al (2015) Investigation of molecular mechanisms and regulatory pathways of pro-angiogenic nanorods. Nanoscale 7:9760–9770. https://doi.org/10.1039/c5nr01327e

Nethi SK, P NAA, Rico-Oller B, Rodriguez-Dieguez A, Gomez-Ruiz S, Patra CR (2017) Design, synthesis and characterization of doped-titanium oxide nanomaterials with environmental and angiogenic applications. Sci Total Environ 599–600:1263–1274. https://doi.org/10.1016/j.scitotenv.2017.05.005

Nikolov IG, Joki N, Vicca S, Patey N, Auchère D, Benchitrit J, Flinois JP, Ziol M, Beaune P, Drüeke TB, Lacour B (2010) Tissue accumulation of lanthanum as compared to aluminum in rats with chronic renal failure–possible harmful effects after long-term exposure. Nephron Exp Nephrol 115:E112–E121. https://doi.org/10.1159/000313492

Niu J, Azfer A, Rogers LM, Wang X, Kolattukudy PE (2007) Cardioprotective effects of cerium oxide nanoparticles in a transgenic murine model of cardiomyopathy. Cardiovasc Res 73:549–559. https://doi.org/10.1016/j.cardiores.2006.11.031

Niu J, Wang K, PE K (2011) Cerium oxide nanoparticles inhibits oxidative stress and nuclear factor-κB activation in H9c2 cardiomyocytes exposed to cigarette smoke extract. J Pharmacol Exp Ther 338:53–61

Nohynek GJ, Lademann J, Ribaud C, Roberts MS (2007) Grey goo on the skin? Nanotechnology, cosmetic and sunscreen safety. Crit Rev Toxicol 37:251–277. https://doi.org/10.1080/10408440601177780

Oberdorster G (2010) Safety assessment for nanotechnology and nanomedicine: concepts of nanotoxicology. J Intern Med 267:89–105. https://doi.org/10.1111/j.1365-2796.2009.02187.x

Oberdorster G, Oberdorster E, Oberdorster J (2005) Nanotoxicology: an emerging discipline evolving from studies of ultrafine particles. Environ Health Perspect 113:823–839

Oliveira MS, Duarte I, Paiva AV, Yunes SN, Almeida CE, Mattos RC, Sarcinelli PN (2014) The role of chemical interactions between thorium, cerium, and lanthanum in lymphocyte toxicity. Arch Environ Occup Health 69:40–45

Oro D et al (2016) Cerium oxide nanoparticles reduce steatosis, portal hypertension and display anti-inflammatory properties in rats with liver fibrosis. J Hepatol 64:691–698. https://doi.org/10.1016/j.jhep.2015.10.020

Palasz A, Czekaj P (2000) Toxicological and cytophysiological aspects of lanthanides action. Acta Biochim Pol 47:1107–1114

Patil SN, Paradeshi JS, Chaudhari PB, Mishra SJ, Chaudhari BL (2016) Bio-therapeutic potential and cytotoxicity assessment of pectin-mediated synthesized nanostructured cerium oxide. Appl Biochem Biotechnol 180:638–654. https://doi.org/10.1007/s12010-016-2121-9

Patra CR, Bhattacharya R, Patra S, Basu S, Mukherjee P, Mukhopadhyay D (2007) Lanthanide phosphate nanorods as inorganic fluorescent labels in cell biology research. Clin Chem 53:2029–2031. https://doi.org/10.1373/clinchem.2007.091207

Patra CR et al (2008) Pro-angiogenic properties of europium(III) hydroxide nanorods. Adv Mater 20:753–756

Patra CR et al (2009) In vivo toxicity studies of europium hydroxide nanorods in mice. Toxicol Appl Pharmacol 240:88–98. https://doi.org/10.1016/j.taap.2009.07.009

Patra CR et al (2011) Reactive oxygen species driven angiogenesis by inorganic nanorods. Nano Lett 11:4932–4938. https://doi.org/10.1021/nl2028766

Pawar K, Kaul G (2013) Toxicity of europium oxide nanoparticles on the Buffalo (Bubalus bubalis) spermatozoa DNA damage. Adv Sci Eng Med 5:11–17

Peng H, Stich MI, Yu J, Sun LN, Fischer LH, Wolfbeis OS (2010) Luminescent europium(III) nanoparticles for sensing and imaging of temperature in the physiological range. Adv Mater 22:716–719. https://doi.org/10.1002/adma.200901614

Peng C, Chen Y, Pu Z, Zhao Q, Tong X, Chen Y, Jiang L (2017) CeO$_2$ nanoparticles alter the outcome of species interactions. Nanotoxicology 11:1–12. https://doi.org/10.1080/17435390.2017.1340527

Perera TSH, Han Y, Lu X, Wang X, Dai H, Li S (2015) Rare earth doped apatite nanomaterials for biological application. J Nanomater 2015:5

Petros RA, DeSimone JM (2010) Strategies in the design of nanoparticles for therapeutic applications. Nat Rev Drug Discov 9:615–627. https://doi.org/10.1038/nrd2591

Portioli C et al (2013) Short-term biodistribution of cerium oxide nanoparticles in mice: focus on brain parenchyma. Nanosci Nanotechnol Lett 5:1174–1181. https://doi.org/10.1166/nnl.2013.1715

Pourkhalili N et al (2011) Biochemical and cellular evidence of the benefit of a combination of cerium oxide nanoparticles and selenium to diabetic rats. World J Diabetes 2:204–210. https://doi.org/10.4239/wjd.v2.i11.204

Pourkhalili N et al (2012) Improvement of isolated rat pancreatic islets function by combination of cerium oxide nanoparticles/sodium selenite through reduction of oxidative stress. Toxicol Mech Methods 22:476–482. https://doi.org/10.3109/15376516.2012.673093

Preaubert L et al (2016) Cerium dioxide nanoparticles affect in vitro fertilization in mice. Nanotoxicology 10:111–117. https://doi.org/10.3109/17435390.2015.1030792

Pulido-Reyes G et al (2015) Untangling the biological effects of cerium oxide nanoparticles: the role of surface valence states. Sci Rep 5:15613. https://doi.org/10.1038/srep15613

Qie Y et al (2016) Surface modification of nanoparticles enables selective evasion of phagocytic clearance by distinct macrophage phenotypes. Sci Rep 6:26269. https://doi.org/10.1038/srep26269

Renna M, Jimenez-Sanchez M, Sarkar S, Rubinsztein DC (2010) Chemical inducers of autophagy that enhance the clearance of mutant proteins in neurodegenerative diseases. J Biol Chem 285:11061–11067. https://doi.org/10.1074/jbc.R109.072181

Rim KT, Koo KH, Park JS (2013a) Toxicological evaluations of rare earths and their health impacts to workers: a literature review. Saf Health Work 4:12–26

Rim KT, Song SW, Kim HY (2013b) Oxidative DNA damage from nanoparticle exposure and its application to workers' health: a literature review. Saf Health Work 4:177–186. https://doi.org/10.1016/j.shaw.2013.07.006

Rocha U et al (2014) Neodymium-doped LaF(3) nanoparticles for fluorescence bioimaging in the second biological window. Small 10:1141–1154. https://doi.org/10.1002/smll.201301716

Rubio L, Annangi B, Vila L, Hernandez A, Marcos R (2016) Antioxidant and anti-genotoxic properties of cerium oxide nanoparticles in a pulmonary-like cell system. Arch Toxicol 90:269–278. https://doi.org/10.1007/s00204-015-1468-y

Sahu SC, Casciano DA (2009) Nanotoxicity: from in vivo and in vitro models to health risks, ISBN: 978-0-470-74137-5

Sancey L et al (2014) Laser spectrometry for multi-elemental imaging of biological tissues. Sci Rep 4:6065. https://doi.org/10.1038/srep06065

Schulz J, Hohenberg H, Pflucker F, Gartner E (2002) Distribution of sunscreens on skin. Adv Drug Deliv Rev 54(Suppl 1):S157–S163

Shanta Singh N, Kulkarni H, Pradhan L, Bahadur D (2013) A multifunctional biphasic suspension of mesoporous silica encapsulated with YVO4:Eu3+ and Fe₃O₄ nanoparticles: synergistic effect towards cancer therapy and imaging. Nanotechnology 24:065101. https://doi.org/10.1088/0957-4484/24/6/065101

Sharma G, Sharma AR, Nam JS, Doss GP, Lee SS, Chakraborty C (2015) Nanoparticle based insulin delivery system: the next generation efficient therapy for type 1 diabetes. J Nanobiotechnol 13:74. https://doi.org/10.1186/s12951-015-0136-y

Shen J, Sun LD, Yan CH (2008) Luminescent rare earth nanomaterials for bioprobe applications. Dalton Trans 42:5687–5697. https://doi.org/10.1039/b805306e

Siafaka PI, Ustundag Okur N, Karavas E, Bikiaris DN (2016) Surface modified multifunctional and stimuli responsive nanoparticles for drug targeting: current status and uses. Int J Mol Sci 17:1440. https://doi.org/10.3390/ijms17091440

Singh AK (2016) Engineered nanoparticles: structure, properties and mechanisms of toxicity. Elsevier Inc., Boston. https://doi.org/10.1016/C2013-0-18974-X

Singh LP, Singh N, Srivastava SK (2015) Terbium doped SnO₂ nanoparticles as white emitters and SnO₂: 5Tb/Fe ₃O₄ magnetic luminescent nanohybrids for hyperthermia application and biocompatibility with HeLa cancer cells. Dalton Trans 44:6457–6465

Sisler JD et al (2016) Differential pulmonary effects of CoO and La₂O₃ metal oxide nanoparticle responses during aerosolized inhalation in mice. Part Fibre Toxicol 13:42. https://doi.org/10.1186/s12989-016-0155-3

Smijs TG, Bouwstra J (2010) Focus on skin as a possible port of entry for solid nanoparticles and the toxicological impact. J Biomed Nanotechnol 6:469–484

Son A, Dosev D, Nichkova M, Ma Z, Kennedy IM, Scow KM, Hristova KR (2007) Quantitative DNA hybridization in solution using magnetic/luminescent core shell nanoparticles. Anal Biochem 370:186–194

Son A, Nichkova M, Dosev D, Kennedy IM, Hristova KR (2008) Luminescent lanthanide nanoparticles as labels in DNA microarrays for quantification of methyl tertiary butyl ether degrading bacteria. J Nanosci Nanotechnol 8:2463–2467

Stern ST, McNeil SE (2008) Nanotechnology safety concerns revisited. Toxicol Sci 101:4–21. https://doi.org/10.1093/toxsci/kfm169

Stipic F, Pletikapic G, Jaksic Z, Frkanec L, Zgrablic G, Buric P, Lyons DM (2015) Application of functionalized lanthanide-based nanoparticles for the detection of okadaic acid-specific immunoglobulin G. J Phys Chem B 119:1259–1264. https://doi.org/10.1021/jp506382w

Tan H, Zhang L, Ma C, Song Y, Xu F, Chen S, Wang L (2013) Terbium-based coordination polymer nanoparticles for detection of ciprofloxacin in tablets and biological fluids. ACS Appl Mater Interfaces 5:11791–11796. https://doi.org/10.1021/am403442q

Taylor NS, Merrifield R, Williams TD, Chipman JK, Lead JR, Viant MR (2016) Molecular toxicity of cerium oxide nanoparticles to the freshwater alga Chlamydomonas reinhardtii is associated with supra-environmental exposure concentrations. Nanotoxicology 10:32–41. https://doi.org/10.3109/17435390.2014.1002868

Thill A, Zeyons O, Spalla O, Chauvat F, Rose J, Auffan M, Flank AM (2006) Cytotoxicity of CeO_2 nanoparticles for Escherichia coli, physic chemical insight of the cytotoxicity mechanism. Environ Sci Technol 40:6151–6156

Turfus SC, Delgoda R, Picking D, Gurley BJ (2017) Pharmacokinetics pharmacognosy. Academic, Boston, pp 495–512. https://doi.org/10.1016/B978-0-12-802104-0.00025-1

van De Rijke F, Zijlmans H, Li S, Vail T, Raap AK, Niedbala RS, Tanke HJ (2001) Up-converting phosphor reporters for nucleic acid microarrays. Nat Biotechnol 19:273–276. https://doi.org/10.1038/85734

Victor SP, Paul W, Vineeth VM, Komeri R, Jayabalan M, Sharma CP (2016) Neodymium doped hydroxyapatite theranostic nanoplatforms for colon specific drug delivery applications. Colloids Surf B Biointerfaces 145:539–547. https://doi.org/10.1016/j.colsurfb.2016.05.067

Wadkins T, Benz J, Briner W (1998) The effect of lanthanum administration during neural tube formation on the emergence of swimming behavior. Metal Ions Biol Med Int Symp 5:168–171

Wang L, Li P, Wang L (2009) Luminescent and hydrophilic LaF3-polymer nanocomposite for DNA detection. Luminescence 24:39–44. https://doi.org/10.1002/bio.1061

Wang G, Peng Q, Li Y (2011) Lanthanide-doped nanocrystals: synthesis, optical-magnetic properties, and applications. Acc Chem Res 44:322–332. https://doi.org/10.1021/ar100129p

Wang K, Wu Y, Li H, Li M, Guan F, Fan H (2014) A hybrid antioxidizing and antibacterial material based on $Ag-La_2O_3$ nanocomposites. J Inorg Biochem 141:36–42. https://doi.org/10.1016/j.jinorgbio.2014.08.009

Wang L, Hu C, Shao L (2017) The antimicrobial activity of nanoparticles: present situation and prospects for the future. Int J Nanomedicine 12:1227–1249. https://doi.org/10.2147/IJN.S121956

Wei PF et al (2014) Accelerating the clearance of mutant huntingtin protein aggregates through autophagy induction by europium hydroxide nanorods. Biomaterials 35:899–907. https://doi.org/10.1016/j.biomaterials.2013.10.024

Wei PF et al (2015) Differential ERK activation during autophagy induced by europium hydroxide nanorods and trehalose: maximum clearance of huntingtin aggregates through combined treatment. Biomaterials 73:160–174. https://doi.org/10.1016/j.biomaterials.2015.09.006

Wu Y et al (2013) Core-shell structured luminescent and mesoporous beta-NaYF4:Ce3+/Tb3+@mSiO2-PEG nanospheres for anti-cancer drug delivery. Dalton Trans 42:9852–9861. https://doi.org/10.1039/c3dt50658d

Xiang J et al (2016) Cerium oxide nanoparticle modified scaffold interface enhances vascularization of bone grafts by activating calcium channel of mesenchymal stem cells. ACS Appl Mater Interfaces 8:4489–4499. https://doi.org/10.1021/acsami.6b00158

Xiong L, Yang T, Yang Y, Xu C, Li F (2010) Long-term in vivo biodistribution imaging and toxicity of polyacrylic acid-coated upconversion nanophosphors. Biomaterials 31:7078–7085. https://doi.org/10.1016/j.biomaterials.2010.05.065

Xu YJ et al (2016) Lanthanide co-doped paramagnetic spindle-like mesocrystals for imaging and autophagy induction. Nanoscale 8:13399–13406. https://doi.org/10.1039/c6nr03171d

Yacobi NR et al (2010) Mechanisms of alveolar epithelial translocation of a defined population of nanoparticles. Am J Respir Cell Mol Biol 42:604–614. https://doi.org/10.1165/rcmb.2009-0138OC

Yang D, Dai Y, Ma P, Kang X, Cheng Z, Li C, Lin J (2013a) One-step synthesis of small-sized and water-soluble NaREF4 upconversion nanoparticles for in vitro cell imaging and drug delivery. Chemistry 19:2685–2694. https://doi.org/10.1002/chem.201203634

Yang L et al (2013b) Intrinsically radiolabeled multifunctional cerium oxide nanoparticles for in vivo studies. J Mater Chem B 1:1421–1431. https://doi.org/10.1039/c2tb00404f

Yi G, Lu H, Zhao S, Ge Y, Yang W, Chen D, Guo LH (2004) Synthesis, characterization, and biological application of size-controlled nanocrystalline NaYF4:Yb, Er infrared-to-visible upconversion phosphors. Nano Lett 4:2191–2196

Zeng HH, Qiu WB, Zhang L, Liang RP, Qiu JD (2016) Lanthanide coordination polymer nanoparticles as an excellent artificial peroxidase for hydrogen peroxide detection. Anal Chem 88:6342–6348

Zhai JH, Wu Y, Wang XY, Cao Y, Xu K, Xu L, Guo Y (2016) Antioxidation of cerium oxide nanoparticles to several series of oxidative damage related to type II diabetes mellitus in vitro. Med Sci Monit 22:3792

Zhang Z, Ma X, Geng Z, Wang K, Wang Z (2015) One-step synthesis of carboxyl-functionalized rare-earth fluoride nanoparticles for cell imaging and drug delivery. RSC Adv 5:33999–34007

Zhao X, He S, Tan MC (2016a) Design of infrared-emitting rare earth doped nanoparticles and nanostructured composites. J Mater Chem C 4:8349–8372

Zhao H et al (2016b) Lanthanide hydroxide nanoparticles induce angiogenesis via ROS-sensitive signaling. Small 12:4404–4411. https://doi.org/10.1002/smll.201600291

Zheng W, Huang P, Tu D, Ma E, Zhu H, Chen X (2015) Lanthanide-doped upconversion nano-bioprobes: electronic structures, optical properties, and biodetection. Chem Soc Rev 44:1379–1415. https://doi.org/10.1039/c4cs00178h

Zou P, Yu Y, Zheng N, Yang Y, Paholak HJ, Yu LX, Sun D (2012) Applications of human pharmacokinetic prediction in first-in-human dose estimation. AAPS J 14:262–281. https://doi.org/10.1208/s12248-012-9332-y

Nanomedicine for Hepatic Fibrosis

Ezhilarasan Devaraj and S. Rajeshkumar

Abstract

Hepatic fibrosis is a wound-healing response and commonly proceeded by chronic liver injury. Phenotypic activation of hepatic stellate cells (HSCs) plays a significant role in the progression of hepatic fibrosis; thus, they are the target cells of antifibrotic therapy. Many drugs show promising antifibrotic effects *in vitro* and *in vivo* studies, and they often exhibit a poor effect in clinical translation due to an insufficient amount of drug accumulation around the target cells (HSCs, hepatocytes, Kupffer cells, etc.) responsible for hepatic fibrosis. Nanomedicines used as theranostic agents can provide novel therapeutic opportunities to deliver antifibrotic compounds with poor water solubility and bioavailability. In recent years, nanoparticle-based antifibrotic therapy has emerged as one of the strategies to suppress the HSC activation and to resolve hepatic fibrosis. The inorganic and organic nanoparticles laden with poorly soluble herbal and synthetic drugs, siRNA with the decoration of HSC-specific molecules, i.e., retinol or receptors, have been studied as the therapeutic strategies to deliver the drugs precisely into HSCs. This review highlights various nano-based HSC targets used in the treatment of liver fibrosis.

Keywords

Extracellular matrix · Hepatic fibrosis · Hepatic stellate cells · Nanomedicine · Nanoparticles

E. Devaraj (✉)
Department of Pharmacology, Saveetha Dental College (SDC), Saveetha Institute of Medical and Technical Sciences (SIMATS), Chennai, Tamil Nadu, India

Biomedical Research Unit and Laboratory Animal Centre, SDC, SIMATS, Chennai, Tamil Nadu, India

S. Rajeshkumar
Department of Pharmacology, Saveetha Dental College (SDC), Saveetha Institute of Medical and Technical Sciences (SIMATS), Chennai, Tamil Nadu, India

© Springer Nature Singapore Pte Ltd. 2020
A. K. Shukla (ed.), *Nanoparticles and their Biomedical Applications*,
https://doi.org/10.1007/978-981-15-0391-7_2

2.1   Introduction

The liver is vulnerable to many forms of injuries from drugs, chemicals, herbal medicines due to its unique anatomic location and function. The acute liver injury is often reversible while chronic liver injury is responsible for various pathological manifestations like hepatic inflammation, fibrosis, cirrhosis, portal hypertension, and hepatocellular carcinoma (Ezhilarasan 2018). Chronic liver injury may lead to sustained scarring response which gradually disrupts the liver vascular architecture owing to an accumulation of extracellular matrix (ECM) in perisinusoidal space that eventually causes liver failure (Higashi et al. 2017). Among chronic liver diseases, fibrosis is regarded as a common pathway that represents a convergent point from many etiologies, most prominently viral hepatitis, nonalcoholic steatohepatitis, and alcohol.

2.2   Hepatic Fibrosis

Hepatic fibrosis represents 45% of all mortality which makes one of the largest unmet needs in clinical medicine (Friedman 2015). Liver fibrosis is a highly orches- trated process characterized by the net accumulation of ECM resulting from the wound-healing response to chronic liver injury of any etiology (Yoon et al. 2016). Liver fibrosis commonly precedes cirrhosis, and it is the major cause of significant morbidity and mortality of patients with chronic liver diseases (CLD) (Schuppan et al. 2018). Liver fibrogenesis is initiated as a result of chronic insults from exces- sive alcohol consumption, viral hepatitis, hepatotoxic drugs, hepatotoxins, nonalco- holic steatohepatitis (NASH), and autoimmune diseases such as primary biliary cirrhosis, primary sclerosing cholangitis, and metabolic disorders (Ezhilarasan et al. 2018). Generally, in a pathophysiological point of view, the scar formation/ECM synthesis by HSCs after the parenchymal liver injury is often beneficial to limit injury. However, when the liver is injured chronically, the ongoing HSC activation is persistent, and it synthesizes an enormous amount of ECM that can lead to fibro- sis, cirrhosis, and its complications including encephalopathy, portal hypertension, coagulopathy, variceal bleeding, liver failure, and death.

2.3   Orchestrating Role of HSCs in Hepatic Fibrosis

Hepatic stellate cells (HSCs), portal fibroblasts, cholangiocytes, and macrophages are central drivers of hepatic fibrosis (Schuppan et al. 2018). Among them, HSCs play a pivotal role in the progression of hepatic fibrosis. HSCs are non-parenchymal and resident perisinusoidal cells and account for 5−8% (quiescent HSCs) of the cells in the liver (Puche et al. 2013). In normal liver, HSCs contribute liver tissue regeneration, homeostasis of ECM synthesis and degradation, retinoid metabolism, endothelial cell-mediated vasoregulation, secretion of growth factors and cytokines, immunoregulation, lipid metabolism, detoxification, etc. (Wallace et al. 2015).

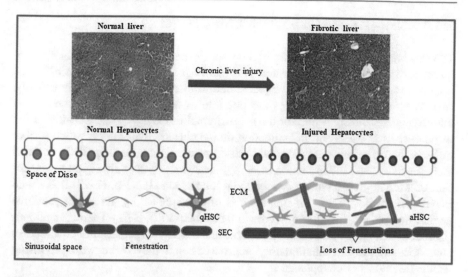

Fig. 2.1 Progression of hepatic fibrosis after chronic liver injury. *qHSCs* quiescent hepatic stellate cells, *aHSCs* activated hepatic stellate cells, *ECM* extracellular matrix, *SECs* sinusoidal endothelial cells

Further, HSCs also play a pivotal role in the scarring process. When there is any toxic insult to liver parenchyma, HSCs are activated through autocrine or paracrine signaling results in excessive ECM synthesis to protect hepatic parenchyma (Tsuchida and Friedman 2017). During CLD, HSCs are activated and acquire contractile, α-smooth muscle actin-positive myofibroblasts (MFBs)-like phenotype and are responsible for an excess synthesis and accumulation of ECM in the perisinusoidal space of Disse leading to portal hypertension and hindrance of hepatic metabolism (Thomson et al. 2017) (Fig. 2.1). Besides, platelet-derived growth factor (PDGF)-mediated HSC proliferation and transforming growth factor-β (TGF-β)-mediated ECM synthesis further aggravate the fibrosis progression (Lee et al. 2015).

2.4   Antifibrotic Strategies

Ever since the discovery of HSC role in the progression of hepatic fibrosis, several studies have been conducted with the objective to control the HSC activation for the purpose of fibrosis regression in injured liver (deLeeuw et al. 1984; Friedman et al. 1985; Ezhilarasan et al. 2012, 2014, 2016, 2017). When viewed over the last three decades, significant experimental and clinical advancements have been made in understanding the principles underlying the development and mechanism of fibrosis progression; reversal and specific antifibrotic targets have matured toward clinical translation (Friedman 1990, 2008, 2015; deLeeuw et al. 1984; Friedman and Bissell 1990).

2.4.1 Non-HSC-Mediated Antifibrotic Targets

The key issues behind the hepatic fibrosis are the persistent liver injury and subsequent activation of HSCs. Therefore, the paradigm of HSCs activation offers an important template for defining targets of antifibrotic therapy (Fig. 2.2). Practically, two types of treatment strategies are possible in hepatic fibrosis, (i) removal of underlying cause responsible for the parenchymal tissue injury and (ii) attenuation of parenchymal stress and inflammation due to chronic liver injury, which further reduces the fibrosis progression, and these are considered as non-HSC-mediated therapeutic targets. Inflammation often aggravates the fibrogenic signal via secretion of various proinflammatory mediators interleukin (IL) 1-β, IL-13, IL-17, and PDGF-BB, while chronic inflammation is often regulated by the potent profibrogenic cytokine, i.e., transforming growth factor-beta 1 (TGF-β1). Therefore, attenuation of inflammatory responses considered an attractive target, and previous studies have focused on anti-inflammatory approaches and they came with promising results (Mehal and Schuppan 2015).

2.4.2 HSC-Mediated Antifibrotic Targets

HSC-mediated targets are (i) inhibition of HSC proliferation, (ii) inhibition of profibrogenic cytokine and growth factors secretion (TGF-β, PDGF-BB, etc.), (iii) inhibition of fibrogenesis (ECM synthesis), (iv) induction of ECM degradation, (v)

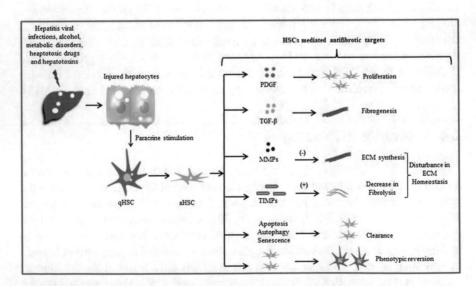

Fig. 2.2 Schematic representation of various antifibrotic targets. *qHSCs* quiescent hepatic stellate cells, *aHSCs* activated hepatic stellate cells, *PDGF* platelet-derived growth factor, *TGF-β* transforming growth factor-β, *MMPs* matrix metalloproteinases, *TIMPs* tissue inhibitors of metalloproteinases, *ECM* extracellular matrix

induction of aHSC apoptosis and senescence, and (vi) phenotypic reversion of aHSCs (MFBs) into qHSCs, are considered as HSC-mediated antifibrotic targets (Ezhilarasan et al. 2018). Novel antifibrotic agents are being developed with a combination of aforementioned targets (Fig. 2.3). Since the proliferation of activated HSCs plays a vital role in the progression of hepatic fibrosis, several studies have concentrated on these lines, and they were able to show the fibrosis regression through the anti-proliferative effect *in vitro* (Ezhilarasan et al. 2016, 2017). Inhibitors of TGF-β, PDGF-BB, connective tissue growth factor (CTGF), tumor necrosis factor α (TNF-α), and epidermal growth factor have been tested successfully *in vitro* and *in vivo* experiments and came out with promising suppression of hepatic fibrosis and they have reached clinical trials (Yoon et al. 2016; Nakamura et al. 2014). In fibrosis, the main problem is the accumulation of ECM. There is a five- to tenfold increase especially in the fibril-forming collagens (types I and III), and other ECM components such as elastin, laminin, and proteoglycans were reported experimentally as well as in human fibrosis (Mehal and Schuppan 2015; Schuppan 1990). Studies have also shown that activation of matrix metalloproteinases (MMPs) and inhibition of tissue inhibitors of metalloproteinases (TIMPs) could promote the degradation and clearance of ECM from the fibrotic liver (Jiang et al. 2013; Ramachandran et al. 2012; Hemmann et al. 2007). For instance, lysyl oxidase-like molecule 2 (LOXL2) mediates collagen crosslinking and fibrotic matrix stabilization during liver fibrosis; therefore, promotion of ECM degradation may be possible by targeting LOXL2. Accordingly, experimental studies have selectively targeted LOXL2 in fibrotic conditions. The inhibition of LOXL2 caused

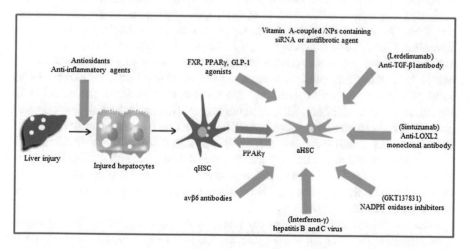

Fig. 2.3 Strategies were used to prevent HSC activation and regression of fibrosis. *qHSCs* quiescent hepatic stellate cells, *aHSC* activated hepatic stellate cells, *HSP-47* heat shock protein-47, *siRNA* small interfering RNA, *LOXL2* lysyl oxidase-like 2, *NADPH* nicotinamide adenine dinucleotide phosphate, *FXR* farnesoid X receptor, *PPARγ* peroxisome proliferator-activated receptor gamma, *GLP-1* glucagon-like peptide-1, *TGF-β* transforming growth factor-β, *MMPs* matrix metalloproteinases, *TIMPs* tissue inhibitors of metalloproteinases, *ECM* extracellular matrix, *NPs* nanoparticles

significant reduction in ECM components (Ikenaga et al. 2017). In a recent clinical trial, simtuzumab (formerly GS-6624A), a monoclonal antibody, was tested against LOXL2 in clinical subjects with liver fibrosis. However, in a phase 2b trial of patients with bridging fibrosis or cirrhosis associated with NASH, simtuzumab was found ineffective in reducing fibrosis. (Clinicaltrials.gov no: NCT01672866 and NCT01672879) (Harrison et al. 2018). Further, phenotypic reversion of MFBs into qHSCs is also employed as an approach to regress the fibrosis. During qHSC transformation into MFBs, the cannabinoid receptors 1 and 2 (CB 1 and 2) are activated. The activation of both CB 1 and 2 receptors was shown to have profibrotic and antifibrotic effects, respectively. Therefore, CB 1 receptor antagonist and agonist for CB 2 receptors have been considered as one of the therapeutic strategies (Muñoz-Luque et al. 2008; Kisseleva et al. 2012). Clearance of activated MFBs from the injured liver by apoptosis and halting their cell cycle via induction of senescence has also been tried previously (Kisseleva et al. 2012).

2.5   Challenges in Antifibrotic Therapy

Several key issues related to hepatic fibrosis targets have been addressed. To date there is no single therapeutic drug available for the treatment of hepatic fibrosis. However, combination therapies are being tried to provide symptomatic relief or regress the fibrosis. Like simtuzumab several antifibrotic agents have shown promising effect in experimental models; however, they failed in clinical translation. Possibly, the answer to such failure is *in vitro* experimental models; the antifibrotics evaluated only against HSCs. However, *in vivo*, the scenario is completely different, and there are several cell types (hepatocytes, macrophages, cholangiocytes, portal fibroblasts) involved in the progression of hepatic fibrosis. Thus far, the targeted antifibrotics have not contained multiple effects against the pleiotropic potential of MFBs (i.e., activation, proliferation, secretion of profibrogenic cytokines, ECM synthesis) and are targeted against a single mode of actions, for instance, as PDGF inhibitors, TGF-β inhibitors, PPARγ agonists, etc. Numerous herbal-derived compounds have been shown to have beneficial effects on experimental liver fibrosis (Ezhilarasan et al. 2014; Ezhilarasan and Karthikeyan 2016). However, their poor water solubility and bioavailability and targeted delivery are still subject to debate. Several newly introduced synthetic compounds show promising effects; however, they lack specificity targeting the MFBs in the fibrotic liver, and their extra hepatic toxicity also cannot be ruled out.

Nanomedicines can provide novel therapeutic opportunities to deliver antifibrotic compounds with poor water solubility and bioavailability and are also used as theranostic agents. In recent years, nanoparticle-based antifibrotic therapy has emerged as one of the strategies to (i) deliver the retinol-decorated antifibrotic compounds into activated HSCs, (ii) suppress the HSCs activation, and (iii) resolve hepatic fibrosis. Nano based and MFB-associated decoration of antifibrotics can provide a novel therapeutic approach to target activated HSCs in the fibrotic liver.

Hence, the present review further overviews nanoformulation-based novel targeting of MFB in the injured liver for the regression of hepatic fibrosis.

2.6 Nanoformulation for the Therapy of Hepatic Fibrosis

Several plant-derived phytocompounds offer poor aqueous solubility resulting in poor bioavailability (Ezhilarasan et al. 2014). Therefore, to circumvent the suboptimal bioavailability of these compounds nanoformulations are prepared (Cengiz et al. 2015). Various novel particulate preparations of herbal drugs in the form of liposomes, niosomes, nanoparticles (NPs), micelles, nanosuspensions, nanocapsulation, nanoemulsion, etc., can be used to enhance the targeting efficiency of the drug delivery systems (DDSs) for receptors present on HSCs, hepatocytes, macrophages, etc. In experimental studies, as nanoformulations, several synthetic and plant-derived compounds have been successfully tested against various drug and chemical-induced hepatic fibrosis.

2.6.1 Plant-Derived Antifibrotic Nanoformulations

2.6.1.1 Silymarin

Until now silymarin (SIL) from *Silybum marianum* is used as a standard hepatoprotective agent in experimental models and also used clinically in patients with various forms of liver diseases (Li et al. 2018; Kieslichova et al. 2018). Though SIL is reported to have a hepatoprotective effect, it has poor water solubility and oral bioavailability that limits hepatoprotective efficacy. Therefore, alternative preparations of SIL as nanoformulation are constantly on the rise (Shangguan et al. 2014). Incorporation of SIL in the liposomal carrier system increased oral bioavailability and exhibited better hepatoprotective and anti-inflammatory effects when compared with conventional SIL suspension (Kumar et al. 2014a). SIL-loaded Eudragit®RS100 NPs (mean particle size of 632.28 ± 12.15 nm with mean drug encapsulation efficiency (EE) of $89.47 \pm 1.65\%$) have been shown to improve bioavailability and excellent antifibrotic properties against bile duct-ligated (BDL) rats. SIL as NPs improved the cholestasis-induced hepatic fibrosis by restoring hepatic regenerative capabilities via mitigation of serum tumor necrosis factor-α (TNF-α), TGF-β1, hydroxyproline level; downregulation of the hepatic expression of TIMP-1 and cytokeratin-19; and upregulation of MMP-2 (Younis et al. 2016).

2.6.1.2 Curcumin

Curcumin has myriad of beneficial properties against liver diseases. However, its bioavailability is low due to limited intestinal uptake and rapid metabolism. Therefore, studies have come up with different nanoformulations with a focus on improving the bioavailability of curcumin (Jamwal 2018). Curcumin as micronized powder and liquid micelles significantly improved solubility, stability, and the slowdown of the first-pass metabolism (Schiborr et al. 2014). In a carbon tetrachloride

(CCl$_4$)-induced fibrosis model, polymeric curcumin (NanoCurc™) nano preparation has better solubility and bioavailability with sustained intrahepatic curcumin delivery to hepatocytes and HSCs (Bisht et al. 2011). Curcumin-encapsulated hyaluronic acid-polylactide NPs (60–70 nm, CEHPNPs) significantly reduced serum marker enzymes of hepatotoxicity and liver collagen levels in thioacetamide (TAA)-induced murine model of hepatic fibrosis (Chen et al. 2016). *In vitro* settings, the above nanoformulations of curcumin-based DDS induced significant aHSC cell death via apoptosis without affecting qHSCs and parenchymal cells (Bisht et al. 2011; Chen et al. 2016). It has to underline that the efficacy of CEHPNPs was approximately 1/30 that of the free drug-treated group *in vitro* (Chen et al. 2016). Phosphatidylserine-decorated curcumin NP system was studied against CCl$_4$-induced hepatic fibrosis, in which phosphatidylserine-modified nanostructured lipid carriers containing curcumin (mean particle size of 204.6 ± 1.97 nm with EE of 89.06 ± 0.47%) reduced pro-inflammatory cytokines, collagen, and α-SMA and enhanced collagenase activities (Wang et al. 2018).

2.6.1.3 Salvianolic Acid

The salvianolic acid B (SAB) from the herbal plant *Salvia miltiorrhiza* (SM) has been shown to have antifibrotic effects (Tsai et al. 2010; Hou et al. 2011; Qiang et al. 2014) however; it has poor water solubility and bioavailability that limits hepatoprotective efficacy against hepatic fibrosis (Gao et al. 2009). Therefore, nano based DDS has been tried in previous studies. The rhodamine B (RhB) covalently grafted SBA-15-structured mesoporous silica NPs (sized 400 nm, MSNs-RhB) have been developed as a DDS for SAB (SAB@MSNs-RhB). This nanoformulation improved SAB uptake by cells, bioaccessibility, and antifibrotic efficacy via antioxidative mechanisms and further, *in vitro* this NP system, showed the endocytosis-mediated sustained release of drug in LX-2 cells (He et al. 2010). Tanshinone (TA) IIA from SM has been tested in the form of TA IIA-loaded globin NPs (360 nm in size) against TAA-induced hepatic fibrosis in rats. Incorporation of liver digesting, biodegradable globin to TA IIA caused a maximum release into the liver and subsequently reduced fibrosis score and its progression (Meng et al. 2015).

2.6.1.4 Miscellaneous

Carvacrol is a monoterpenoid phenol found in several aromatic plants, including oregano (*Origanum vulgare*), pepperwort (*Lepidium flavum*), thyme (*Thymus vulgaris*), and wild bergamot. Like other plant-derived compounds, carvacrol is also highly lipophilic with low solubility in water and has poor bioavailability properties. Therefore, nano encapsulated and nanoemulsion form of carvacrol was synthesized. In the TAA-induced liver fibrosis model of rats, the nano encapsulated form (mean EE of 76.4 ± 4.2) of carvacrol was shown to have a more prominent antifibrotic effect than the nanoemulsion form (mean EE of 49.3 ± 4.5). The nano encapsulated form of carvacrol significantly mitigated the oxidative stress, inflammation, apoptosis, and hydroxyproline (Hussein et al. 2017).

Similarly, green tea extracts encapsulated with chitosan NPs reduced the liver collagen accumulation in CCl$_4$-induced fibrotic liver in rats (Safer et al. 2015a).

They have also demonstrated the antifibrotic efficacy of green tea nanoformulation (160 nm) against CCl$_4$ and ethanol-induced hepatic fibrosis in rats. In both studies, as nanoformulations, green tea extract significantly ameliorated the fibrotic changes induced by the above hepatotoxins (Safer et al. 2015b).

2.6.2 Synthetic Antifibrotic Nanoformulations

2.6.2.1 Sorafenib

Sorafenib is a tyrosine kinase inhibitor, an approved drug widely used for liver cancer. It has an inhibitory effect on vascular endothelial growth factor receptor (VEGF) and PDGF and was shown to exert antifibrotic activity in preclinical and clinical studies (Wang et al. 2010; Pinter et al. 2012). However, two major challenges limit the preventative, chronic use of sorafenib as anti-fibrotic in patients. Firstly, studies have shown that sorafenib can induce paradoxical activation of the mitogen-activated protein kinase (MAPK) pathway in both malignant and normal stromal cells (Duncan et al. 2012; Chen et al. 2017). In liver fibrosis context, MAPK, a potent mitogen, is responsible for the activation of HSCs during the progression of fibrosis (Sung et al. 2018). Secondly, as a result, off-target uptake of sorafenib by normal tissues often causes side effects such as hand-foot syndrome, diarrhea, and hypertension (Sung et al. 2018). On account of the above challenges, sorafenib/MEK inhibitor-loaded CXCR4-targeted NPs (140 nm) have developed. Interestingly, sorafenib in combination with a MEK inhibitor suppressed paradoxical MAPK-induced HSC activation *in vitro* and alleviated liver fibrosis in a CCl$_4$-induced murine model (Sung et al. 2018). Furthermore, sorafenib-loaded poly(lactic-co-glycolic acid (PLGA) NPs (100–300 nm with EE > 82%) treatment significantly decreased α-SMA and collagen content in the fibrotic liver of CCl$_4$-treated mice. Interestingly, increasing the PLGA content in the PEGPLGA/PLGA mixture led to increasing in the particle size and EE of sorafenib into the NPs and a decrease in the drug release rate (Lin et al. 2016).

2.6.2.2 Paclitaxel

Taxol® (paclitaxel), an anticancer drug, has been shown to have an antifibrotic effect by targeting the TGF-β pathway (Zhou et al. 2010; Wang et al. 2013). However, off-target adverse effects induced by Taxol® such as neutropenia limit its potential clinical applications to treat liver fibrosis (Cella et al. 2003). Therefore, in a recent study, carboxymethyl cellulose-docetaxel-conjugated NPs (Cellax, an albumin-bound DDS) have been developed to selectively target aHSCs. Cellax NPs (120 nm) significantly alleviated the *in vitro* profibrogenic potential of HSCs and CCl$_4$-induced liver fibrosis in mice (Chang et al. 2018).

2.6.2.3 Hyaluronic Acid

Hyaluronic acid (HA) is a biocompatible, biodegradable, non-toxic, and linear polysaccharide and it is one of the important components of ECM (Sudha and Rose 2014). CD44 expression is reported to increase with hepatic fibrosis and has an important role in the HSCs activation and migration (Kikuchi et al. 2005). Hence, CD44 has been targeted for HA receptor-mediated DDS, and to achieve this micelles have developed with HA polymer back bone to deliver losartan (angiotensin II receptor antagonist) via a CD44 receptor. These losartan-loaded HA micelles (300 nm) accumulated in the liver and offered significant antifibrotic effect via reduction of α-SMA expression *in vitro* and *in vivo* against TAA and ethanol-induced fibrosis; therefore, losartan-loaded HA micelles could be an attractive option for antifibrotic therapy (Thomas et al. 2015).

2.6.2.4 Miscellaneous

Hedgehog (Hh) and peroxisome proliferator-activated receptor gamma (PPAR-γ) are major signaling pathways involved in the pathogenesis of liver fibrosis (Hsu et al. 2013; Yang et al. 2014). Since Hh inhibitor, vismodegib (GDC), and PPAR-γ agonist, rosiglitazone (RSG), have poor water solubility, a previous study was formulated methoxy-polyethylene-glycol-b-poly(carbonate-colactide)(mPEG-b-p(CB-co-LA)), biodegradable polymeric NPs (120–130 nm with EE of GDC (98%) and RSG (95%)) for treating liver fibrosis. Intravenous tail vein injection of NPs encapsulating GDC and RSG provided hepatoprotection by reducing Hh pathway ligands and increasing PPAR-γ activity in BDL rats (Kumar et al. 2014b). Oxidative stress plays a pivotal role in HSC activation and subsequent progression of hepatic fibrosis (Ezhilarasan 2018). Therefore, oral administration of redox NPs (poly(ethylene glycol)-b-poly[4-(2,2,6,6-tetramethylpiperidine-1-oxyl)aminomethylstyrene] (MeO-PEG-b-PMNT)) used as a novel treatment approach for the therapy of hepatic fibrosis (Eguchi et al. 2015). It is noteworthy to mention that most of the orally administered NP size ranging between 10 and 100 nm is not usually absorbed from the gastrointestinal tract. However, redox polymers are absorbed after their oral administration, and this kind of redox NP preparation could be an ideal oral medication for oxidative stress-induced hepatic fibrosis (Eguchi et al. 2015). Zinc oxide NPs have reduced oxidative stress, hydroxyproline level, and α-SMA expression in TAA-induced hepatic fibrosis in rats (Bashandy et al. 2018). Similarly, intravenous injection of gold NPs (5–10 nm) ameliorated ethanol and methamphetamine-induced activation of Kupffer cells and HSCs, oxidative stress, and fibrosis through modulation of signaling pathways of protein kinase B (AKT)/ phosphoinositide 3-kinase and MAPK (de Carvalho et al. 2018). Cerium oxide NPs (4–20 nm) were concentrated in the diseased liver and mitigated the expression of genes responsible for oxidative and endoplasmic reticulum stress, macrophage infiltration, α-SMA expression, hepatic steatosis, inflammatory cytokines, and portal pressure in CCl_4-induced hepatic fibrosis in rats (Oró et al. 2016). Oral administration of citrate-functionalized manganese tetroxide NPs (mean size of 6.19 ± 0.05 nm) significantly ameliorated the oxidative stress and improved fibrosis than that of conventional SIL preparation against CCl_4-induced fibrotic liver (Adhikari et al. 2016).

2.7 Antifibrotic Therapy with NP-Laden Small Interfering RNA (siRNA)

RNA interference is a sequence-specific manner to inhibit the expression of homologous genes that results in gene silencing. The siRNA has high efficiency and specificity, providing new avenues for the gene therapy of hepatic fibrosis (Schuppan et al. 2018). However, the gene-silencing effect of siRNA is seriously hindered by off-target effects. To address these problems, the delivery system-based cationic lipids and polymers have been widely studied. For instance, positively charged cationic liposomes and stable nucleic acid-lipid NPs were combined with negatively charged siRNA, thereby increasing the entrapment and transfection efficiency and siRNA delivery safely and effectively (Mussi and Torchilin 2013). Heat shock protein 47 (HSP47) is a collagen-specific molecular chaperone in the endoplasmic reticulum, and its expression is dramatically upregulated in the pathological process of hepatic fibrosis (Zhao et al. 2017). Therefore, in a recent study, polypeptide pPB-modified stable nucleic acid-lipid NPs (pPB-SNALPs) were prepared (50 nm) to selectively deliver siRNAs against HSP 47 to the liver. The pPB-SNALPs loaded siRNA system used as the targeted delivery system for activated HSCs by specific receptor and exhibited good targeting ability *in vitro* against LX-2 cells and mouse primary HSCs and also *in vivo* against TAA-induced hepatic fibrosis in mice (Jia et al. 2018). Chemokine receptor type 4 (CXCR4)-targeted NPs were formulated to deliver siRNAs against VEGF into fibrotic livers to block angiogenesis. AMD3100, a CXCR4 antagonist, was incorporated into the NPs (mean size of 81.62 ± 20.26 nm with EE—80%). This NP system served dual functions by acting as a targeting moiety and suppressed the progression of fibrosis by inhibiting the proliferation and activation of HSCs. The CXCR4-targeted NPs delivered VEGF siRNAs to fibrotic livers, subsequently, decreased EGF expression, angiogenesis and normalized the distorted vessels in the fibrotic livers in CCl_4-induced mouse model (Kong et al. 2013).

Cationic solid lipid NPs (mean size of 106.2 ± 5.4 nm, CSLNs) reconstituted from natural low-density lipoprotein were designed and targeted with specific systemic delivery of CTGF siRNA (siCTGF) for the treatment of hepatic fibrosis. In an *N*-nitrosodimethylamine-induced fibrosis model, intravenous injection of CSLN/siCTGF complex specifically delivered into the liver and resulted in a significant reduction of collagen, profibrogenic and proinflammatory cytokines such as TGF-β, and CTGF, TNF-α, and IL-6 (Liu et al. 2016). The cationic lipid NPs (80 nm) loaded with siRNA to the procollagen α1(I) gene (LNP-siCol1a1) have been shown to reduce collagen production in mouse models of liver fibrosis (Jiménez Calvente et al. 2015). In a similar study, using cationic nanohydrogel particles (40 nm) have demonstrated as well-defined *in vivo* model of anti-Col1α1 siRNA delivery to HSCs without detectable toxicity. The intravenously administered siRNA-loaded nanogel particles exclusively concentrated in the liver and 50% of which taken particularly by MFBs. The anti-Col1α1 siRNA carriers specifically reduced the collagen deposition in the fibrotic liver (Kaps et al. 2015).

The sterically stabilized phospholipid NPs (SSLNPs) have developed with different nitrogen to phosphate (N/P) ratios (30, 20, and 10), and this nanocarrier is capable of incorporating siRNA to CTGF in its core through self-association with a novel cationic lipid composed of naturally occurring phospholipids and amino acids. Interestingly, EE for siRNA was proportionately increased with N/P ratios, and SSLNPs with N/P = 30 have shown maximum siRNA EE of 85 ± 16% and mean particle size of 83 ± 13 nm. Galactosamine (GalN) is known to target asialo-glycoprotein receptors, expressed on the surface of hepatocytes (D'Souza and Devarajan 2015); therefore, these receptors were targeted by attaching the GalN ligand to the nanocarriers (SSLNP-GalN encapsulated siRNA), which enhances the uptake of NPs by hepatocytes (one of the main sources of CTGF) during fibrosis. The above nanoconstruct reduced collagens 1 and III and α-SMA expressions in HSCs (Khaja et al. 2016). In an attempt to deliver siRNA to HSCs, several types of proton-activated lipid-like materials (ssPalms) that contain myristic acid (mean size of 161 ± 4 nm, ssPalmM), hydrophobic vitamins A (mean size of 182 ± 13 nm, ssPalmA) and E (mean size of 167 ± 10 nm, ssPalmE) as hydrophobic scaffolds were used. Among them, ssPalmA lipid NP system showed a significant inhibitory effect on collagen production in a CCl_4-induced fibrosis model of mice (Toriyabe et al. 2017).

2.8   HSC Targeted NP Delivery: Decoration and Drug-Laden Strategies

NPs synthesized or fabricated with the affinity toward aHSC-specific molecules or receptors to ensure the targeted drug delivery. Decoration with HSC-specific molecules or receptors has been tried for targeted drug delivery. In a previous study, the NP drug carrier facilitates the formation of corona composed of native transport proteins. The corona protein formed on the drug vehicles, which was considered as a new strategy for the design of smart vehicles for targeted drug delivery. Retinol, one of the HSC-specific molecules, conjugated with low-molecular-weight polyethylenimine (PEI), which further combined nucleotides to form NPs. The introduction of retinol specifically recruited retinol-binding protein 4 (RBP) in the corona that successfully directed the drug-laden particles into the HSCs in the liver. Retinol-conjugated polyetherimine (RcP) NP system was selectively recruited RBP in its corona components. RBP was found to bind retinol and direct the antisense oligonucleotide (ASO)-laden RcP carrier to HSCs. This NP system has been tested against CCl_4 and BDL model of hepatic fibrosis, in which the ASO-laden RcP particles effectively suppressed the expression of type I collagen and consequently ameliorated hepatic fibrosis (Zhang et al. 2015). Cationic lipid NPs are recognized as being easily internalized into the cells because of the negatively charged phospholipids on the cell membrane; however, in contrast, cationic retinol-loaded silibinin (SBN) nanostructured lipid carriers (mean size of 183.1 ± 2 nm with SBN mean

EE of 98.9 ± 0.2%) showed a slower SBN release as compared to the anionic nano-carriers (mean size of 261.2 ± 3.9 nm with SBN EE of 97.9 ± 0.7%). Further, retinol loading assisted active delivery of nanocarriers to the liver, and in fact, the incorporation of retinol in nanoformulation did not effectively increase the uptake by HSCs due to an insufficient amount of retinol in the formulations (0.02%) as suggested by authors. In light of the above report, it is to suggest that to achieve effective NP uptake by HSCs, the concentration of retinol needs to be higher than 0.02%. However, SBN nanostructured lipid carriers suppressed the α-SMA and cleared the activated HSCs by apoptosis (Pan et al. 2016).

Nitric oxide (NO) plays an imperative role in inhibiting the development of hepatic fibrosis and its ensuing complication of portal hypertension by inhibiting HSC activation (Iwakiri 2015). Therefore, gold (<5 nm) and silica NP-mediated DDS containing NO donors (S-nitroso-N-acetyl-DL-penicillamine, glyco-SNAP, 3-morpholino-sydnonimine, S-nitrosoglutathione) have developed. Both gold and silica nanoconjugates with NO donors significantly inhibited the HSC proliferation and its vascular tube formation ability *in vitro* (Das et al. 2010). Later, this strategy has been applied *in vivo*, in which polymeric NPs are designed to transport and deliver NO into HSCs for the treatment of liver fibrosis and portal hypertension. S-nitrosoglutathione, a NO donor, was incorporated with the NPs and is designed for liver delivery, minimizing systemic delivery of NO. The NPs are also decorated with retinol to specifically target HSCs. This NO-releasing retinol-decorated NP significantly downregulated the expression of profibrotic genes *in vitro* and *in vivo*. Further, this NP system also attenuated the HSC contraction via downregulation of endothelin-1 (ET-1) and accurately alleviated hemodynamic disorders in BDL-induced portal hypertension as evidenced by decreasing portal pressure (20%) and without changing mean arterial pressure (Duong et al. 2015).

The fibroblast growth factor-inducible 14 (Fn14), a membrane receptor highly and specifically expressed only in aHSCs, is reported to be the key-driven factor of hepatic fibrosis, and thus, it has a great potential as a novel target for the development of effective treatment (Wilhelm et al. 2016). Huang et al. identified a d-enantiomeric peptide ligand of Fn14 through a mirror-image mRNA display. The d-peptide ligand showed strong binding to Fn14 while maintaining high proteolytic resistance. As a targeting moiety, this d-peptide successfully mediated high selectivity of activated HSCs for liposomal vehicles compared to that of other major cell types in the liver and significantly enhanced the accumulation of liposomes in the hepatic fibrosis region of a CCl4-induced mouse model. Moreover, in combination with curcumin as an encapsulated load, a liposomal formulation conjugated with this d-peptide showed powerful inhibition of the proliferation of aHSCs and reduced the hepatic fibrosis significantly *in vivo*. Therefore, aHSC-specific targets like Fn14 may provide a promising approach to targeted drug delivery for hepatic fibrosis treatment (Ezhilarasan et al. 2018; Huang et al. 2017). The general schematic representation of various HSC-specific targets has been presented in Fig. 2.4.

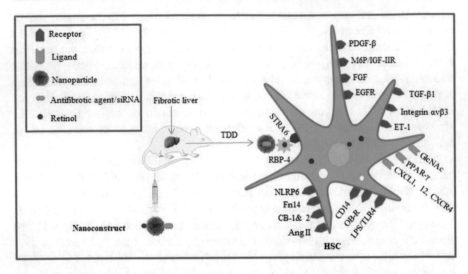

Fig. 2.4 Possible nanoparticle-based antifibrotic targets with various receptors and ligands activated during HSC activation. *siRNA* interfering RNA, *TDD* targeted drug delivery, *RBP-4* Retinol-Binding Protein 4, *PDGF* platelet-derived growth factor, *M6P/IGF-IIR* mannose 6-phosphate/insulin-like growth factor II receptor, *FGF* fibroblast growth factor, *EGFR* epidermal growth factor receptor, *TGF-β1* transforming growth factor-beta1, *Ang II* angiotensin-II, *ET-1* endothelin-1, *GlcNAc* N-acetylglucosamine, *Fn14* fibroblast growth factor-inducible 14, *NLRPs* nod-like receptor pyrin domain-containing proteins, *CB-1&2* cannabinoid receptor-1&2, *CXCL1* chemokine (C-X-C motif) ligand 1, *OB-R* leptin receptor

2.9   Advantages and Limitations of NP-Based HSC Targets

Most of the antifibrotic agents studied have poor efficacy because of their poor water solubility and bioavailability. These antifibrotic agents have better bioavailability than their raw drugs when prepared as a nanoformulation. The optimized size, charged particle surface, and lipophilic nature of the NPs make drugs soluble and improve their bioavailability and accessibility to the liver by passive targeting. Further, NPs are useful as a theranostic agent, for instance, antifibrotic agents with NP can be decorated with HSC-specific molecules or receptors present in the cells responsible for hepatic fibrosis. The specific decoration ensures the targeted drug delivery. Bioimaging tools are also available to monitor the organ distribution of NPs which may further ensure targeted drug delivery and extra hepatic toxicity (Fig. 2.5).

Inorganic NPs are commonly used to load therapeutic drugs, and they have little toxic effect; therefore, nanomaterial-induced toxicity, especially for inorganic NPs, should be considered before using as DDS. Interestingly, silica NPs are used as a model to induce hepatic fibrosis in experimental animals (Yu et al. 2017). The toxicity risks associated with organic NPs are discussed to be less as they made from natural or highly biocompatible polymers like PEG. In the hepatic fibrosis context, most of the NPs studied are intravenously injected because of their poor GIT

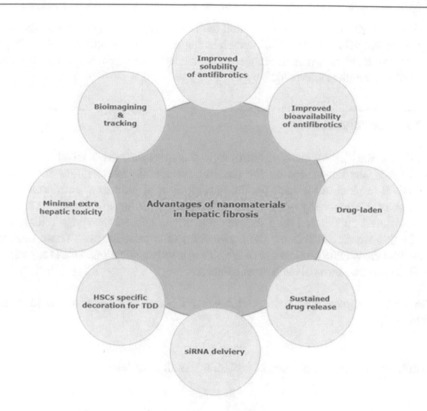

Fig. 2.5 Advantages of nanoparticles for the treatment of hepatic fibrosis

absorption. These intravenously injected NPs can be cleared by the resident macrophages such as Kupffer cells in the liver. The NP access to the other liver cells, including hepatocytes, was possible after extravasating liver sinusoidal fenestration. Therefore, as suggested by Almeida et al. larger NPs generally may have efficient hepatic uptake than the smaller ones (Almeida et al. 2011).

2.10   Future Directions

A Couple of decades before most of the experimentally studied antifibrotic drugs were generally tested as therapeutics without targeting any specific cell types responsible for hepatic fibrosis. Now, the trend has changed and most of the studies are currently focusing on the activation of HSCs and their control in fibrotic liver. However, hepatic fibrosis is a complex process which involves multiple cell type, viz., resident hepatocytes, HSCs, Kupffer cells, sinusoidal endothelial cells, and portal fibroblasts. So far, only a few studies have targeted the aHSCs or MFB with nanomaterial-laden drugs containing HSC-specific decoration such as retinol. These decorations for targeted drug delivery showed promising effect, and therefore, in the future, more such studies are warranted with different targets against the other cell type involved in hepatic fibrosis. Further, as suggested by Giannitrapani et al. (2014) the development of a nanovigilance or regulatory framework based on objective scientific research is warranted.

2.11 Conclusion

Until now the development of antifibrotic therapies remains elusive; the greater advance of our knowledge on the possible therapeutic interventions for hepatic fibrosis is in need of the hour for the development of promising antifibrotic drug candidates. Undoubtedly, most of the NPs tailored for hepatic fibrotic treatments were focused on targeted drug delivery and which ensures the drugs release into HSCs that are responsible for the progression of hepatic fibrosis. These findings may pave the way for clinical use of the NPs as a safe medication of CLD associated with fibrosis and cirrhosis in human subjects.

Financial Support The authors received no financial support to produce this manuscript.

Conflict of Interest The authors declare no conflicts of interest.

Authors' Contribution ED solely determined the theme of this review manuscript; ED and SR reviewed the literature, drafted the manuscript, designed the figures, and submitted the manuscript.

References

Adhikari A, Polley N, Darbar S, Bagchi D, Pal SK (2016) Citrate functionalized Mn_3O_4 in nanotherapy of hepatic fibrosis by oral administration. Future Sci OA 2(4):FSO146

Almeida JP, Chen AL, Foster A, Drezek R (2011) *In vivo* biodistribution of nanoparticles. Nanomedicine (Lond) 6(5):815–835

Bashandy SAE, Alaamer A, Moussa SAA, Omara EA (2018) Role of zinc oxide nanoparticles in alleviating hepatic fibrosis and nephrotoxicity induced by thioacetamide in rats. Can J Physiol Pharmacol 96(4):337–344

Bisht S, Khan MA, Bekhit M et al (2011) A polymeric nanoparticle formulation of curcumin (NanoCurc™) ameliorates CCl_4-induced hepatic injury and fibrosis through reduction of pro-inflammatory cytokines and stellate cell activation. Lab Investig 91(9):1383–1395

Cella D, Peterman A, Hudgens S, Webster K, Socinski MA (2003) Measuring the side effects of taxane therapy in oncology: the functional assesment of cancer therapy-taxane (FACT-taxane). Cancer 98(4):822–831

Cengiz M, Kutlu HM, Burukoglu DD, Ayhancı A (2015) A comparative study on the therapeutic effects of silymarin and silymarin-loaded solid lipid nanoparticles on D-GaIN/TNF-α-induced liver damage in Balb/c mice. Food Chem Toxicol 77:93–100

Chang CC, Yang Y, Gao DY et al (2018) Docetaxel-carboxymethylcellulose nanoparticles ameliorate CCl₄-induced hepatic fibrosis in mice. J Drug Target 26(5–6):516–524

Chen YN, Hsu SL, Liao MY et al (2016) Ameliorative effect of curcumin-encapsulated hyaluronic acid-PLA nanoparticles on thioacetamide-induced murine hepatic fibrosis. Int J Environ Res Public Health 14(1):11. pii: E11

Chen Y, Liu YC, Sung YC et al (2017) Overcoming sorafenib evasion in hepatocellular carcinoma using CXCR4-targeted nanoparticles to co-deliver MEK-inhibitors. Sci Rep 7:44123

D'Souza AA, Devarajan PV (2015) Asialoglycoprotein receptor mediated hepatocyte targeting – strategies and applications. J Control Release 203:126–139

Das A, Mukherjee P, Singla SK et al (2010) Fabrication and characterization of an inorganic gold and silica nanoparticle mediated drug delivery system for nitric oxide. Nanotechnology 21(30):305102

de Carvalho TG, Garcia VB, de Araújo AA et al (2018) Spherical neutral gold nanoparticles improve anti-inflammatory response, oxidative stress and fibrosis in alcohol-methamphetamine-induced liver injury in rats. Int J Pharm 548(1):1–14

deLeeuw AM, McCarthy SP, Geerts A, Knook DL (1984) Purified rat liver fat-storing cells in culture divide and contain collagen. Hepatology 4:392–403

Duncan JS, Whittle MC, Nakamura K et al (2012) Dynamic reprogramming of the kinome in response to targeted MEK inhibition in triple-negative breast cancer. Cell 149(2):307–321

Duong HT, Dong Z, Su L et al (2015) The use of nanoparticles to deliver nitric oxide to hepatic stellate cells for treating liver fibrosis and portal hypertension. Small 11(19):2291–2304

Eguchi A, Yoshitomi T, Lazic M et al (2015) Redox nanoparticles as a novel treatment approach for inflammation and fibrosis associated with nonalcoholic steatohepatitis. Nanomedicine (Lond) 10(17):2697–2708

Ezhilarasan D (2018) Oxidative stress is bane in chronic liver diseases: clinical and experimental perspective. Arab J Gastroenterol 19(2):56–64

Ezhilarasan D, Karthikeyan S (2016) Silibinin alleviates N-nitrosodimethylamine-induced glutathione dysregulation and hepatotoxicity in rats. Chin J Nat Med 14(1):40–47

Ezhilarasan D, Karthikeyan S, Vivekanandan P (2012) Ameliorative effect of silibinin against N-nitrosodimethylamine-induced hepatic fibrosis in rats. Environ Toxicol Pharmacol 34(3):1004–1013

Ezhilarasan D, Sokal E, Karthikeyan S, Najimi M (2014) Plant derived antioxidants and antifibrotic drugs: past, present and future. J Coast Life Med 2(9):738–745

Ezhilarasan D, Evraerts J, Brice S et al (2016) Silibinin inhibits proliferation and migration of human hepatic stellate LX-2 cells. Clin Exp Hepatol 6(3):167–174

Ezhilarasan D, Evraerts J, Sid B et al (2017) Silibinin induces hepatic stellate cell cycle arrest via enhancing p53/p27 and inhibiting Akt downstream signaling protein expression. Hepatobiliary Pancreat Dis Int 16(1):80–87

Ezhilarasan D, Sokal E, Najimi M (2018) Hepatic fibrosis: it is time to go with hepatic stellate cell-specific therapeutic targets. Hepatobiliary Pancreat Dis Int 17:192–197

Friedman SL (1990) Cellular sources of collagen and regulation of collagen production in liver. Semin Liver Dis 10(1):20–29

Friedman SL (2008) Hepatic fibrosis – overview. Toxicology 254(3):120–129

Friedman SL (2015) Hepatic fibrosis: emerging therapies. Dig Dis 33(4):504–507

Friedman SL, Bissell DM (1990) Hepatic fibrosis: new insights into pathogenesis. Hosp Pract (Off Ed) 25(5):43–50

Friedman SL, Roll FJ, Boyles J, Bissell DM (1985) Hepatic lipocytes: the principal collagen-producing cells of normal rat liver. Proc Natl Acad Sci U S A 82:8681–8685

Gao DY, Han LM, Zhang LH, Fang XL, Wang JX (2009) Bioavailability of salvianolic acid B and effect on blood viscosities after oral administration of salvianolic acids in beagle dogs. Arch Pharm Res 32(5):773–779

Giannitrapani L, Soresi M, Bondì ML, Montalto G, Cervello M (2014) Nanotechnology applications for the therapy of liver fibrosis. World J Gastroenterol 20(23):7242–7251

Harrison SA, Abdelmalek MF, Caldwell S et al (2018) Simtuzumab is ineffective for patients with bridging fibrosis or compensated cirrhosis caused by nonalcoholic steatohepatitis. Gastroenterology 155:1140–1153. S0016-5085(18):34758-9

He Q, Zhang J, Chen F et al (2010) An anti-ROS/hepatic fibrosis drug delivery system based on salvianolic acid B loaded mesoporous silica nanoparticles. Biomaterials 31(30):7785–7796

Hemmann S, Graf J, Roderfeld M, Roeb E (2007) Expression of MMPs and TIMPs in liver fibrosis – a systematic review with special emphasis on anti-fibrotic strategies. J Hepatol 46(5):955–975

Higashi T, Friedman SL, Hoshida Y (2017) Hepatic stellate cells as key target in liver fibrosis. Adv Drug Deliv Rev 121:27–42

Hou J, Tian J, Jiang W, Gao Y, Fu F (2011) Therapeutic effects of SMND-309, a new metabolite of salvianolic acid B, on experimental liver fibrosis. Eur J Pharmacol 650(1):390–395

Hsu WH, Lee BH, Hsu YW, Pan TM (2013) Peroxisome proliferator-activated receptor-γ activators monascin and rosiglitazone attenuate carboxymethyllysine-induced fibrosis in hepatic stellate cells through regulating the oxidative stress pathway but independent of the receptor for advanced glycation end products signaling. J Agric Food Chem 61(28):6873–6879

Huang L, Xie J, Bi Q et al (2017) Highly selective targeting of hepatic stellate cells for liver fibrosis treatment using a d-Enantiomeric peptide ligand of Fn14 identified by Mirror-image mRNA display. Mol Pharm 14(5):1742–1753

Hussein J, El-Banna M, Mahmoud KF et al (2017) The therapeutic effect of nano-encapsulated and nano-emulsion forms of carvacrol on experimental liver fibrosis. Biomed Pharmacother 90:880–887

Ikenaga N, Peng ZW, Vaid KA et al (2017) Selective targeting of lysyl oxidase-like 2 (LOXL2) suppresses hepatic fibrosis progression and accelerates its reversal. Gut 66(9):1697–1708

Iwakiri Y (2015) Nitric oxide in liver fibrosis: the role of inducible nitric oxide synthase. Clin Mol Hepatol 21(4):319–325

Jamwal R (2018) Bioavailable curcumin formulations: a review of pharmacokinetic studies in healthy volunteers. J Integr Med 16(6):367–374. pii: S2095-4964(18)30077-3

Jia Z, Gong Y, Pi Y et al (2018) pPB peptide-mediated siRNA-loaded stable nucleic acid lipid nanoparticles on targeting therapy of hepatic fibrosis. Mol Pharm 15(1):53–62

Jiang H, Xia LZ, Li Y, Li X, Wu J (2013) Effect of Panax notoginseng saponins on expressions of MMP-13 and TIMP-1 in rats with hepatic fibrosis. Zhongguo Zhong Yao ZaZhi 38(8):1206–1210

Jiménez Calvente C, Sehgal A et al (2015) Specific hepatic delivery of procollagen α1(I) small interfering RNA in lipid-like nanoparticles resolves liver fibrosis. Hepatology 62(4):1285–1297

Kaps L, Nuhn L, Aslam M et al (2015) In vivo gene-silencing in fibrotic liver by siRNA-loaded cationic nanohydrogel particles. Adv Healthc Mater 4(18):2809–2815

Khaja F, Jayawardena D, Kuzmis A, Önyüksel H (2016) Targeted sterically stabilized phospholipid siRNANanomedicine for hepatic and renal fibrosis. Nanomaterials (Basel) 6(1):pii: E8

Kieslichova E, Frankova S, Protus M et al (2018) Acute liver failure due to Amanita phalloides poisoning: therapeutic approach and outcome. Transplant Proc 50(1):192–197

Kikuchi S, Griffin CT, Wang SS, Bissell DM (2005) Role of CD44 in epithelial wound repair: migration of rat hepatic stellate cells utilizes hyaluronic acid and CD44v6. J Biol Chem 280(15):15398–15404

Kisseleva T, Cong M, Paik Y et al (2012) Myofibroblast revert to an inactive phenotype during regression of liver fibrosis. Proc Natl Acad Sci U S A 109(24):9448–9453

Kong WH, Park K, Lee MY et al (2013) Cationic solid lipid nanoparticles derived from apolipoprotein-free LDLs for target specific systemic treatment of liver fibrosis. Biomaterials 34(2):542–551

Kumar N, Rai A, Reddy ND et al (2014a) Silymarin liposomes improves oral bioavailability of silybin besides targeting hepatocytes, and immune cells. Pharmacol Rep 66(5):788–798

Kumar V, Mundra V, Mahato RI (2014b) Nanomedicines of hedgehog inhibitor and PPAR-γ agonist for treating liver fibrosis. Pharm Res 31(5):1158–1169

Lee YA, Wallace MC, Friedman SL (2015) Pathobiology of liver fibrosis: a translational success story. Gut 64(5):830–841

Li Y, Mu M, Yuan L, Zeng B, Lin S (2018) Challenges in the early diagnosis of patients with acute liver failure induced by amatoxin poisoning: two case reports. Medicine (Baltimore) 97(27):e11288

Lin TT, Gao DY, Liu YC et al (2016) Development and characterization of sorafenib-loaded PLGA nanoparticles for the systemic treatment of liver fibrosis. J Control Release 221:62–70

Liu CH, Chan KM, Chiang T et al (2016) Dual-functional nanoparticles targeting CXCR4 and delivering AntiangiogenicsiRNA ameliorate liver fibrosis. Mol Pharm 13:2253–2262

Mehal WZ, Schuppan D (2015) Antifibrotic therapies in the liver. Semin Liver Dis 35(2):184–198

Meng Z, Meng L, Wang K et al (2015) Enhanced hepatic targeting, biodistribution and antifibrotic efficacy of tanshinone IIA loaded globin nanoparticles. Eur J Pharm Sci 73:35–43

Muñoz-Luque J, Ros J, Fernández-Varo G et al (2008) Regression of fibrosis after chronic stimulation of cannabinoid CB2 receptor in cirrhotic rats. J Pharmacol Exp Ther 324(2):475–483

Mussi SV, Torchilin VP (2013) Recent trends in the use of lipidic nanoparticles as pharmaceutical carriers for cancer therapy and diagnostics. J Mater Chem B 1(39):5201

Nakamura I, Zakharia K, Banini BA et al (2014) Brivanib attenuates hepatic fibrosis in vivo and stellate cell activation in vitro by inhibition of FGF, VEGF and PDGF signaling. PLoS One 9(4):e92273

Oró D, Yudina T, Fernández-Varo G et al (2016) Cerium oxide nanoparticles reduce steatosis, portal hypertension and display antiinflammatory properties in rats with liver fibrosis. J Hepatol 64(3):691–698

Pan TL, Wang PW, Hung CF et al (2016) The impact of retinol loading and surface charge on the hepatic delivery of lipid nanoparticles. Colloids Surf B Biointerfaces 141:584–594

Pinter M, Sieghart W, Reiberger T et al (2012) The effects of sorafenib on the portal hypertensive syndrome in patients with liver cirrhosis and hepatocellular carcinoma – a pilot study. Aliment Pharmacol Ther 35:83–91

Puche JE, Saiman Y, Friedman SL (2013) Hepatic stellate cells and liver fibrosis. Compr Physiol 3(4):1473–1492

Qiang G, Yang X, Xuan Q et al (2014) Salvianolic acid a prevents the pathological progression of hepatic fibrosis in high-fat diet-fed and streptozotocin-induced diabetic rats. Am J Chin Med 42(5):1183–1198

Ramachandran P, Pellicoro A, Vernon MA et al (2012) Differential Ly-6C expression identifies the recruited macrophage phenotype, which orchestrates the regression of murine liver fibrosis. Proc Natl Acad Sci U S A 109(46):E3186–E3195

Safer AM, Hanafy NA, Bharali DJ, Cui H, Mousa SA (2015a) Effect of green tea extract encapsulated into chitosan nanoparticles on hepatic fibrosis collagen fibers assessed by atomic force microscopy in rat hepatic fibrosis model. J Nanosci Nanotechnol 15(9):6452–6459

Safer AM, Sen A, Hanafy NA, Mousa SA (2015b) Quantification of the healing effect in hepatic fibrosis induced by chitosan nano-encapsulated green tea in rat model. J Nanosci Nanotechnol 15(12):9918–9924

Schiborr C, Kocher A, Behnam D et al (2014) The oral bioavailability of curcumin from micronized powder and liquid micelles is significantly increased in healthy humans and differs between sexes. Mol Nutr Food Res 58(3):516–527

Schuppan D (1990) Structure of the extracellular matrix in normal and fibrotic liver: collagens and glycoproteins. Semin Liver Dis 10(1):1–10

Schuppan D, Ashfaq-Khan M, Yang AT, Kim YO (2018) Liver fibrosis: direct antifibrotic agents and targeted therapies. Matrix Biol 68–69:435–451

Shangguan M, Lu Y, Qi J et al (2014) Binary lipids-based nanostructured lipid carriers for improved oral bioavailability of silymarin. J Biomater Appl 28(6):887–896

Sudha PN, Rose MH (2014) Beneficial effects of hyaluronic acid. Adv Food Nutr Res 72:137–176

Sung YC, Liu YC, Chao PH et al (2018) Combined delivery of sorafenib and a MEK inhibitor
    using CXCR4-targeted nanoparticles reduces hepatic fibrosis and prevents tumor development.
    Theranostics 8(4):894–905
Thomas RG, Moon MJ, Kim JH, Lee JH, Jeong YY (2015) Effectiveness of losartan-loaded
    Hyaluronic Acid (HA) micelles for the reduction of advanced hepatic fibrosis in C3H/HeN
    mice model. PLoS One 10(12):e0145512
Thomson J, Hargrove L, Kennedy L, Demieville J, Francis H (2017) Cellular crosstalk during
    cholestatic liver injury. Liver Res 1(1):26–33
Toriyabe N, Sakurai Y, Kato A et al (2017) The delivery of small interfering RNA to hepatic
    stellate cells using a lipid nanoparticle composed of a vitamin A-scaffold lipid-like material.
    J Pharm Sci 106(8):2046–2052
Tsai MK, Lin YL, Huang YT (2010) Effects of salvianolic acids on oxidative stress and hepatic
    fibrosis in rats. Toxicol Appl Pharmacol 242(2):155–164
Tsuchida T, Friedman SL (2017) Mechanisms of hepatic stellate cell activation. Nat Rev
    Gastroenterol Hepatol 14(7):397–411
Wallace MC, Friedman SL, Mann DA (2015) Emerging and disease-specific mechanisms of
    hepatic stellate cell activation. Semin Liver Dis 35(2):107–118
Wang Y, Gao J, Zhang D et al (2010) New insights into the antifibrotic effects of sorafenib on
    hepatic stellate cells and liver fibrosis. J Hepatol 53:132–144
Wang C, Song X, Li Y et al (2013) Low-dose paclitaxel ameliorates pulmonary fibrosis by sup-
    pressing TGF-β1/Smad3 pathway via miR-140 upregulation. PLoS One 8(8):e70725
Wang J, Pan W, Wang Y et al (2018) Enhanced efficacy of curcumin with phosphatidylserine-
    decorated nanoparticles in the treatment of hepatic fibrosis. Drug Deliv 25(1):1–11
Wilhelm A, Shepherd EL, Amatucci A et al (2016) Interaction of TWEAK with Fn14 leads to the
    progression of fibrotic liver disease by directly modulating hepatic stellate cell proliferation.
    J Pathol 239(1):109–121
Yang JJ, Tao H, Li J (2014) Hedgehog signaling pathway as key player in liver fibrosis: new
    insights and perspectives. Expert Opin Ther Targets 18(9):1011–1021
Yoon YJ, Friedman SL, Lee YA (2016) Antifibrotic therapies: where are we now? Semin Liver Dis
    36(1):87–98
Younis N, Shaheen MA, Abdallah MH (2016) Silymarin-loaded Eudragit(®) RS100 nanoparticles
    improved the ability of silymarin to resolve hepatic fibrosis in bile duct ligated rats. Biomed
    Pharmacother 81:93–103
Yu Y, Duan J, Li Y et al (2017) Silica nanoparticles induce liver fibrosis via TGF-β1/Smad3 path-
    way in ICR mice. Int J Nanomedicine 12:6045–6057
Zhang Z, Wang C, Zha Y et al (2015) Corona-directed nucleic acid delivery into hepatic stellate
    cells for liver fibrosis therapy. ACS Nano 9(3):2405–2419
Zhao Y, Dang Z, Xu S, Chong S (2017) Heat shock protein 47 effects on hepatic stellate cell-
    associated receptors in hepatic fibrosis of Schistosomajaponicum-infected mice. Biol Chem
    398(12):1357–1366
Zhou J, Zhong DW, Wang QW, Miao XY, Xu XD (2010) Paclitaxel ameliorates fibrosis in hepatic
    stellate cells via inhibition of TGF-beta/Smad activity. World J Gastroenterol 16(26):3330–3334

Biomedical Applications of Zinc Oxide Nanoparticles Synthesized Using Eco-friendly Method

3

S. Rajeshkumar and D. Sandhiya

Abstract

Nanotechnology is an emerging area of research and plays a vital role in various fields of application. Consequently, it mainly focused on synthesis of nanoparticles using novel approaches. Among this, synthesis of zinc oxide using biological method plays a unique role in research, such as cost-effective and environment-friendly method. In this review paper, we mainly focused on synthesis of zinc oxide nanoparticles using biological methods such as plant-mediated, bacterial-mediated, fungal-mediated, and algal-mediated method. These biological materials are enriched with biomolecules, and they play a major role in reduction of metals. Based on this, bioreduction capacity of various biological materials used to synthesize zinc oxide nanoparticles under different conditions is also provided in this review. Various instrumental techniques such as Fourier transform infrared (FT-IR), scanning electron microscopy (SEM), and X-ray diffraction (XRD) are used to characterize the size and functional group present in the nanoparticles, and some other biological techniques are also used to identify the effectiveness of novel-mediated zinc oxide nanoparticles. Finally, this review provides enough detail about the biological-mediated zinc oxide nanoparticles and its functional groups, and biological application; it helps researcher to identify previous results of the study and helps to pave new way for research.

Keywords

Green synthesis · zinc oxide nanoparticles · Medicine · Antimicrobial · Characterization

S. Rajeshkumar (✉) · D. Sandhiya
Nanobiomedicine Lab, Department of Pharmacology, Saveetha Dental College, Saveetha Institute of Medical and Technical Sciences (SIMATS), Chennai, Tamil Nadu, India

3.1   Introduction

Nanotechnology In rising generation, nanotechnology can lead to a drastic change in every field of science. This nanotechnology was used with combination of different fields such as electronics, optics, biomedical, and material science. Therefore, researchers are finding a new way to introduce this innovative method (Rico et al. 2011; Sabir et al. 2014).

Nanotechnology is a field which deals with nanoparticles, ranging from 1 to 100 nm in size. The atomic and molecular aggregates of the nanoparticles were characterized by this size. We can modify the atomic and molecular properties using base elements. Nanoparticles have numerous advantages because of their beneficial properties (Sabir et al. 2014; Daniel and Astruc 2004; Kato 2011; Cauerhff and Castro 2013).

Role of Zinc Oxide Nanoparticles Zinc oxide is an inorganic compound. It appears as white powder and doesn't dissolve in water. It has wurtzite (B4) crystal structure. The powder form of ZnO which was used to make different products includes glass, ceramics, cement, rubber, paints, ointments, lubricants, adhesives, plastics, sealants, pigments, foods which are rich in Zn nutrients, batteries, ferrites, and fire retardants (Sabir et al. 2014; Wang et al. 2004; Do Kim et al. 2007; Hamminga et al. 2004).

Zinc oxide with less particle size, especially like ultrafine ZnO, has a very good application like paints, cement, etc., (Do Kim et al. 2007; Hamminga et al. 2004), since it has been used in the preparation of gas sensor, solar cells, and chemical absorbent (Do Kim et al. 2007; Wang et al. 2001; Lin et al. 1998; Turton et al. 2004). Zinc oxide is an n-type semiconductor, and it has wide band gap of 3.37 eV with large exciton energy of ~60 meV. It is considered as an active photocatalyst, because of its environmental sustainability and cost-effectiveness (Yan et al. 2017; Kayaci et al. 2014). In many cases, we obtain a large crystal of ZnO with less specific area owing to its low crystallization temperature and fast growth rate. Due to this reason, it does not produce photocatalyst activity (Manna et al. 2015).

3.2   Different Methods Used to Synthesize Nanoparticles

Different methods were used to synthesize nanoparticle such as physical, chemical, and biological methods, and hybrid techniques (Patra and Baek 2014; Mohanpuria et al. 2008; Tiwari et al. 2008; Luechinger et al. 2010). The production of physical and chemical methods releases some toxic by-products, which are very hazardous in environment. Especially, in clinical field, chemical particles cause health-related issues (Parashar et al. 2009a, b). Some of the physical

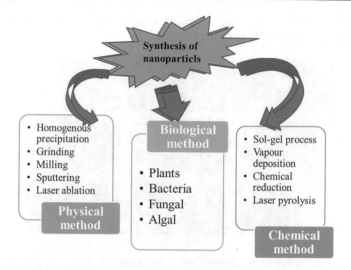

Fig. 3.1 The image of different methods to synthesize nanoparticles is shown

and chemical methods are homogeneous precipitation, hydrothermal synthesis, sol-gel process, electrodeposition, and so on (Do Kim et al. 2007; Kim et al. 2003, 2006; Li et al. 2001; Kim and Kim 2002; O'Regan and Schwartz 2000). During the last few decades, some of the researchers showed their interest toward nanotechnology and nanoscience such as biological synthesis of metal oxide nanoparticles (Li et al. 2011). This method helps to reduce the toxicity of metal during biological reduction, as shown inTherefore, figure below (Fig. 3.1).

3.3   Green Synthesis of Nanoparticles

In the recent years, scientists mainly focused to synthesize the nanoparticle which doesn't produce toxic products in the manufacturing process (Daniel and Astruc 2004; Patra and Baek 2014; Joerger et al. 2000). This process can be obtained with the help of biological process. The biological process using biotechnological tools should be safe to use for nanofabrication. Therefore, this type of process is called green technology or green nanotechnology. The green synthesis of nanoparticles or nanomaterials using green routes or biological routes involves microorganisms, plants, algae, and viruses (biproducts of proteins and lipids), which was produced by the help of techniques in biotechnology. The advantages of green synthesis of nanoparticles are that it consumes inexpensive chemicals, less energy, and eco-friendly by-products (Patra and Baek 2014; Joerger et al. 2000; Narayanan and Sakthivel 2011).

Bottom-up approach method is used to synthesize biological-based synthesis of nanoparticles. This process is involved in reducing agents or stabilizing agents. Nowadays, researchers have started introducing more biological compounds to synthesize bionanomaterial, which is mainly used in medical field (Patra and Baek 2014; Mohanpuria et al. 2008; Ahmed et al. 2017).

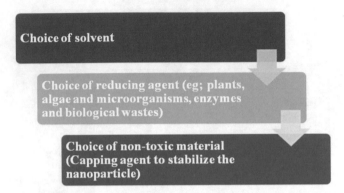

Steps for the synthesis of nanoparticles using biological method

3.4   Biosynthesis of Zinc Oxide Nanoparticles

Among different metal oxide nanoparticles, zinc oxides are the most widely used nanoparticles because of their huge applications such as optics, piezoelectric, magnetic, and gas-sensing process. They employ various plant extracts, algal extract microorganisms, and other biological by-products (Gunalan et al. 2012a; Suresh et al. 2015a, b, c, d; Udayabhanu et al. 2015; Raja Naika et al. 2015; Nethravathi et al. 2015; Pavan Kumar et al. 2015). From recent years, research shows their huge interest toward novel method like green synthesis using plant extracts, plant latex, and fruit juices as a fuel to produce zinc oxide nanoparticles (Pavithra et al. 2017; Aruna and Mukasyan 2008). Green synthesis method helps to reduce or eliminate the toxic and hazardous products, and also very effective in synthesis of NPs (Nava et al. 2017; Thakkar et al. 2010; Elumalai et al. 2015a; Iravani 2011). Some of the green synthesized ZnO NPs are listed or tabulated in this review article such as plant-mediated synthesis of ZnO NPs, bacterial-mediated synthesis of ZnO NPs, algal synthesis of ZnO NPs, and fungal-mediated synthesis of ZnO NPs.

Among different metal oxide nanoparticles, most of the researchers show their interest toward ZnO NPs due to their wide properties such as antimicrobial, catalytic, and optical properties. The zinc oxide nanoparticles act as reducing agents using biological method (Ahmed et al. 2017; Madhumitha et al. 2016). The biological compounds are made up of phytochemical constituents, such as phenolic, carbonyl, amine groups, proteins, pigments, flavonoids, terpenoids, alkaloids, and also other reducing agents, which is responsible for reduction process (Vijayaraghavan and Ashokkumar 2017; Asmathunisha and Kathiresan 2013). Therefore, this present review is considered about biological synthesis of zinc oxide nanoparticles such as plants, bacterial, fungal, and algal. It may help researcher in future to go through the whole biological synthesis of zinc oxide nanoparticles in this chapter (Fig. 3.2).

Fig. 3.2 The images of biosynthesis of ZnO NPs are shown

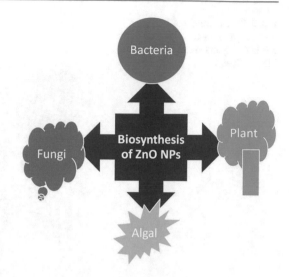

3.4.1   Plant-Mediated Synthesis of ZnO Nanoparticles and Its Biological Application

This review, briefly explains the advantage by synthesis of plant-mediated ZnO NP that acts as a reducing or capping agent (Gunalan et al. 2012a; Shamaila et al. 2016; Matinisea et al. 2017; Ramesha et al. 2015) which is reliable method (Shamaila et al. 2016), no additional chemical (Shamaila et al. 2016), cost effective (Gunalan et al. 2012a; Suresh et al. 2015a; Ramesha et al. 2015; Santhoshkumar et al. 2017; Dobrucka and Dugaszewska 2016), simple (Suresh et al. 2015a; Santhoshkumar et al. 2017; Dobrucka and Dugaszewska 2016), environmental friendly (Nava et al. 2017; Gandhi et al. 2017; Vijayakumar et al. 2016a; Elumalai and Velmurugan 2015), biocompatible reagents (Sundrarajan et al. 2015), and also represented in Fig. 3.3. Additionally, we successfully tabulated the green synthesis of ZnO NPs by employing various plant extracts in Table 3.1.

Here, we tabulated the synthesis of ZnO NPs using plant extracts of *Moringa oleifera* (Matinisea et al. 2017), *Solanum nigrum* (Ramesha et al. 2015), *Aloe vera* (Gunalan et al. 2012a), *Passifloraceae caerulea* L. (Santhoshkumar et al. 2017), *Buchanania lanzan (leaves)* (Suresh et al. 2015a), *Trifolium pretense* (Dobrucka and Dugaszewska 2016), *Momordica charantia Linn* (Gandhi et al. 2017), *Camellia sinensis* (Nava et al. 2017), *Laurus nobilis* (Vijayakumar et al. 2016a), *Azadirachta indica* (Elumalai and Velmurugan 2015), *Pongamia pinnata* (Sundrarajan et al. 2015), *Eucalyptus globulus* (Siripireddy and Mandal 2017), *Acalypha indica* (Karthik et al. 2017), *Suaeda aegyptiaca* (Rajabia et al. 2017), *Nyctanthus* (Jamdagni et al. 2016), *Cassia fistula* (Suresh et al. 2015d), *Ginger* (Chinnammal Janaki and Sailatha 2015), *P. niruri* (Anbuvannana et al. 2015), *Glycosmispentaphyll* (Vijayakumara

Fig. 3.3 The image of
advantage of plant-
mediated synthesis of ZnO
NPs is shown

et al. 2018), *V. trifolia* (Elumalai et al. 2015b), *Aloe vera* (Gunalan et al. 2012b), *Adhatoda vasica* (Sonia et al. 2017), *Cochlospermum religiosum* (Mahendra et al. 2017), *P. hysterophorus* L. (Rajiv et al. 2013), *P. niruri* (Anbuvannan et al. 2015), *Calotropis procera* (Salem et al. 2015), *Green tea* (Dhanemozhi et al. 2017), *Gossypium hirsutum* L. (Venkatachalam et al. 2017), *Punica granatum* (Kaviya et al. 2017), *Date seed* (El-Naggar et al. 2018), and *Carica papaya latex* (Chandrasekaran et al. 2016). The size of the zinc oxide NPs was analyzed by scanning electron microscopy, transmission electron microscopy. Mostly, the size of zinc oxide nanoparticles is 2–200 nm in range. The possible biomolecules present in the plant extracts is responsible for bioreduction of ZnO NPs. It comes under the bond vibration peaks, which is identified from the wave number of FT-IR techniques (Ramesha et al. 2015). Probably, the nanoparticles are surrounded by proteins, amino acids, terpenoids, alkaloids, and flavonoids of organic compound, which is detected by using FT-IR spectroscopy with its range. In this review, mostly ZnO NPs are consist of aromatic stretching of various bioactive compounds, N=C=S stretching vibration and C=O, C=H, C=N, NH aromatic stretching vibration, O-H stretching, aldehydic C-H stretching, symmetric stretching of the carboxyl side groups, C-N stretching vibration of amine groups, stretching vibration, hydroxyl group, -C-O, -C-O-C stretching mode, C=C stretching mode, etc. (Matinisea et al. 2017; Ramesha et al. 2015; Santhoshkumar et al. 2017; Dobrucka and Dugaszewska 2016).

The plant-mediated ZnO NPs possess more biological activities like antibacterial, antifungal, antidiabetic (Gunalan et al. 2012a), anticancer activity (Suresh et al. 2015a), acaricidal activity (Dobrucka and Dugaszewska 2016), pediculicidal activity (Dobrucka and Dugaszewska 2016), and larvicidal activity (Dobrucka and

Table 3.1 Different characteristics of ZnO NPs

S.no	Plants (scientific name)	Common name	Parts used for synthesis	Size (nm)	Structure of nanoparticles	Functional group present in these nanoparticles	Biological activities	Remarks	References
1	*Moringa oleifera*	Moringa	Leaves	6–10 nm (HRTEM)	Anisctropy	Aromatic stretching of various bioactive compounds, N=C=S stretching vibration and C=O, C=H, C=N, NH aromatic stretching vibration	–	–	Matinisea et al. (2017)
2	*Solanum nigrum*	Black nightshade, black-berry	Leaves	20–30 nm (FE-SEM), 29.79 nm (TEM)	Quasi-spherical	O-H stretching, aldehydic C-H stretching, symmetric stretching of the carboxyl side groups, C-N stretching vibration of amine groups.	Antibacterial activity	XPS-one symmetrical peak and two strong peaks PL analysis shows five emission properties	Ramesha et al. (2015)

(continued)

Table 3.1 (continued)

S.no	Plants (scientific name)	Common name	Parts used for synthesis	Size (nm)	Structure of nanoparticles	Functional group present in these nanoparticles	Biological activities	Remarks	References
3	*Aloe vera*	Indian Aloe	Leaf	–	–		Antibacterial and antifungal activity	–	Gunalan et al. (2012a)
4	*Passifloraceae caerulea* L.	Blue crown passion flower	Leaf	70 nm (SEM)	Spherical	O-H stretching vibration, C=C stretch in aromatic ring and C=O stretch in polyphenols, C-N stretch of amide-I in proteins, C-O stretch in amino group, C-N stretching, C-H bending, C- alkyl chloride, hexagonal phase ZnO	Antimicrobial activity	–	Santhoshkumar et al. (2017)
5	*Buchanania lanzan* (*leaves*)	Almondette tree	Leaves	100 nm (TEM)	Agglomerated or foam (like bunch of particles)	–	Antibacterial activity and antioxidant activity	–	Suresh et al. (2015a)

	Plant	Common name	Part	Size	Shape	FTIR groups	Activity	Characterization	References
6	*Trifolium pratense*	Clover	Flower	100–190 nm (SEM)	Agglomerated	Stretching vibration, hydroxyl group, -C-O, -C-O-C stretching mode and C=C stretching mode	Antimicrobial activity	TXRF strongly suggest than ZnO nanoparticles is the major elements in *Trifolium pratense*	Dobrucka and Dugaszewska (2016)
7	*Momordica charantia Linn*	English African cucumber	Leaf	–	Spherical shape	OH, vinyl ethers, aldehydes, beta lactones, and aliphatic amines groups	Acaricidal activity, pediculicidal activity and larvicidal activity	GC-MS Z-10-methyl-11-tetradecen-1-ol propionate, 1-heptadecyne, N-hexadecanoic acid, heptadecane, 2,6,10,15-tetramethyl-, heptacosane, octacosane, nonacosane, 2-hexanone, 3-cyclohexylidene-4-ethyl- compounds identified	P. Rajiv Gandhi et al. (2017)
8	*Camellia sinensis*	Tea plant	Leaves	–	–	C=C in aromatic rings, C=O stretch in polyphenols, C-C stretch in aromatic ring.	–	Catalytic activity shows 37% degradation in 120 min (good improvement in dye degradation)	Nava et al. (2017)

(continued)

Table 3.1 (continued)

S.no	Plants (scientific name)	Common name	Parts used for synthesis	Size (nm)	Structure of nanoparticles	Functional group present in these nanoparticles	Biological activities	Remarks	References
9	*Laurus nobilis*	**Bay laurel**, Sweet bay	Leaves	20–50 nm (SEM)	Agglomerated (flower like structure)	OH stretching vibration, C-C stretching of aromatic groups, C-H bend in alkane group,	Antibacterial activity, anticancer activity	Antibiofilm assay- *Laurus nobilis* ZnO NPs inhibited the biofilm formation	Vijayakumar et al. (2016a)
10	*Azadirachta indica*	Neem	Leaves	–	Spherical shape	O-H and C-H stretching of polyols, C-C stretching vibration of aromatic rings, O-H and C-OH vibration of polyols, C-N, and N-H of amines, C=O stretching vibration of carboxylic acid	Antibacterial activity, antifungal activity, MBC, MIC, MFC	–	Elumalai and Velmurugan (2015)

11	*Pongamia pinnata*	Indian beach	Leaves	100 nm (SEM)	Spherical shape	C=O bond of nonionic carboxylic group, C-O-H bending mode, hexagonal phase Zn-O vibration	Antibacterial activity	DLS measures size distribution of prepared ZnO NPs	Sundrarajan et al. (2015)
12	*Eucalyptus globulus*	Eucalyptus	Leaves	~11.6 nm (PSA),	Spherical and hexagonal	C-OH stretch of tertiary alkyl group, O-H stretching in polyphenols, C-O stretching vibration of carboxylic acids, C-Cl stretching of alkyl halide	Antioxidant activity	Particle size analyzer shows average distribution of particle size, Raman spectroscopy study shows structural determinants, and disorder in nanostructure material	Siripireddy and Mandal (2017)
13	*Acalypha indica*	Indian Copperleaf	Leaves	–	–	–	Antibacterial activity		Karthik et al. (2017)

(continued)

Table 3.1 (continued)

S.no	Plants (scientific name)	Common name	Parts used for synthesis	Size (nm)	Structure of nanoparticles	Functional group present in these nanoparticles	Biological activities	Remarks	References
14	*Suaeda aegyptiaca*	Lower Jordan Valley	(Leaves)	<80 nm	Spherical	Alcohol, amine and aldehyde	Antibacterial activity, minimal inhibitory concentration, minimal bactericidal concentration, antifungal activity, and antioxidant activity	Determination of total phenolic and flavonoid compound using Folin-Ciocalteau reagent, DNA interaction study was done by using gel electrophoresis, antioxidant activity was done using DPPH analysis	Rajabia et al. (2017)
15	*Nyctanthus*	Coral Jasmine	Flower	12–32 nm (TEM)	Agglomerates	-NH vibration stretch of protein amide linkage, C-N stretch of aliphatic amines, C-H bond alkynes	Antifungal activity, minimal inhibitory concentration	Dynamic light scattering used to measure the size of nanoparticle	Jamdagni et al. (2016)

			Part	Size	Shape		Activity		Reference
16	*Cassia fistula*	Indian laburnum	Leaves	–	–	–	Antibacterial activity, antioxidant activity, flavonoid assay, polyphenol assay	–	Suresh et al. (2015d)
17	*Zingiber officinale*	Ginger	Plant	23–25 nm (SEM)	Spherical	Alkaloids 6-gingerol, 6-Shogal,a – Zingeberene	Antibacterial activity, antifungal activity	–	Chinnammal Janaki and Sailatha (2015)
18	*P. niruri*	Stonebreaker	Leaves	–						Anbuvannana et al. (2015)
19	*Glycosmispentaphyll*	Gin Berry,	Leaves	32–40 nm (SEM)	Spherical	—OH stretching vibration, alkane –CH stretching vibration, C-H stretching vibration, C=O stretching bond, C-H bending, C=C bending and C-O stretching	Antibacterial activity and antifungal activity		Vijayakumara et al. (2018)

(continued)

Table 3.1 (continued)

S.no	Plants (scientific name)	Common name	Parts used for synthesis	Size (nm)	Structure of nanoparticles	Functional group present in these nanoparticles	Biological activities	Remarks	References
20	*V. trifolia*	Hand of Mary	Leaves	15–46 nm (SEM)	Spherical and less aggregated	O-H stretching in alcohol and phenol group, C-H stretching in alkanes, C=C and N-H bend 10 amines, C-C and O-H stretching in aromatic group, C=O and C-H stretching alkane, C-O in alcohol, ester, C-N aliphatic and C-O in aromatic groups	Antibacterial activity, antifungal activity, minimal inhibitory concentration	Photocatalytic activity and GC-MS spectroscopy	Elumalai et al. (2015b)

21	*Aloe vera*	Indian Aloe	Leaves	–	–	–	Antibacterial activity, antifungal activity, minimal inhibitory concentration, minimal bactericidal concentration and minimal fungicidal concentration	–	Gunalan et al. (2012b)
22	*Adhatoda vasica*	Adusa	Leaves	10–12 nm (TEM)	Spherical and hexagonal	C=O group, C-O stretching, C-N stretching, C-Cl stretching vibration	Antibacterial activity, antifungal activity, antioxidant activity, minimal inhibitory concentration, minimal bactericidal concentration, minimal fungicidal concentration	–	Sonia et al. (2017)

(continued)

Table 3.1 (continued)

S.no	Plants (scientific name)	Common name	Parts used for synthesis	Size (nm)	Structure of nanoparticles	Functional group present in these nanoparticles	Biological activities	Remarks	References
23	*Cochlospermum religiosum*	Buttercup	Leaves		Hexagonal	C-H stretch of alkanes, C=C stretch of aromatic ring, carbonyl group, aromatic C-C, C-H alkanes	Antibacterial activity, minimal inhibitory concentration	Dynamic light scattering, live and dead cell analysis, antimitotic activity	Mahendra et al. (2017)
24	*P. hysterophorus L*	Santa-Maria	Leaves	27–84 nm (TEM)	Spherical, hexagonal and some are agglomerated	N-H bending, NH stretching vibration, M-O vibration band	Antifungal activity		Rajiv et al. (2013)
25	*P. niruri*	Stonebreaker	Leaves	25.61 nm	Rectangle, triangle, radial hexagonal, spherical and rod shape	C-H stretching vibration, C-O and C-H vibration mode of starch, O-H group	–	Photoluminescence property	Anbuvannan et al. (2015)
26	*Caltropis procera*	Rubber bush, apple of Sodom	Leaves and fruits	–	–	–	Antimicrobial activity, MIC, growth curve	Inductively coupled plasma mass spectrometry (ICP-MS), static biofilm assay, SDS-PAGE immunoblot analysis	Salem et al. (2015)

No.	Scientific name	Common name	Plant part	Characterization	Morphology	FTIR	Activity	Method	References
27	*Camellia sinensis*	Green tea	Leaf	–	–	Stretching vibration in O-H groups of water, N-H stretching amines, C=C stretching in aromatic ring, C=O stretch in polyphenol, C-N stretch of amide-I in protein, C-O stretch in amino acid	–	Electrochemical impedance spectroscopy (ESI), cyclic voltammetry (CV)	Dhanemozhi et al. (2017)
28	*Gossypium hirsutum L.*	Upland cotton	Roots	–	–	–	Antioxidant enzyme activity	Superoxide dismutase enzyme activity, detection of melanodialdehyde contents, catalase, and peroxidase enzyme activity	Venkatachalam et al. (2017)
29	*Punica granatum*	Pomegranate	Peel	SEM (20 nm)	Spherical	–	Antibacterial activity	–	Kaviya et al. (2017)
30	*Phoenix dactylifera*	Date seed	Seed	–	Agglomerated and spherical	–	Antibacterial activity	Cytotoxicity assay	El-Naggar et al. (2018)
31	*Carica papaya latex*	Papaya	Latex	–	Agglomerated and spherical	N=H stretching vibration of amine group NH$_2$, aromatic C-C bending of proteins	Antibacterial activity	–	Chandrasekaran et al. (2016)

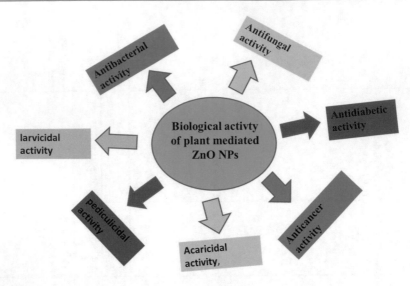

Fig. 3.4 Biomedical applications of plant mediated ZnONPs

Dugaszewska 2016), as shown in Fig. 3.4. Other activities were also done to confirm the effectiveness of ZnO NPs such as minimal inhibitory concentration, minimal bactericidal concentration, minimal antifungicidal concentration, and antioxidant activity. Table 3.1 shows that the other activities of ZnO NPs like catalytic activity and dynamic light scattering are used to analyze the particle size of the nanoparticles, photocatalytic activity, gas chromatography (GC)–mass spectrometry (MS), live and dead cell analysis were used to distinguish the cells, antimitotic activity was used to identify chromosomal abnormality, inductively coupled plasma mass spectrometry (ICP-MS), static biofilm assay, sodium dodecyl sulfate (SDS)-polyacrylamide gel electrophoresis (PAGE) immunoblot analysis, electrochemical impedance spectroscopy (ESI), cyclic voltammetry (CV) are used to enhance specific capacitance and electrochemical stability of metal oxide (Elumalai et al. 2015b; Mahendra et al. 2017; Salem et al. 2015).

3.4.2 Bacterial-Mediated Synthesis of ZnO Nanoparticles and Its Biological Application

Nowadays, microbial polysaccharides and their derivatives are used to produce ZnO NPs, due to their chemical constituents and their complex structure. Additionally, microbial polysaccharide has higher reducibility agent (Vijayakumar et al. 2016b; Thaya et al. 2016; Ismail and Nampoothiri 2010; Abinaya et al. 2018). Therefore, researchers are focusing on the synthesis of microbial exopolysaccharide (EPS), due to their nontoxic, biocompatible, easily biodegradable nature, abundant presence in natural sources, and also their use as a natural polymer to stabilize and reduce the toxicity of the nanoparticles (Vijayakumar et al. 2016c; Suganya et al.

2017). Table 3.2 shows the bacterial-mediated ZnO NPs. Some of the microbes such as *B. licheniformis* (Abinaya et al. 2018), *B. megaterium* (Saravanan et al. 2018), *Serratia ureilytica* (Dhandapani et al. 2014), *B. cereus* (Hussein et al. 2009), and *R. pyridinivorans* (Kundu et al. 2014) were used for the synthesis of ZnO NPs. It also possesses some biological activity such as antibacterial, antifungal, and antioxidant activities. The size of bacterial-mediated ZnO NPs is in the range 20–120 nm.

Table 3.2 shows the bacterial-mediated synthesis of ZnO nanoparticles.

3.4.3   Fungal-Mediated Synthesis of ZnO Nanoparticles

Table 3.3 shows that the biosynthesis of fungal-mediated ZnO nanoparticles is used to synthesize steroidal pyrazolines (Shamsuzzaman et al. 2017), suitable for plant growth (Raliya and Tarafdar 2013), which is cost-effective, nontoxic, and eco-friendly.

Table 3.3 shows the fungal mediated synthesis of ZnO nanoparticles.

3.4.4   Algal-Mediated Synthesis of ZnO Nanoparticles

Nanotechnology and nanoscience are a great promise of marine environment. Biosynthesis of ZnO NPs is nontoxic, biosafe, and biocompatible (Rosi and Mirkin 2005; Nagarajan and Arumugam Kuppusamy 2013). Seaweeds are widely present in the marine environment and also used for effective synthesis of nanoparticles. This marine macroalgae consists of protein, sugar, amino acids, and fat, which are used for the development of novel drugs (Manilal et al. 2011; Molinski et al. 2009; Murugan et al. 2017). In this review, we tabulated algal-mediated synthesis of ZnO nanoparticles, as shown in Table 3.4. In this, *Ulva lactuca* contain sugar ring vibration and possess biological activity, which is tested against thermal burns (Dumbrava et al. 2018); *Padina tetrastromatica* contains C-H wagging, aging sulfate groups, which is tested against BET analysis (Pandimurugan and Thambidurai 2017); *Sargassum muticum* and *Sargassum wightii* possess Larvicidal and pupicidal activity, which is used for treating Scrofula, goiter, tumor, edema, testicular pain, and swelling (Azizi et al. 2014; Kandale et al. 2011). *Caulerpa peltata, red Hypnea Valencia*, and *brown Sargassum myriocystum* contain weak bands and hydroxyl group; it is effective against antimicrobial activity (Nagarajan and Arumugam Kuppusamy 2013).

3.5   Conclusion

Therefore, the main objective of the review has been focused on the synthesis of zinc oxide nanoparticles using biological methods such as plant mediated, bacterial mediated, fungal mediated, and algal mediated, as was tabulated. These biological materials are enriched with biomolecules, which plays a major role in reduction of

Table 3.2 Bacterial mediated ZnO NPs

S.no.	Bacteria	Family	Size (nm)	Structure of nanoparticles	Functional group present in these nanoparticles	Biological activities	Remarks	References
1	B. lichenformis	Bacillaceae	SEM (45–95 nm)	Rod and cubic	C-H and O-H stretch, C=C stretch, C-OH stretch, C-O stretch, C-C stretch, C=O stretch, C-N stretch	Antibacterial, antifungal, antioxidant activity and minimal inhibitory concentration	Mosquito larvicidal and antibiofilm activity	Abinaya et al. (2018)
2	B. megaterium	Bacillaceae	–	Agglomerated and hexagonal	O-H hydroxyl group, C-O carbonyl, and –N-H stretching vibration, C-H stretching vibration, COO- carboxyl group, O-H, CH, and C-N stretching vibration of aromatic, aliphatic amino group	Antibacterial activity, Minimal inhibitory concentration	Alamar blue and LDH assay, detection of ROS	Vijayakumar et al. (2016c)
3	Serratia ureilytica	Enterobacteriaceae	–	Hexagonal	–	Antibacterial activity	–	Suganya et al. (2017)
4	B. cereus	Bacillaceae	20–30 nm	Raspberry and plate-like structure	–	–	–	Saravanan et al. (2018)
5	R. pyridinivorans	Nocardiaceae	100–120 nm	Quasi-spherical and hexagonal	O-H groups in alkanes, H-H groups in alkanes, alkenes, strong alkynes, alcohol, acetate, ethers, carboxyl compounds, lactanes, lactoms, nitro compounds	Antibacterial activity	Cytotoxicity assay, photocatalytic activity	Dhandapani et al. (2014)

Table 3.3 The fungal mediated synthesis of ZnO nanoparticles

S.no	Fungal name	Family	Size (nm)	Structure of nanoparticle	Functional group present in these nanoparticles	Antibacterial activity	Remarks	References
1	*C. albicans*	Saccharomycetaceae	SEM (15 and 25 nm)	Quasi-spherical	–	–	Photoluminescence	Shamsuzzaman et al. (2017)
2	*Aspergillus fumigatus*	Trichocomaceae	SEM (1.2–6.8 nm)	Spherical	–	–	Dynamic light scattering (DLS)	Raliya and Tarafdar (2013)

Table 3.4 Algae-mediated ZnO NPs

S.no.	Algae	Family	Size (nm)	Structure of nanoparticles	Functional group present in these nanoparticles	Biological activities	Remarks	References
1	*Ulva lactuca*	Ulvaceae	XRD (39.31 nm and 58.78 nm)	Rod-like structure, cubic, spherical	Sugar ring vibration, sulfate esters	Antioxidant activity	Tested against thermal burns	Dumbrava et al. (2018)
2	*Padina tetrastromatica*	Dictyotaceae	XRD (28 and 24 nm)	Hexagonal)	O-H, C-H, C-O, C=O stretching, C-H wagging, aging sulfate groups, O-H vibration	Antibacterial activity	BET analysis	Pandimurugan and Thambidurai (2017)
3	*Sargassum muticum*	Sargassaceae	SEM (30–57 nm)	Agglomerated with hexagonal	C-O hydroxyl group, C=O amine group,	–	–	Azizi et al. (2014)
4	*Sargassum wightii*	Sargassaceae	SEM (20–62 nm)	Spherical	C=O stretching vibration of diisobutyl phthalate, hexahydrofarnesyl acetone, tannin, flavonoid, C=O of amide group	–	Larvicidal and pupicidal activity	Murugan et al. (2017)
5	*Padina tetrastromatica*	Dictyotaceae	–	Spherical, rectangular	O-H and C-H stretching vibration, O-H-O carboxylate group, C-O and C-C stretching vibration of pyranose ring, O=S=O deformation of sulfates	Antimicrobial analysis	BET analysis	Pandimurugan and Thambidurai (2016)

| 6 | Ulva lactuca | Ulvaceae | TEM (10–50 nm) | Hexagonal, triangle, rod and rectangle | H bonded in alcohol and phenol groups, C-C stretching in aromatic rings, O-H stretching | Antibacterial analysis | Photocatalytic assay, biofilm efficacy, larvicidal activity, histopathological and microscopic analysis of larvicidal activity | Aishwarya et al. (2018) |
| 7 | Caulerpa peltata, red Hypnea valencia, and brown Sargassum myriocystum | | SEM (96–120 nm) | Spherical, rod, rectangle, triangle, radial hexagonal | Weak bands and hydroxyl group | Antimicrobial activity | Reactive oxygen species | Nagarajan and Arumugam Kuppusamy (2013) |

metals. Based on this, bioreduction capacity of various biological materials is used to synthesize zinc oxide nanoparticles under different condition. The functional group present in the nanoparticles and some other biological techniques are also been used to identify the effectiveness of novel-mediated zinc oxide nanoparticles.

References

Abinaya M, Vaseeharan B, Divya M, Sharmili A, Govindarajan M, Alharbi NS, Kadaikunnan S, Khaled JM, Benelli G (2018) Technical note bacterial exopolysaccharide (EPS)-coated ZnO nanoparticles showed high antibiofilm activity and larvicidal toxicity against malaria and Zika virus vectors. J Trace Elem Med Biol 45:93–103

Ahmed S, Annu, Chaudhry SA, Ikram S (2017) Biogenic synthesis of ZnO nanoparticles using plant extracts and microbes: a prospect towards green chemistry. J Photochem Photobiol B: Biology 166:272–284

Anbuvannan M, Ramesh M, Viruthagiri G, Shanmugam N, Kannadasan N (2015) Synthesis, characterization and photocatalytic activity of ZnO nanoparticles prepared by biological method. Spectrochim Acta A Mol Biomol Spectrosc 143:304–308

Anbuvannana M, Ramesh M, Viruthagiri G, Shanmugam N, Kannadasan N (2015) Synthesis, characterization and photocatalytic activity of ZnO nanoparticles prepared by biological method. Mol Biomol Spectrosc 143:304–308

Aruna ST, Mukasyan AS (2008) Combustion synthesis and nanomaterials. Curr Opinion Solid State Mater Sci 12:44

Asmathunisha N, Kathiresan K (2013) A review on biosynthesis of nanoparticles by marine organisms. Colloids Surf B: Bionterfaces 103:283–287

Azizi S, Ahmad MB, Namvar F, Mohamad R (2014) Green biosynthesis and characterization of zinc oxide nanoparticles using brown marine macroalga Sargassum muticum aqueous extract. Mater Lett 116:275–277

Cauerhff A, Castro GR (2013) Bionanoparticles, a green nanochemistry approach. Electron J Biotechnol 16(3)

Chandrasekaran R, Gnanasekar S, Seetharaman P, Keppanan R, Arockiaswamy W, Sivaperumal S (2016) Formulation of Carica papaya latex-functionalized silve rnanoparticles for its improved antibacterial and anticancer applications. J Mol Liq 219:232–238

Chinnammal Janaki E, Sailatha SG (2015) Synthesis, characteristics and antimicrobial activity of ZnO nanoparticles. Spectrochim Acta A Mol Biomol Spectrosc 144:17–22

Daniel M-C, Astruc D (2004) Gold nanoparticles: assembly, supramolecular chemistry, quantum-size-related properties, and applications toward biology, catalysis, and nanotechnology. Chem Rev 104(1):293–346

Dhandapani P, Siddarth AS, Kamalasekaran S, Maruthamuthu S, Rajagopal G (2014) Bio-approach: Ureolytic bacteria mediated synthesis of ZnO nanocrystals on cotton fabric and evaluation of their antibacterial properties. Carbohydr Polym 103:448–455

Dhanemozhi AC, Rajeswari V, Sathyajothi S (2017) Green synthesis of zinc oxide nanoparticle using green tea leaf extract for Supercapacitor application. Mater Today: Proc 4:660–667

Do Kim K, Choi DW, Choa Y-H, Kim HT (2007) Optimization of parameters for the synthesis of zinc oxide nanoparticles by Taguchi robust design method. Colloids Surf A: Physicochem Eng Asp 311:170–173

Dobrucka R, Dugaszewska J (2016) Biosynthesis and antibacterial activity of ZnO nanoparticles using Trifolium pratense flower extract. Saudi J Biol Sci 23:517–523

Dumbrava A, Berger D, Matei C, Radu MD, Gheorghe E (2018) Characterization and applications of a new composite material obtained by green synthesis, through deposition of zinc oxide onto calcium carbonate precipitated in green seaweeds extract. Ceram Int 44:4931–4936

El-Naggar ME, Shaarawy S, Hebeish AA (2018) Multifunctional properties of cotton fabrics coated with in situ synthesis of zinc oxide nanoparticles capped with date seed extract. Carbohydr Polym 181:307–316

Elumalai K, Velmurugan S (2015) Green synthesis, characterization and antimicrobial activities of zinc oxide nanoparticles from the leaf extract of *Azadirachta indica*, (L.). Appl Surf Sci 345:329–333

Elumalai K, Velmurugan S, Ravi S, Kathiravan V, Ashokkumar S (2015a) Green synthesis of zinc oxide nanoparticles using Moringa Oleifera leaf extract and evaluation of its antimicrobial activity. Spectrochim Acta A 143:158e164

Elumalai K, Velmurugan S, Ravi S, Kathiravan V, Adaikala Raj G (2015b) Plant mediated synthesis of ZnO nanoparticles and their catalytic reduction of methylene blue and antimicrobial activity. Adv Powder Technol 26:1639–1651

Gandhi PR, Jayaseelan C, Mary RR, Mathivanan D, Suseem SR (2017) Acaricidal, pediculicidal and larvicidal activity of synthesized ZnO nanoparticles using Momordica charantia leaf extract against blood feeding parasites. Exp Parasitol 181:47e56

Gunalan S, Sivaraj R, Rajendran V (2012a) Green synthesized ZnO nanoparticles against bacterial and fungal pathogens. Prog Nat Sci: Mater Int 22(6):693–700

Gunalan S, Sivaraj R, Rajendran V (2012b) Green synthesized ZnO nanoparticles against bacterial and fungal pathogens. Mater Int 22(6):693–700

Hamminga GM, Mul G, Moulijin JA (2004) Effects of impregnation with styrene and nano-zinc oxide on fire-retarding, physical, and mechanical properties of poplar wood. Chem Eng Sci 59:5479

Hussein MZ, Azmin WHWN, Mustafa M, Yahaya AH (2009) Bacillus cereus as a biotemplating agent for the synthesis of zinc oxide with raspberry- and plate-like structures. J Inorg Biochem 103:1145–1150

Iravani S (2011) Green synthesis of metal nanoparticles using plants. Green Chem 13:2638e2650

Ishwarya R, Vaseeharan B, Kalyani S, Banumathi B, Govindarajan M, Alharbid NS, Kadaikunnan S, Al-anbrd MN, Khaled JM, Benelli G (2018) Facile green synthesis of zinc oxide nanoparticles using *Ulva lactuca* seaweed extract and evaluation of their photocatalytic, antibiofilm and insecticidal activity. J Photochem Photobiol B Biol 178:249–258

Ismail B, Nampoothiri KM (2010) Production, purification and structural characterization of an exopolysaccharide produced by a probiotic Lactobacillus plantarum MTCC 9510. Arch Microbiol 192:1049–1057

Jamdagni P, Khatri P, Rana JS (2016) Green synthesis of zinc oxide nanoparticles using flower extract of Nyctanthes arbor-tristis and their antifungal activity. J King Saud Univ Sci 30(2):168–175

Joerger R, Klaus T, Granqvist CG (2000) Biologically produced silver carbon composite materials for optically functional thin film coatings. Adv Mater 12(6):407–409

Kandale A, Meena AK, Rao MM, Panda P, Mangal AK, Reddy G et al (2011) Marine algae: anintroduction, foodvalue andmedicinaluses. J Pharm Res 4(1):219e221

Karthik S, Siva P, Balu KS, Suriyaprabha R, Rajendran V, Maaza M (2017) *Acalypha indica*–mediated green synthesis of ZnO nanostructures under differential thermal treatment: effect on textile coating, hydrophobicity, UV resistance, and antibacterial activity. Adv Powder Technol 28:3184–3194

Kato H (2011) In vitro assays: tracking nanoparticles inside cells. Nat Nanotechnol 6(3):139–140

Kaviya S, Kabila S, Jayasree KV (2017) Hexagonal bottom-neck ZnO nano pencils: a study of structural, optical and antibacterial activity. Mater Lett 204:57–60

Kayaci F, Vempati S, Donmez I, Biyikli N, Uyar T (2014) Nanoscale 6:10224–10234

Kim KD, Kim HT (2002) Preparation of silica nanoparticles determination of the optimal synthesis conditions for small and uniform particles. Colloids and surfaces a physicochemical and engineering aspects. J Sol Gel Sci Technol 25:183

Kim KD, Lee TJ, Kim HT (2003) Dynamics of population code for working memory in the prefrontal cortex. Colloids Surf A Physicochem Eng Asp 224:1

Kim KD, Han DN, Lee JB, Kim HT (2006) New and future developments in catalysis: solar pho-
    tocatalysis. Scripta Mater 54:143
Kundu D, Hazra C, Chatterjee A, Chaudhari A, Mishra S (2014) Extracellular biosynthesis of zinc
    oxide nanoparticles using Rhodococcus pyridinivorans NT2: multifunctional textile finishing,
    biosafety evaluation and in vitro drug delivery in colon carcinoma. J Photochem Photobiol B
    Biol 140:194–204
Li WJ, Shi EW, Zheng YQ, Yin ZW (2001) Hydrothermal preparation of nanometer ZnO powders.
    J Mater Sci Lett 20:1381
Li X, Xu H, Chen Z-S, Chen G (2011) Biosynthesis of nanoparticles by microorganisms and their
    applications. J Nanomater 2011, Article ID 270974, 16 pages
Lin HM, Tzeng SJ, Hsiau PJ, Tsai WL (1998) Low temperature electronics and low temperature
    cofired ceramic. Nanostruct Mater 10:465
Luechinger NA, Grass RN, Athanassiou EK, Stark WJ (2010) Bottom-up fabrication of metal/
    metal nano composites from nanoparticles of immiscible metals. Chem Mater 22(1):155–160
Madhumitha G, Elango G, Roopan SM (2016) Biotechnological aspects of ZnOnanoparticles:
    overview on synthesis and its applications. Appl Microbiol Biotechnol 100:571–581
Mahendra C, Murali M, Manasa G, Ponnamma P, Abhilash MR, Lakshmeesha TR, Satish A,
    Amruthesh KN, Sudarshana MS (2017) Antibacterial and antimitotic potential of bio-fabricated
    zinc oxide nanoparticles of Cochlospermum religiosum (L.). Microb Pathog 110:620e629
Manilal A, Thajuddin N, Selvin J, Idhayadhulla A, Kumar RS, Sujith S (2011) In vitro mosquito
    larvicidal activity of marine algae against the human vectors, Culex quinquefasciatus (Say) and
    Aedes aegypti (Linnaeus) (Diptera: Culicidae). Int J Zool Res 7(3):272e278
Manna J, Goswami S, Shilpa N, Sahu N, Rana RK (2015) Biomimetic method to assemble
    nanostructured ag@ZnO on cotton fabrics: application as self-cleaning flexible materi-
    als with visible-light Photocatalysis and antibacterial activities. ACS Appl Mater Interfaces
    7:8076–8082
Matinisea N, Fukua XG, Kaviyarasua K, Mayedwa N, Maaza M (2017) ZnO nanoparticles via
    Moringa oleifera green synthesis: physical properties & mechanism of formation. Appl Surf
    Sci 406:339–347
Mohanpuria P, Rana NK, Yadav SK (2008) Biosynthesis of nanoparticles: technological concepts
    and future applications. J Nanopart Res 10(3):507–517
Molinski TF, Dalisay DS, Lievens SL, Saludes JP (2009) Drug development from marine natural
    products. Nat Rev Drug Discov 8(1):69e85
Murugan K, Roni M, Panneerselvam C, Aziz AT, Suresh U, Rajaganesh R, Aruliah R, Mahyoub
    JA, Trivedi S, Rehman H, Al-Aoh HAN, Kumar S, Higuchi A, Vaseeharan B, Wei H, Senthil-
    Nathan S, Canale A, Benelli G (2017) Sargassum wightii-synthesized ZnO nanoparticles
    reduce the fitness and reproduction of the malaria vector Anopheles stephensi and cotton boll-
    worm Helicoverpa armigera. Phys Mol Plant Pathol 101(2018):202–213
Nagarajan S, Arumugam Kuppusamy K (2013) Extracellular synthesis of zinc oxide nanoparticle
    using seaweeds of gulf of Mannar, India. J Nanobiotechnol 11:39
Narayanan KB, Sakthivel N (2011) Green synthesis of biogenic metal nanoparticles by terrestrial
    and aquatic phototrophic and heterotrophic eukaryotes and biocompatible agents. Adv Colloid
    Interf Sci 169(2):59–79
Nava OJ, Luque PA, Gomez-Gutierrez CM, Vilchis-Nestor AR, Castro-Beltran A, Mota-Gonzalez
    ML, Olivas A (2017) Influence of Camellia sinensis extract on zinc oxide nanoparticle green
    synthesis. J Mol Struct 1134:121e125
Nethravathi PC, Shruthi GS, Suresh D, Udayabhanu, Nagabhushana H, Sharma SC (2015)
    Garcinia xanthochymus mediated green synthesis of ZnO nanoparticles: photoluminescence,
    photocatalytic and antioxidant activity studies. Ceram Int 41:8680–8687
O'Regan B, Schwartz DT (2000) Electrodeposited nanocomposite n–p heterojunctions for solid-
    state dye-sensitized photovoltaics. Adv Mater 12:1263
Pandimurugan R, Thambidurai S (2016) Novel seaweed capped ZnO nanoparticles for effective
    dye photodegradation and antibacterial activity. Adv Powder Technol 27:1062–1072

Pandimurugan R, Thambidurai S (2017) UV protection and antibacterial properties of seaweed capped ZnO nanoparticles coated cotton fabrics. Int J Biol Macromol 105:788–795

Parashar UK, Saxena PS, Srivastava A (2009a) Bioinspired synthesis of silver nanoparticles. Dig J Nanomater Biostruct 4(1):159–166

Parashar V, Parashar R, Sharma B, Pandey AC (2009b) Partenium leaf extract mediated synthesis of silver nanoparticles: a novel approach towards weed utilization. Dig J Nanomater Biostruct 4(1):45–50

Patra JK, Baek K-H (2014) Green nanobiotechnology: factors affecting synthesis and characterization techniques. J Nanomater 2014, Article ID 417305, 12 pages

Pavan Kumar MA, Suresh D, Nagabhushana H, Sharma SC (2015) Beta vulgaris aided green synthesis. of ZnO nanoparticles and their luminescence, photo-. catalytic and antioxidant properties. Eur Phys J Plus 130:109–116

Pavithra NS, Lingaraju K, Raghu GK, Nagaraju G (2017) Citrus maxima (Pomelo) juice mediated eco-friendly synthesis of ZnO nanoparticles: Applicationsto photocatalytic, electrochemical sensor and antibacterial activities. Mol Biomol Spectrosc 185:11–19

Raja Naika H, Lingaraju K, Manjunath K, Kumar D, Nagaraju G, Suresh D, Nagabhushana H (2015) Green synthesis of CuO nanoparticles using Gloriosa. superba L. extract and their antibacterial activity. J Taibah Univ Sci 9:7–12

Rajabia HR, Naghiha R, Kheirizadeh M, Sadatfaraji H, Mirzaei A, Alvand ZM (2017) Microwave assisted extraction as an efficient approach for biosynthesis of zinc oxide nanoparticles: synthesis, characterization, and biological properties. Mater Sci Eng C 78:1109–1118

Rajiv P, Rajeshwari S, Venckatesh R (2013) Spectrochimica Acta Part A: bio-fabrication of zinc oxide nanoparticles using leaf extract of Parthenium hysterophorus L. and its size-dependent antifungal activity against plant fungal pathogens. Mol Biomol Spectrosc 112:384–387

Raliya R, Tarafdar JC (2013) ZnO nanoparticle biosynthesis and its effect on phosphorous-mobilizing enzyme secretion and gum contents in clusterbean (Cyamopsis tetragonoloba L.). Agric Res 2(1):48–57

Ramesha M, Anbuvannan M, Viruthagiri G (2015) Green synthesis of ZnO nanoparticles using Solanum nigrum leaf extract and their antibacterial activity. Mol Biomol Spectrosc 136 (864–870

Rico CM, Majumdar S, Duarte-Gardea M, Peralta-Videa JR, Gardea-Torresdey JL (2011) Interaction of nanoparticles with edible plants and their possible implications in the food chain. J Agric Food Chem 59(8):3485–3498

Rosi NL, Mirkin CA (2005) Nanostructures in biodiagnostics. Chem Rev 105:1547–1562

Sabir S, Arshad M, Chaudhari SK (2014) Zinc oxide nanoparticles for revolutionizing agriculture: synthesis and applications. Sci World J 925494:8

Salem W, Leitner DR, Zingl FG, Schratter G, Prassl R, Goessler W, Reidl J, Schild S (2015) Antibacterial activity of silver and zinc nanoparticles against Vibrio cholera and enterotoxic Escherichia coli. Int J Med Microbiol 305:85–95

Santhoshkumar J, Kumar SV, Rajeshkumar S (2017) Synthesis of zinc oxide nanoparticles using plant leaf extract against urinary tract infection pathogen. Res Effic Technol 3:459–465

Saravanan M, Gopinath V, Chaurasia MK, Syed A, Ameen F, Purushothaman N (2018) Green synthesis of anisotropic zinc oxide nanoparticles with antibacterial and cytofriendly properties. Microb Pathog 115:57–63

Shamaila S, Sajjad AKL, Ryma N-u-A, Farooqi SA, Jabeen N, Majeed S, Farooq I (2016) Review advancements in nanoparticle fabrication by hazard free eco-friendly green routes. Appl Mater Today 5:150–199

Shamsuzzaman AM, Khanam H, Aljawfi RN (2017) Biological synthesis of ZnO nanoparticles using C. albicans and studying their catalytic performance in the synthesis of steroidal pyrazolines. Arab J Chem 10:S1530–S1536

Siripireddy B, Mandal BK (2017) Facile green synthesis of zinc oxide nanoparticles by Eucalyptus globulus and their photocatalytic and antioxidant activity. Adv Powder Technol 28:785–797

Sonia S, Linda Jeeva Kumari H, Ruckmani K, Sivakumar M (2017) Antimicrobial and antioxidant potentials of biosynthesized colloidal zinc oxide nanoparticles for a fortified cold cream formulation: a potent nanocosmeceutical application. Mater Sci Eng C 79:581–589

Suganya P, Vaseeharan B, Vijayakumar S, Balan B, Govindarajan M, Alharbi NS, Kadaikunnan S, Khaled JM, Benelli G (2017) Biopolymer zein-coated gold nanoparticles: synthesis, antibacterial potential, toxicity and histopathological effects against the Zika virus vector Aedes aegypti. J Photochem Photobiol B 173:404–411

Sundrarajan M, Ambika S, Bharathi K (2015) Plant-extract mediated synthesis of ZnO nanoparticles using *Pongamia pinnata* and their activity against pathogenic bacteria. Adv Powder Technol 26:1294–1299

Suresh D, Nethravathi PC, Udayabhanu MAPK, Raja Naika H, Nagabhushana H, Sharma SC (2015a) Chironji mediated facile green synthesis of ZnO nanoparticles and their photoluminescence, photodegradative, antimicrobial and antioxidant activities. Mater Sci Semicond Process 40:759–765760

Suresh D, Udayabhanu PCN, Lingaraju K, Rajanaika H, Sharma SC, Nagabhushana H (2015b) EGCG assisted green synthesis of ZnO nanopowders: Photodegradative, antimicrobial and antioxidant activities. Spectrochim Acta A 136:1467–1474

Suresh D, Shobharani RM, Nethravathi PC, Pavan Kumar MA, Nagabhushana H, Sharma SC (2015c) Artocarpus gomezianus aided green synthesis of ZnO nanoparticles: luminescence, photocatalytic and antioxidant properties. Spectrochim Acta A 141:128–134

Suresh D, Nethravathi PC, Udayabhanu, Rajanaika H, Nagabhushana H, Sharma SC (2015d) Green synthesis of multifunctional zinc oxide (ZnO) nanoparticles using Cassia fistula plant extract and their photodegradative, antioxidant and antibacterial activities. Mater Sci Semicond Process 31:446–454

Thakkar KN, Mhatre SS, Parikh RY (2010) Biological synthesis of metallic nanoparticles. Nanomedicine 6:257e262

Thaya R, Malaikozhundan B, Vijayakumar S, Sivakamavalli J, Jeyasekar R, Shanthi S, Vaseeharan B, Ramasamy P, Sonawane A (2016) Chitosan coated Ag/ZnO nanocomposite and their antibiofilm, antifungal and cytotoxic effects on murine macrophages. Microb Pathog 100:124–132

Tiwari DK, Behari J, Sen P (2008) Time and dose-dependent antimicrobial potential of Ag nanoparticles synthesized by top down approach. Curr Sci 95(5):647–655

Turton R, Berry DA, Gardner TH, Miltz A (2004) Polymer nanocomposite for electro-optics: perspectives on processing technologies. Indian Eng Chem Res 43:1235

Udayabhanu PC, Nethravathi MA, Pavan Kumar D, Suresh K, Lingaraju H, Rajanaika H, Nagabhushana SCS (2015) Green synthesis of copper oxide (CuO) nanoparticles using banana peel extract and their photocatalytic activities. Mater Sci Semicond Process 33:81–88

Venkatachalam P, Priyank N, Manikandan K, Ganeshbabu I, Indiraarulselvi P, Geeth N, Muralikrishna K, Bhattacharya RC, Tiwari M, Sharma N, Sahi SV (2017) Enhanced plant growth promoting role of phycomolecules coated zinc oxide nanoparticles with P supplementation in cotton (Gossypium hirsutum L.). Plant Physiol Biochem 110:118e127

Vijayakumar S, Vaseeharan B, Malaikozhundan B, Shobiya M (2016a) *Laurus nobilis* leaf extract mediated green synthesis of ZnO nanoparticles: characterization and biomedical applications. Biomed Pharmacother 84:1213–1222

Vijayakumar S, Malaikozhundan B, Gobi N, Vaseeharan B, Murthy C (2016b) Protective effects of chitosan against the hazardous effects of zinc oxide nanoparticle in freshwater crustaceans Ceriodaphnia cornuta and Moina micrura. Limnologica 61:44–51

Vijayakumar S, Malaikozhundan B, Ramasamy P, Vaseeharan B (2016c) Assessment of biopolymer stabilized silver nanoparticle for their ecotoxicity on Ceriodaphnia cornuta and antibiofilm activity. J Environ Chem Eng 4:2076–2083

Vijayakumara S, Krishnakumar C, Arulmozhi P, Mahadevan S, Parameswari N (2018) Biosynthesis, characterization and antimicrobial activities of zinc oxide nanoparticles from leaf extract of *Glycosmis pentaphylla* (Retz.) DC. Microb Pathog 116:44–48

Vijayaraghavan K, Ashokkumar T (2017) Plant-mediated biosynthesis of metallic nanoparticles: a review of literature, factors affecting synthesis, characterization techniques and applications. J Environ Chem Eng 5:4866–4883

Wang ZS, Huang CH, Huang YY, Hou YJ, Xie PH, Zhang BW, Cheng HM (2001) Synthesis and physicochemical characterization of ZnOPorphyrin based hybrid materials for potential photovoltaic applications. Chem Mater 13:678

Wang X, Ding Y, Summers CJ, Wang ZL (2004) Large scale synthesis of six-nanometer-wide ZnO nanobelts. J Phys Chem B 108(26):8773–8777

Yan X, Shumin W, Lib X, Meng H, Zhanga X, Wang Z, Han Y (2017) Ag nanoparticle-functionalized ZnO micro-flowers for enhanced photodegradation of herbicide derivatives. Chem Phys Lett 679:119–126

Potential Applications of Greener Synthesized Silver and Gold Nanoparticles in Medicine

4

Naumih Noah

Abstract

Silver and gold nanoparticles have changed the medical field in various ways. Due to their very small sizes, high surface area, and physical and chemical properties, they have found widespread applications in drug delivery, imaging, diagnosis, and therapeutics. They usually respond significantly to the magnetic field which varies with time, and hence they can transfer enough thermal energy. Their unique physicochemical properties have led to the development of biosensors for point-of-care disease diagnosis. Greener synthesized gold and silver nanoparticles have also revealed anticancer activity toward numerous cancer cells. In this book chapter, we will therefore explore the potential application of greener synthesized silver and gold nanoparticles in the medical field.

Keywords

Gold nanoparticles · Green nanotechnology · Greener synthesized nanoparticles in medicine · Silver nanoparticles

4.1 Introduction

The field of nanotechnology which studies the management of matter on atomic and molecular levels and involves connecting atoms and molecules to yield particles and structures with functions which are from the same material at the bulk form (Bagherzade et al. 2017; Noah 2018) is advancing rapidly. Nanotechnology also involves fabrication and use of the physical, chemical, and biological systems at the

N. Noah (✉)
School of Pharmacy and health Sciences, United States International University-Africa, Nairobi, Kenya
e-mail: mnoah@usiu.ac.ke

nanometer scale (Noah 2018). These materials are well-known as nanoparticles or nanomaterials and are changing the scientific world mainly because of their extraordinary physical, chemical, and biological properties, in comparison to their bulk counterparts (Gatebe 2012; Noah 2018). The nanotechnology is progressing very fast, fabricating an unlimited development in various fields (Bagherzade et al. 2017; Noah 2018) such as field of biotechnology, biomedical, optical, medical imaging, catalysis, and electronics (Bagherzade et al. 2017; Makarov et al. 2014; Noah 2018; Usman et al. 2019; Verma and Mehata 2016). Metal nanoparticles such as silver and gold nanoparticles show extraordinary physical and chemical properties that are different from those of the bulk metals (Rodriguez-Lorenzo and Alvarez-Puebla 2014). Due to their very small sizes, high surface area, and physical and chemical properties, these nanoparticles have found widespread applications in drug delivery, imaging, diagnosis, and therapeutics (Khan et al. 2014; Noah 2018). They usually respond significantly to the magnetic field which varies with time, and hence they can transfer enough toxic thermal energy (Khan et al. 2014; Noah 2018), and thus, they show remarkable potential applications. For example, silver nanoparticles have found applications in high sensitivity biomolecular detection, catalysis biosensors and medicine, due to their incomparable optical, electrical, thermal properties (Ahmed et al. 2016b; Noah 2018) and they also display expansive spectrum of antibacterial activity (Bagherzade et al. 2017). Gold nanoparticles on the other hand have been extensively used in biotechnology and biomedical fields due to their large surface area and high electron conductivity and have proved to be safe and much less toxic agents for drug distribution in the body (Alaqad and Saleh 2016; Khan et al. 2017; Noah 2018; Tedesco et al. 2010).

Numerous physical and chemical methods have been developed for the production of silver and gold nanoparticles of ideal shapes and sizes (Logeswari et al. 2013; Makarov et al. 2014; Noah 2018; Verma and Mehata 2016). These methods include decomposition; electrochemical; microwave-assisted processes; gas-phase condensation; vapor-phase synthesis; and colloidal or liquid-phase methods (Afolabi et al. 2011; Logeswari et al. 2013; Mhlanga et al. 2010, 2011; Noah 2018; Scriba et al. 2008; Swihart 2003). However, their limitations and toxic nature of the chemicals used during the synthesis or in the use of the nanoparticles hinder their usage in many applications such as in the biomedical field (Noah 2018; Usman et al. 2019).

This has steered the development of green synthesis of nanoparticles by applying the green chemistry principles in the synthesis of these nanoparticles in order to minimize the hazardous effects faced in the earlier mentioned methods and to maximize the safety and sustainability of the nanoparticle production (Alivisatos 2004; Bo and Ren-Cheng 2017; Noah 2018; Usman et al. 2019). The mechanism of the green synthesis of these nanoparticles has been explained due to the presence of phytochemicals such as flavonoids, alkaloids, glycosides, terpenoids, phenols, etc. which act as capping or reducing agents (Usmani et al. 2018)

4.2   Green Synthesis of Silver and Gold Nanoparticles

The most common green synthetic approaches used in the synthesis of silver and gold nanoparticles as described in literature include the use of microorganism such as bacteria and fungus and plant extracts since they do not contain unsafe chemicals (Logeswari et al. 2013; Noah 2018), and the nanoparticle produced using these green methods has been found to exhibit comparable or slightly higher antibacterial activities as compared to those obtained using chemical methods (Shaik et al. 2016). The green synthesis using microorganisms is also known as "bio-nano factories" due to its rapidity, environmental effectiveness, affordability, and low cost, distinctively structured with high capability of metal uptake, and can maintain safe levels (Menon et al. 2017). These methods use various microbes such as fungi, bacterial, algae, and virus as reducing agents (Du et al. 2007; Dumur et al. 2011; Sarkar et al. 2012; Soltani Nejad et al. 2015; Xuwang et al. 2016) in the synthesis of the nanoparticles making them environmentally friendly since the enzymes present in the microbes can degrade the toxic chemicals produced during the biosynthesis of the nanoparticles (Menon et al. 2017; Noah 2018). As it has been described in literature (Noah 2018), numerous bacterial strains (Fayaz et al. 2011; Husseiny et al. 2007; Luo et al. 2014; Mewada et al. 2017; Mishra et al. 2011; Srinath and Ravishankar Rai 2015; Syed et al. 2016) as well as algal strains, yeast, and fungal strains (Ahmad et al. 2003; Castro et al. 2013; Li and Zhang 2016; Menon et al. 2017; Sarkar et al. 2012; Singaravelu et al. 2007; Singh et al. 2013; Venkatesan et al. 2014) have been successfully used in the biosynthesis of gold and silver nanoparticles of diameter of between 5 and 80 nm.

As it has been reported in literature by several researchers, the biomolecules present in plant extracts are said to act both as a reducing and stabilizing agents in a one-step green synthesis process of nanoparticles (Mhlanga et al. 2010; Noah 2018; Sadeghi and Gholamhoseinpoor 2015). The reducing and stabilizing agents involved comprise of various water-soluble plant metabolites such as alkaloids, phenolic compounds, and terpenoids and coenzymes (Noah 2018). The synthesis is usually rapid, readily conducted at ambient temperatures and pressure, cheap, scalable, ecologically safe, and harmless for clinical research (Ikram and Ahmed 2015; Noah 2018). Silver nanoparticles and gold nanoparticles have been of focus in these plant-based syntheses owing to their strong antibacterial activity (Noah 2018). Extracts of several plant species (Ali et al. 2015b; Bagherzade et al. 2017; Lim et al. 2016; Mittal et al. 2013; Singh and Srivastava 2015; Verma and Mehata 2016) have been successfully used in the synthesis silver and gold nanoparticles of diameter between 6 and 80 nm. Agricultural waste has also been explored in the biosynthesis of these nanoparticles (Noah 2018). For example, Ndikau et al. (2017) have reported the synthesis of 17.96 nm gold nanoparticles using the rind extract of watermelon which is usually thrown away as waste, hence minimizing waste to the environment (Ndikau et al. 2017). Mango peel extract has been used to synthesize 6.03–18 nm spherical gold nanoparticles with an in vitro cytotoxic effect on two normal cells (Yang et al. 2014).

4.3 Application of Greener Synthesized Silver and Gold Nanoparticle in Medicine

Due to eco-friendly, cost-effective, high product yielding properties, greener synthesized silver and gold nanoparticles have changed the medical field of medicine in various ways (Chintamani et al. 2018; Noah 2018). They have been increasingly used because of their several advantages, which include small size, high stability, loading capacity (Katas et al. 2018b), physical and chemical properties, high surface area, tunable optical, and non-cytotoxicity (Khan et al. 2014). Owing to their high biocompatibility, chemical stability, convenient surface bioconjugation with molecular probes, and excellent surface plasmon resonance, these nanoparticles have widespread biomedical applications such as directed drug delivery, biosensing and imaging, cancer treatment, DNA-RNA analysis, gene therapy, diagnosis, antibacterial agents, and therapeutics among others (Kumar and Yadav 2009; Paciotti et al. 2006; Tang et al. 2006; Tuhin Subhra Santra et al. 2014). Additionally, these nanoparticles are synthesized using green and economical methods. Figure 4.1 shows some potential applications of gold nanoparticles in the biomedical field. These applications can be classified into two broad categories (diagnostics and therapeutics) (Kotcherlakota et al. 2018) whose recent advancements are discussed in the following text.

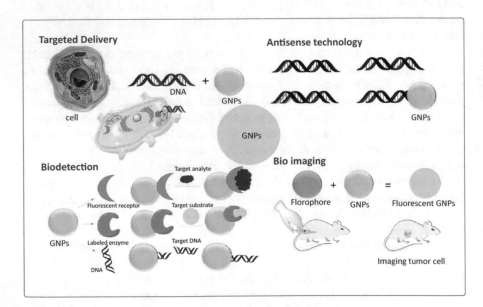

Fig. 4.1 Some potential applications of gold nanoparticles in the biomedical field. (Reproduced from Alaqad and Saleh (2016, p. 384) An open-access article)

4.3.1 Diagnostic and Imaging Applications of Greener Synthesized Gold and Silver Particles

Due to their unique physicochemical properties of silver and gold nanoparticles, they have been used as diagnostic agents especially in the development of biosensors for point-of-care disease diagnosis (Noah 2018). The interparticle plasmon coupling of the nanoparticles causes color changes which have been widely used in biosensors based on aggregation of the nanoparticles. For example, the color of gold nanoparticle is red when the nanoparticles are very small and well dispersed, but it turns blue or purple when the nanoparticles aggregate (Lin et al. 2013), while silver nanoparticle is yellowish brownish when dispersed but turns black when the nanoparticles aggregate (Gonçalo et al. 2012). This intrinsic property of the gold and silver nanoparticles has been used in a series of biosensors based on predictable color changes such as biosensors for detection of α-1-fetoprotein (Che et al. 2010), anti-hepatitis B virus antibodies in human serum (de la Escosura-Muñiz et al. 2010), breast cancer biomarkers (Ambrosi et al. 2010), serum p53 protein in head and neck squamous cell carcinoma (Zhou et al. 2011), mycobacteria of the *Mycobacterium tuberculosis* complex (Costa et al. 2010), and human immunodeficiency virus type 1 DNA (Wabuyele and Vo-Dinh 2005). In addition, the SPR peaks and line widths are said to be sensitive to the size and shape of the nanoparticles making these nanoparticles very sensitive to be used in bio-imaging (Abou El-Nour et al. 2010). For example, silver nanoparticles have been used as agents for photo thermal therapy for ablation of breast cancer cells (Loo et al. 2005)

Gold nanoparticles have a high-binding affinity to analytes which changes their physicochemical properties such as conductivity, redox behavior, and SPR. This makes them to form detectable signals enabling them to be used as diagnostic agents (Katas et al. 2018b). They also act as probes in the microscopy examination of cancer cells since they can accumulate and exert optical scattering effect in tumor cells, and hence they can be used in cancer diagnosis (Cai and Chen 2007; Tomar and Garg 2013). Gold nanoparticles are also said to permit in vitro detection and act as a diagnostic agent for diseases by readily conjugating with biomarkers such as oligonucleotides or antibodies to detect the target biomolecules (Huo et al. 2011).

Graphene sheets decorated with greener synthesized gold nanoparticles have been reported for label free electrochemical impedance hybridization sensing (Hu et al. 2011; Tuhin Subhra Santra et al. 2014). Hu et al. (2011) reported that functionalization of the graphene sheets with 3, 4, 9, 10-perylene tetra carboxylic acid (PTCA) can separate them and introduce more negative –COOH and enable the decoration of graphene with gold nanoparticles. After the application of amine-terminated ionic liquid (NH2-IL) $HAuCl_4$ is reduced to gold nanoparticles, it then aids the immobilization of the DNA probes via electrostatic interaction and adsorption effect due to graphene sheet and NH_2-IL protected gold nanoparticles. Their results indicated that for label-free DNA detection, electrochemical impedance values increased after DNA probes immobilization. They further concluded that the sensor can successfully detect the sequence of pol gene of human immunodeficiency virus 1(Hu et al. 2011; Tuhin Subhra Santra et al. 2014).

Due to their high surface area to volume ratio, gold nanoparticles have very high plasmon resonance and can be used to detect biomolecules (Tuhin Subhra Santra et al. 2014). As a result, biosynthesized gold nanoparticles from *C. nudiflora* plant extract have been used in the detection of HCG hormone in pregnant women on both pregnancy positive and negative urine samples (Kuppusamy et al. 2014; Tuhin Subhra Santra et al. 2014). By mixing 500 µl of biosynthesized gold nanoparticle solution with the same volume of the test sample and testing the solution using a pregnancy test strip, the authors found that the gold nanoparticles changed color into pink when pregnancy was positive and gray when negative. They further claimed that the method was 100% accurate for pregnancy diagnosis and can be used as an alternative method for urine pregnancy test (Kuppusamy et al. 2014; Tuhin Subhra Santra et al. 2014).

Gold nanoparticles synthesized from patuletin isolated from *Tagetes patula* which was used as a capping and reducing agent as reported by Muhammad et al. was used as a chemosensor for piroxicam. In their work, they conjugated the gold nanoparticles with the patuletin and the conjugate was found to be 63.2 by weight. They then examined the conjugate as a potential chemosensor with different drugs, but only one drug, piroxicam, was found to quench luminescence which followed Beer's law in a concentration range of 20–60 µM. The quenching was also found to be stable at different pH, elevated temperatures, or addition of other drugs, and hence they concluded that it could be important for molecular recognition applications (Ateeq et al. 2015).

Biosynthesized silver nanoparticles from the aqueous solution of polysaccharide of guar gum (*Cyamopsis tetragonoloba*) plants (Kotcherlakota et al. 2018; Pandey et al. 2012) acting as a reducing agent with a uniform size of <10 nm and characterized using X-ray diffraction (XRD), scanning electron microscopy (SEM), and transmission electron microscopy (TEM) displayed exceptional optical property toward ammonia with a very short response time of between 2 and 3 s and a detection limit of 1 ppm at room temperature (Kotcherlakota et al. 2018; Pandey et al. 2012). Pandey et al. (2012) explained that these optical properties toward ammonia at room temperature could be used as a sensor for detection of ammonia and further added that the biosensor could detect the ammonia level in biological fluids such as plasma, saliva, cerebrospinal liquid, and sweat suggesting an upcoming application for the ammonia biosensors (Kotcherlakota et al. 2018; Pandey et al. 2012). Moreover, silver nanoparticles could also be used to sense DNA hybridization though the success depends on the salt concentration which should be greater than a minimum threshold (Farkhari et al. 2016).

4.3.2 Therapeutic Applications of Greener Synthesized Gold and Silver Nanoparticles

Recent literature has shown silver and gold nanoparticles to have incredible biomedical properties including anticancer, antibacterial, and antidiabetic agents (Abbasi et al. 2017; Bagherzade et al. 2017; Kamala Priya 2015; Kotcherlakota et al. 2018; Logeswari et al. 2013; Mishra et al. 2011; Nafeesa Khatoon et al. 2017; Sadeghi et al. 2015). These nanoparticles usually respond significantly to the magnetic field which varies with time, and hence they can transfer enough toxic thermal energy to tumor cells as hyperthermic agents (Khan et al. 2014). These properties have led to biosynthesized silver and gold nanoparticles with more therapeutic effect due to their attachment of biological constituents with therapeutic potential (Kotcherlakota et al. 2018). For example, gold nanoparticles are known to bind to a range of organic molecules and hence have been used as therapeutic agents and vaccine carriers into the specific cells increasing drug effectiveness and damage of pathogens (Khan et al. 2014; Noah 2018). The following section discusses various therapeutic applications of greener synthesized silver and gold nanoparticles.

4.3.2.1 Cancer Applications

Alternative treatment strategies of cancer are being pursued due to the growing drug resistance, reduced bioavailability, and the nonspecific toxicity of chemotherapeutic agents which restrict their treatment effects (Kotcherlakota et al. 2018). Recent literature on the anticancer effect of greener synthesized silver and gold nanoparticles recommends their imminent role as therapeutic agents to fight cancer, and several research studies have proved that greener synthesized silver and gold nanoparticles show effective anticancer properties (Kotcherlakota et al. 2018).

Silver nanoparticles biosynthesized using *Bacillus funiculus* culture supernatant were found to exhibit antiproliferative activity in MDA-MB-231 (human breast cancer) cells through production of reactive oxygen species (ROS), leading to apoptosis (Gurunathan et al. 2013; Kotcherlakota et al. 2018). Likewise, protein-capped silver nanoparticles synthesized by Leena et al. (2017) using *Penicillium shearii* AJP05 fungus were proven to have anticancer effect on epithelial (hepatoma) and mesenchymal (osteosarcoma) cells (Kotcherlakota et al. 2018; Leena et al. 2017). The authors found that the generated reactive oxygen species (ROS) was the key cause of the cytotoxic effect of the biosynthesized silver nanoparticles (Kotcherlakota et al. 2018; Leena et al. 2017), and they also claimed that these biosynthesized silver nanoparticles sensitized the cancer cells, making them cisplatin-resistant (Kotcherlakota et al. 2018; Leena et al. 2017). Similarly, silver nanoparticles synthesized from *Alternanthera sessilis* plant extract by Firdhouse and Lalitha (2013) were found to show substantial cytotoxic activity toward prostate cancer cells (PC-3)(Firdhouse and Lalitha 2013; Kotcherlakota et al. 2018).

Biosynthesized silver nanoparticles have also been found to have both antibacterial and anticancer activity. For instance, 136 ± 10.09 nm silver nanoparticles biosynthesized by Sankar et al. (2013) using a the aqueous extract of *Origanum vulgare* (oregano) and characterized using UV-Vis spectroscopy, Fourier-transform infrared spectroscopy (FT-IR), field emission-scanning electron microscopy (FE-SEM), X-ray diffraction (XRD), and dynamic light scattering measurements were found to be valuable for both antibacterial and anticancer activities (Kotcherlakota et al. 2018; Sankar et al. 2013). The authors described a dose-dependent effectiveness of the silver nanoparticles toward the pathogens and human lung cancer cells (A549) (Kotcherlakota et al. 2018; Sankar et al. 2013). Also, 6.9 ± 0.2 nm silver nanoparticles biofabricated by Rajasekharreddy and Rani (2014) using *Sterculia foetida* L. seed extract and characterized using UV-Vis and (TEM) were found to display greater killing ability toward human cervical cancer cell lines (HeLa) as well as an application toward the antiangiogenic activity (Kotcherlakota et al. 2018). Mukherjee's research group (Mukherjee et al. 2014) has proven the synthesis of silver nanoparticles using *Olax scandens* plant extract (Kotcherlakota et al. 2018). They found that the as-synthesized silver nanoparticles showed numerous uses namely biocompatibility, imaging agent, anticancer, and antibacterial agents (Kotcherlakota et al. 2018). They found that the generation of the ROS and activation of the p53 was due to the anticancer mechanism of the nanoparticles' (Mukherjee et al. 2014). Environmentally friendly silver nanoparticles which revealed anticancer activity toward human colon cancer cells (HCT-15) were biosynthesized by Ramar et al. (2015) using ethanolic extract of rose (*Rosa indica*) petals and were found to mitigate toxicity levels while still retaining their anticancer activity (Kotcherlakota et al. 2018). In addition to the anticancer activity, they were also found to have antibacterial and anti-inflammatory activities (Ramar et al. 2015). Greener synthesized silver nanoparticles have also been found to induce cell death and oxidative stress (Noah 2018) of skin carcinoma cells and in human fibrosarcoma (Nafeesa Khatoon et al. 2017), human breast adeno-carcinoma (MCF-7) (Patra et al. 2015; Premkumar et al. 2010; Wang et al. 2016), human lung bronchoalveolar (NCI-H358) (Premkumar et al. 2010), and HEp-2 cancer cell line (Justin Packia Jacob et al. 2012; Nafeesa Khatoon et al. 2017).

Greener synthesized gold nanoparticles have also been found to show anticancer activity toward several cancer cells (Noah 2018). They have been found to induce apoptosis of HL-60 cancerous cells (Geetha et al. 2013), MCF 7 breast cancer cells (Kamala Priya 2015), A549 human lung cancer (Patra et al. 2015; Wang et al. 2016), and human keratinocyte cell line (Wang et al. 2016).

Gold nanoparticles biosynthesized from the flower extract of the pharmacologically importance tree *Couroupita guianensis* and characterized using UV-Vis spectroscopy, FTIR, XRD, SEM, and TEM analysis were found to have a substantial cytotoxicity effect in HL-60 cells in a concentration-dependent trend with the CC_{50} value of 5.14 μM and 113.25 μM for PBMC which was revealed by means of MTT assay, DNA fragmentation, apoptosis by DAPI staining, and comet assay for DNA damage (Geetha et al. 2013). The anticancer properties of greener synthesized gold nanoparticles using *Anacardium occidentale* leaves extract were investigated and found to exhibit a 23.56% viability on MCF-7 cell lines at a maximum concentration of 100 mg/ml (Sunderam et al. 2018).

Stable gold nanoparticles synthesized using aqueous and ethanolic *Taxus baccata* extracts and broadly characterized by UV-Vis spectroscopy, TEM, SEM, AFM, DLS, Zetasizer, EDS, and FT-IR techniques showed a potent, selective, dose- and time-dependent anticancer activity on breast (MCF-7), cervical (HeLa), and ovarian (Caov-4), which was more effective (Kajani et al. 2016). The authors detailed an in vitro study of cell exposure by the synthesized gold nanoparticles using flow cytometry and real-time polymerase chain reaction (RT-PCR) and claimed that it indicated that the *caspase*-independent death program was most feasible anticancer mechanism of the synthesized gold nanoparticles (Kajani et al. 2016). Similarly, gold nanoparticles biosynthesized using the stem bark extract of *Nerium oleander* (commonly known as Karabi) and characterized by surface plasmon resonance spectroscopy (SPR), high-resolution transmission electron microscopy (HRTEM), X-ray diffraction (XRD) studies, and dynamic light scattering (DTS) were found to be very effective for the apoptosis of MCF-7 breast cancer cells using nonradioactive colorimetric assay technique with tetrazolium salt, 3-[4,5-dimethylthiazole-2-yl]-2,5-diphenyl tetrazolium bromide (MTT)(Barai et al. 2018).

Greener synthesized gold nanoparticles have been found to enhance their anticancer activity. For example, a study by Yarramala et al. (2015) biosynthesized luminescent gold nanoparticles capped with apo-α-lactalbumin and established the advantage of the luminescent apo-α-LA-gold nanoparticles in their attack of cancer cells in broad and selective killing of breast cancer cells in particular. They concluded that coating AuNPs with the protein apo-α-lactalbumin improved their anticancer activity by severalfold (Yarramala et al. 2015).

4.3.2.2 Antibacterial Activity

Pathogenic microorganisms have led to antimicrobial resistance and are termed as a major threat to human health (Kotcherlakota et al. 2018; Morens and Fauci 2013; Nanda and Saravanan 2009). Several antibiotics currently used to treat microbial infections have limited use due to the development of antibiotic (Kotcherlakota et al. 2018). The emergency of nanotechnology promises an alternate treatment method to treat microbial infections (Kotcherlakota et al. 2018). As such, silver and gold nanoparticles have been found to have potent antibacterial activity (Abbasi et al. 2017; Ahmed et al. 2016a; Alivisatos 2004; Bagherzade et al. 2017; Bo and Ren-Cheng 2017; Duan et al. 2015; Hebeish et al. 2011; Huang et al. 2007; Kamala Priya 2015; Kaviya et al. 2011; Kim et al. 2009; Logeswari et al. 2013; Masurkar et al. 2011; Mishra et al. 2011; Nabikhan et al. 2010; Nafeesa Khatoon et al. 2017; Nakkala et al. 2014; Narayanan and Park 2014; Prasad and Elumalai 2011; Sadeghi et al. 2015).

A number of researchers have reported remarkable antimicrobial effects of greener synthesized silver nanoparticles. For example, silver nanoparticles synthesized by Singh et al. (2017) using the culture supernatant of endophytic fungus (*Raphanus sativus*) were found to have antibacterial effect on Gram-positive (methicillin-resistant *Bacillus subtilis*, MTCC 441, *Staphylococcus aureus*, MTCC 740) and Gram-negative (*Escherichia coli*, MTCC 443, and *Serratia marcescens*, MTCC 97) bacterial pathogens (Kotcherlakota et al. 2018; Singh et al. 2017). The authors further claimed that the interference of cell membrane and DNA was the main reason for the antibacterial effect of silver nanoparticles (Kotcherlakota et al. 2018; Singh et al. 2017). In another study by Abalkhil et al. (2017), silver nanoparticles synthesized using *Aloe vera, Portulaca oleracea*, and *Cynodon dactylon* were explored for their antibacterial effect against human pathogens (Kotcherlakota et al. 2018). From their work, the SEM analysis of the silver nanoparticles showed that cell wall damage was the key event happening during the antibacterial effect of silver nanoparticles (Abalkhil et al. 2017; Kotcherlakota et al. 2018).

Studies have shown that the development of biofilms with drug-resistant bacteria was the leading challenge for conventional treatments (Kotcherlakota et al. 2018). Biosynthesized silver nanoparticles have been found to have antibacterial effect in bacterial films as demonstrated by Xiang et al. (2013). In their study, they prepared biogenic silver nanoparticles using *Bombyx mori* silk fibroin and established the antibacterial effect in bacterial biofilms (Kotcherlakota et al. 2018). Their results indicated that the silk fibroin-silver nanoparticle composite was effective against methicillin-resistant *Staphylococcus aureus* (*S. aureus*) and consequently repressed the biofilm formation caused by the same bacterium (Xiang et al. 2013). The authors further showed that a maturely formed biofilm formed by methicillin-resistant *Staphylococcus aureus* can be damaged by the silk fibroin-silver nanoparticle composite, hence meeting the mandate of clinical application. They therefore concluded that the silk fibroin-silver nanoparticle composite prepared by that clean and facile method was anticipated to be an effective and inexpensive antimicrobial material in biomedical fields (Xiang et al. 2013). Another study by Nanda and Saravanan (2009) where they synthesized silver nanoparticles using *Staphylococcus aureus* culture supernatant showed antibacterial effect on methicillin-resistant *S. aureus*, *Staphylococcus epidermidis*, and *Streptococcus pyogenes* as well as on *Salmonella typhi* and *Klebsiella pneumoniae* (*Kotcherlakota* et al. 2018; Nanda and Saravanan 2009*).

Studies have also shown that greener synthesized silver nanoparticles can be synergized with antibiotics (Kotcherlakota et al. 2018). This has been illustrated in a study by Railean-Plugaru et al. (2016) where silver nanoparticles synthesized using *Actinobacteria* CGG 11n bacteria were evaluated for their antibacterial activity in combination with kanamycin, ampicillin, neomycin, and streptomycin using flow cytometry (Kotcherlakota et al. 2018). The study showed a more synergistic effect of the combined biogenic silver nanoparticles and antibiotics against *Pseudomonas aeruginosa* representing a prototype of multidrug resistant for which effective therapeutic options are very limited (Railean-Plugaru et al. 2016).

Silver nanoparticles produced using medicinally important plant extracts have been shown to display a potent antibacterial effect (Kotcherlakota et al. 2018). For example, in a study by Singhal et al. (2011), 4–30 nm silver nanoparticles synthesized using *Ocimum sanctum* (Tulsi) leaf extract were assessed for their antibacterial effect on Gram-negative *E. coli* and Gram-positive *S. aureus* (Kotcherlakota et al. 2018) and found that they possessed antimicrobial activity suggesting their probable application in medical industry (Singhal et al. 2011) . Further, silver nanoparticles prepared using *Acalypha indica* leaf extracts by Krishnaraj et al. (2010) exhibited effective inhibitory activity against water-borne pathogens, viz., *Escherichia coli* and *Vibrio cholerae*. From their results, the synthesized silver nanoparticles at 10 μg/ml were recorded as the minimal inhibitory concentration (MIC) against *E. coli* and *V. cholerae*, and the adjustment in membrane permeability and respiration of the silver nanoparticle treated bacterial cells were apparent from the activity of silver nanoparticles (Krishnaraj et al. 2010). Equally, silver nanoparticles synthesized by Kaviya et al. (2011) using *Citrus sinensis* peel extract indicated more potency for antibacterial effects when evaluated for their antibacterial effect on Gram-negative *Escherichia coli* and *Pseudomonas aeruginosa,* and Gram-positive *Staphylococcus aureus* bacteria (Kaviya et al. 2011; Kotcherlakota et al. 2018).

Although gold nanoparticles are not as strong antimicrobial agents as silver nanoparticles, they have been reported to possess antimicrobial activities (Kundu 2017; Lima et al. 2013) indicating that they could be used in biomedical applications. For example, capped gold nanoparticles (C-AuNPs) of ≈20–30 nm synthesized by Rao et al. (2017) using flower and leaf extracts of *Ocimum tenuiflorum* leaves of *Azadirachta indica* and *Mentha spicata* and peel of *Citrus sinensis* plants were found to inhibit 99% growth of *Staphylococcus aureus, Pseudomonas aeruginosa,* and *Klebsiella pneumoniae* antimicrobial strains at 512 and 600 µg ml^{-1}. The authors from their analyses inferred that the synthesized capped gold nanoparticles (C-AuNPS) had interacted with the bacterial cell wall due to the phytochemicals present in the plant extracts and that the C-AuNPs ruptured the cell wall, disturbing the metabolism of the bacteria by inducing chemical activities. They also claimed that the C-AuNPs could have entered inside the pathogenic bacteria to destroy the outer cell wall for their interactions with mitochondria and other organelles of bacteria and concluded that the synthesized gold nanoparticles could be an asset for several other biomedical and bioengineering applications (Rao et al. 2017). Similarly, Naraginti et al. (2017), Nayak et al. (2018), Sunkari et al. (2017), and Yuan et al. (2017) synthesized highly stable gold nanoparticles from *papaya* leaf extract, Kiwi fruit extract, *Citrus maxima* peel extract (a biomass waste), and banana pith extract, respectively, which displayed effective excellent catalytic activity in the reduction of 4-nitrophenol to 4-aminophenol and methylene blue and antimicrobial activity on both Gram-positive and Gram-negative bacteria.

Greener synthesized gold nanoparticles have also been formulated with drugs and their antibacterial activity studies as reported by Shittu et al. (2017). In their work, they synthesized gold nanoparticles using *Piper guineense* aqueous leaf extract and characterized them using UV-Vis spectrophotometer, DLS, TEM/EDS, and FTIR. They used the drug lincomycin to prepare three drug formulations with different compositions. The first one was composed of the polymer (PEG) – greener synthesized gold nanoparticles – drug (lincomycin) and labeled it as PND, while the second one was composed of the polymer (PEG) and the biosynthesized gold nanoparticles and labeled as PN, while the last one was composed of the polymer (PEG) and the drug (lincomycin) and labeled it as PD. They studied the antibacterial activity of the formulations against Gram-positive bacteria *Staphylococcus aureus* and *Streptococcus pyogenes* at different temperatures of 40 °C and 60 °C, respectively. As represented in Fig. 4.2, the antibacterial potential of the nanodrug was seen on the *Staphylococcus aureus* and *Streptococcus pyogenes* with maximum inhibitions of 18 mm (at 40 °C) and 16 mm (at 60 °C) for *Staphylococcus aureus* and 16 mm for *Streptococcus pyogenes* (both at 40 °C and 60 °C). The bacteria growth inhibition was found to continue and lasted for 15 min, while that of non-nanodrug lasted for 9 min with lesser growth inhibition compared to the formulated nanodrug (Fig. 4.2), suggesting that the nanodrug was effective in bacteria growth inhibition.

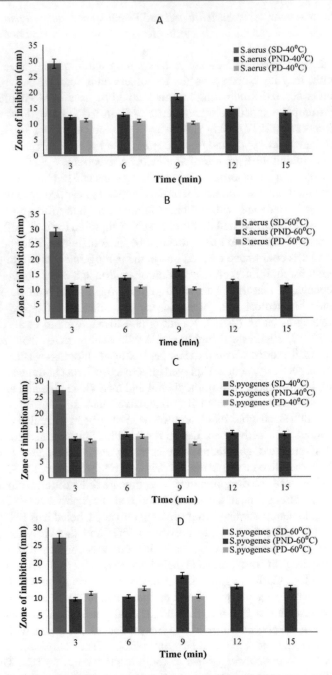

Fig. 4.2 Antimicrobial potential of formulated nanodrugs (PND, PN, PD) where (**a**) *Staphylococcus aureus* at 40 °C, (**b**) *Staphylococcus aureus* at 60 °C, (**c**) *Streptococcus pyogenes* at 40 °C, (**d**) *Streptococcus pyogenes* at 60 °C. (Reproduced from Shittu et al. (2017) An open access article)

Gold nanoparticles biosynthesized from microorganisms have also been found to have antibacterial activity. For example, Thirumurugan et al. (2012) biosynthesized gold nanoparticles using *Bacillus subtilis* supernatant as the reducing agent, and after investigating their antibacterial and antifungal effect, their results indicated an increased antibacterial activity with various antibiotics including norfloxacin ciprofloxacin, nitrofurantoin, and chloramphenicol. They then concluded that these biosynthesized nanoparticles can make a possible platform in future for preparing nanomedicines for bacterial- and fungal-related diseases (Thirumurugan et al. 2012). Similarly, gold nanoparticles synthesized using *Aspergillus foetidus* as source reducing and stabilizing agent showed moderate antimicrobial activity against *Staphylococcus aureus* and *Escherichia coli* strain (Roy 2017).

Further a combination of gold-silver nanoparticles synthesized by Ramasamy et al. (2016) was found to have an enhanced antibacterial activity with an improved therapeutic efficacy against bacterial biofilms. In their work, they synthesized bimetallic gold-silver using γ-proteobacterium and *Shewanella oneidensis* MR-1 and assessed their antibacterial activities on planktonic and biofilm phases of individual and mixed multiculture of Gram-negative (*Escherichia coli* and *Pseudomonas aeruginosa*) and Gram-positive bacteria (*Enterococcus faecalis* and *Staphylococcus aureus*), respectively. Their results indicated a 30–50 μM minimum inhibitory concentration of gold-silver nanoparticles for the tested bacteria. Interestingly, their results showed more effectiveness of the gold-silver nanoparticles in inhibiting bacterial biofilm formation at 10 μM concentration. Additionally, their scanning and transmission electron microscopy results accounted for the impact of gold-silver nanoparticles on biocompatibility and bactericidal effect and that the small size and bio-organic materials covering on gold-silver nanoparticles improved the internalization. They thus concluded that that was what caused bacterial inactivation (Ramasamy et al. 2016). They further claimed that the bacteriogenically synthesized gold-silver nanoparticles appeared to be a promising nano-antibiotic for overcoming the bacterial resistance in the established bacterial biofilms (Ramasamy et al. 2016). A summary of more antimicrobial activity of greener synthesized silver and gold nanoparticles is given in Table 4.1.

4.3.2.3 Drug Delivery Applications

The developments in functionalization of chemistry with advanced nanomaterials and their diverse uses in the treatment of several human diseases have received attention globally (Shittu et al. 2017). Nanotechnology is essential for drug delivery, with many latent uses in clinical medicine and research (Mocan et al. 2010; Shittu et al. 2017). The choice of nanoparticles for drug delivery is greatly preferred due to their exceptional chemical and physical properties that hold support for future advance treatment of diseases offering an efficient strategy in drug delivery system with minimal side effect (Giljohann et al. 2010; Kotcherlakota et al. 2018) since the slow and sustained release of drugs and delivery in targeted sites are the two major criteria for efficient drug delivery systems (Cho et al. 2008; Kotcherlakota et al. 2018) which can be realized through active or passive delivery approaches (Torchilin 2010). Recent literature has demonstrated the potential of several metal

Table 4.1 A summary of a few more antimicrobial activity of greener synthesized silver and gold nanoparticles

Nanoparticle	Green synthetic method	Antibacterial activity	References
Silver	*Punica granatum* L. fruit extract	*Pseudomonas aeruginosa*	Akkiraju et al. (2017)
Silver	Banana plant stems extract	*Escherichia coli and Staphylococcus epidermis*	Dang et al. (2017)
Silver	*Acacia rigidula*	*Escherichia coli, Pseudomonas aeruginosa*, clinical multidrug-resistant strain of *Pseudomonas aeruginosa* and *Bacillus subtilis*	Carlos Enrique Escárcega-González et al. (2018)
Silver	*Eucalyptus globulus* leaf extract	*Staphylococcus aureus*	Ali et al. (2015a)
Silver and gold	Exopolysaccharides and metabolites of *Weissella confusa*	Multidrug resistance (MDR) *Escherichia coli*	Adebayo-Tayo et al. (2019)
Silver, gold, silver-gold	Leaves of *Gmelina arborea* (ROXB) (family *Verbenaceae*) extract	*Bacillus subtilis, Staphylococcus aureus, Escherichia Coli and Pseudomonas aeruginosa*	Khalil et al. (2017)
Gold	*Lignosus rhinocerotis and chitosan*	*Pseudomonas aeruginosa, Escherichia coli, Staphylococcus aureus and Bacillus species*	Katas et al. (2018a)
Gold	*Galaxaura elongata*	*Escherichia coli, Klebsiella pneumoniae, Staphylococcus aureus and Pseudomonas aeruginosa*	Abdel-Raouf et al. (2017)
Gold	Kanamycin, commercial antibiotic	*S. bovis, S. epidermidis, E. aerogenes and Pseudomonas aeruginosa*	Jason et al. (2016)
Gold	*Sambucus ebulus* L leaf extract	*Staphylococcus aureus, Bacillus subtilis, Escherichia coli and Salmonella enteritidis*	Azizian Shermeh et al. (2016)
Gold	*Citrullus lanatus* rind	*Bacillus cereus, Listeria monocytogenes, Staphylococcus aureus, Escherichia coli and Salmonella typhimurium*	Patra and Baek (2015)

nanoparticles for drug delivery applications (Kotcherlakota et al. 2017). For example, as reported by Noah (2018), gold nanoparticles are known to have a strong affinity toward biomolecules such as proteins, peptides, antibodies, oligonucleotides, and pathogens such as bacteria and viruses (Solano-Umaña and Vega-Baudrit 2015). Hence, functionalization of gold nanoparticles with these biomolecules allows them to be used as biomarkers for diseases detection (Lin et al. 2013; Noah 2018) and drug delivery system since they bear high drug load releasing it to the specific sites (Solano-Umaña and Vega-Baudrit 2015) and the gold core of the gold nanoparticles is essentially inert and nontoxic (Connor et al. 2005).

A nanodrug formulated by encapsulating lincomycin with gold nanoparticles biosynthesized using *Piper guineense* aqueous leaf extract and covalently functionalized with polyethylene glycol (PEG) has been reported in literature by Shittu et al. (2017). In their work, they synthesized gold nanoparticles using *Piper guineense* aqueous leaf extract and characterized them using UV-Vis spectrophotometer, DLS, TEM/EDS, and FTIR. They then prepared three formulations with different compositions. The first one was composed of the polymer (PEG) – greener synthesized gold nanoparticles – drug and labeled it as PND, while the second one was composed on the polymer (PEG) and the biosynthesized gold nanoparticles and labeled as PN, while the last one was composed of the polymer (PEG) and the drug and labeled it as PD. They then used in vitro dissolution methods to evaluate the potential performance of the formulated nanodrug by studying the release capability at 40 °C and 60 °C. Their results indicated the maximum release efficiency at the 9th minute (23.4 mg ml^{-1} for 40 °C) and (29.5 mg ml^{-1} for 60 °C) compared with the non-nanodrug as represented in Fig. 4.3. The functionalized drugs (PND, PN, and PD) displayed a progressive increase in the concentration of their drug release, with maximum peak at the 9th minute with PND having the highest standard drug release, as compared to the other formulations. They then concluded that the increase in the rate of the standard drug release was owed to both the PEG and the greener synthesized gold nanoparticles effect and more of the induction was from the nanoparticle rather than the polymer (Shittu et al. 2017).

Greener synthesized silver nanoparticles have also been studied for drug delivery systems. For example, silver nanoparticles synthesized from *Butea monosperma* plant extract by Patra et al. (2015) and characterized using various analytical methods were loaded with the FDA-approved chemotherapeutic drug doxorubicin to prepare drug delivery system (DDS) and proven anticancer efficacy in various cancer cells in vitro (Kotcherlakota et al. 2018; Patra et al. 2015). Their results showed that the DDS system efficiently delivered the doxorubicin into cancer cells and exhibited more cytotoxicity than a pristine drug, and they claimed that they believed that greener synthesized silver nanoparticles could be useful for the development of cancer therapy using nanomedicine approach in the near future (Patra et al. 2015).

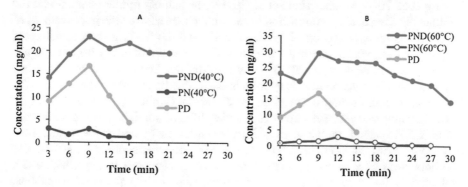

Fig. 4.3 Release efficiency of the formulated drug at different temperatures (**a**) 40 °C and (**b**) 60 °C. (Reproduced from Shittu et al. (2017), an open access article)

4.3.2.4 Wound Healing Applications

Wound healing is an important biological process involving many cell types and can be termed as a restorative operation of tissue injuries (Kotcherlakota et al. 2018), and any cut or blow to the body can be denoted to as a wound (Nguyen et al. 2009; Rieger et al. 2015). Wound healing is a complex process which is generally bothered by many physiological processes such as diabetes and blood loss and needs to be taken care of very carefully (Dong et al. 2014; Kotcherlakota et al. 2018). Angiogenesis plays a crucial role, including a number of growth factors and cytokines to accelerate the process of wound healing (Kotcherlakota et al. 2018; Tonnesen et al. 2000). To improve the process of wound healing, diverse nanomaterials have been incorporated. For example, gold and silver nanoparticles have lately been proven to have anti-inflammatory effects and thus improve wound healing, which has been exploited in developing better dressings for wound and burns (Noah 2018; Solano-Umaña and Vega-Baudrit 2015), and silver nanoparticles in dose-dependent manner is claimed that it can promote wound healing and lessen scar appearance (Shankar et al. 2015).

Green synthesized silver nanoparticles from *Lansium domesticum* fruit peel extract by Shankar et al. (2015) were believed to be have a potential in wound healing due to their antibacterial and antifungal activities of silver nanoparticles (Kotcherlakota et al. 2018; Shankar et al. 2015). In their work, they incorporated the synthesized silver nanoparticles in Pluronic F127 gels as a delivery scheme to evaluate their potential in wound healing. Their results indicated an enhanced wound healing activity of 0.1% w/w from the observed wound tensile strength of 33.41 ± 2.38 N/cm^2 and wound closure time. Also their results did not show any inflammation, under the histopathological and biomedical analysis, which was further corroborated with the high amount of collagen production in the treated groups compared to other groups (Kotcherlakota et al. 2018; Shankar et al. 2015). In another study by Garg et al. (2014) where they synthesized silver nanoparticles using the *Arnebia nobilis* root extract for use as a wound healing agent found that the synthesized silver nanoparticles hydrogel displayed substantial wound healing capability in excision wound models in animals due to their antibacterial property (Garg et al. 2014; Kotcherlakota et al. 2018). The authors further claimed that the epithelialization and the contraction of the wound area showed the improvement of cell proliferation, migration of epithelial cells, and the incrimination in action of myofibroblasts, which proved the wound healing potential of the hydrogel nanoparticles (Kotcherlakota et al. 2018). They also did not find any side effects green synthesis of silver nanoparticles hydrogels suggesting that they could make a useful alternate wound healing agent (Garg et al. 2014). Similarly, (Muhammad et al. 2017) synthesized silver nanoparticles under diffused sunlight using glucuronoxylan (GX) isolated from the seeds of *Mimosa pudica* (MP), and characterized them by UV-is, SEM equipped with STEM, and EDS (Kotcherlakota et al. 2018). The authors prepared a wound dressing from the silver nanoparticle impregnated GX which was found to exhibit a remarkable wound healing potential in rabbits. Likewise, silver nanoparticles synthesized using *Orchidantha chinensis*, an endocytic fungus of a Chinese herb were found to have greater antibacterial properties and wound healing capabilities (Kotcherlakota et al. 2018). The authors recommended that the protein produced by the endocytic fungus having antimicrobial as well as anti-inflammatory activity acted as a capping and stabilizing agent to silver nanoparticles (Kotcherlakota et al. 2018; Wen et al. 2016). Additionally, silver nanoparticles synthesized using linseed hydrogels of by Haseeb et al. (2017) were found to have remarkable healing properties as an antimicrobial dressing (Haseeb et al. 2017; Kotcherlakota et al. 2018) and more greener synthesized silver nanoparticles and their wound healing capabilities have also been reported in literature Al-Shmgani et al. (2017; Ovais et al. 2018; Thanganadar Appapalam and Panchamoorthy 2017).

Gold nanoparticles have also been shown to be helpful in would healing. For example, a study of the effect of a mixture of gold nanoparticles with epigallocatechin gallate (EGCG) in Hs68 and HaCaT cell proliferation and in mouse cutaneous wound healing by Jyh-Gang et al. (2012) was found to significantly increase the wound healing on mouse skin through anti-inflammatory and anti-oxidation effects. The authors proposed that their study might support future studies using other antioxidant agents in the treatment of cutaneous wounds (Jyh-Gang et al. 2012). However, very few reports exist of greener synthesized gold nanoparticles in would healing as compared to silver nanoparticles as earlier reported.

4.3.2.5 Leishmanicidal Agents

The current drugs for leishmaniasis, a disease caused by parasites of the genus *Leishmania* and transmitted to the host by a sand fly vector (McCall et al. 2013), have caused severe toxicity with the parasites developing resistant to the available leishmanicidal agents (Ahmad et al. 2016a; Natera et al. 2007). Biosynthesized gold and silver nanoparticles have been used as a combined therapy of the leishmanicidal agent to overcome the mentioned problem (Katas et al. 2018b). For example, amphotericin B absorbed on the surface of silver nanoparticles synthesized from the aqueous extract of *Isatis tinctoria* (Ahmad et al. 2016b; Katas et al. 2018b) was found to have an enhanced photo-induced anti-leishmanial activity with an IC50 of 2.43 μg/mL. Similarly, silver nanoparticles synthesized using *Anethum graveolens* leaf extract were found to augment anti-leishmanial effect of miltefosine by twofold even though the silver nanoparticles alone did not show any inhibition against the leishmanial parasite (Kalangi et al. 2016). It can be therefore concluded that combining biogenic silver nanoparticles with leishmanicidal agents could offer a safer and more effective alternative treatment for leishmaniasis (Katas et al. 2018b). Gold nanoparticles have also been reported as drug delivery systems for leishmanicidal agents (Katas et al. 2018b). Quercetin-functionalized gold nanoparticles synthesized by Das et al. (2013) were reported to be effective against wild- and resistant-type visceral leishmaniasis. Likewise, gold nanoparticles biosynthesized from an aqueous extract of *Rhazya stricta Decne* as a reducing agent were found to be effective against *Leishmania tropica (HTD7)* (Ahmad et al. 2017)

4.3.2.6 Anti-inflammatory Activity

Anti-inflammation is biological process which produces compounds such as cytokines and interleukins, produced by specific T lymphocytes, B lymphocytes, and macrophages, and it is important for wound healing mechanism (Justin Packia Jacob et al. 2012; Khandel et al. 2018). These anti-inflammatory mediators are secreted from the primary immune organs (Satyavani et al. 2011) and are said to be involved in the biochemical pathways as well as controlling the expansion of diseases (Gurunathan et al. 2009). It has been reported that gold and silver nanoparticles synthesized from plant extracts have significant tissue regeneration through inflammatory functions; thus, they can be used to naturally prevent inflammation (Gurunathan et al. 2009)

Greener synthesized silver nanoparticles using *Salvia officinalis* extract were found to increase IL-8 and TNF-α genes expression by 28.76% and 42%, respectively, but suppressed cyclooxygenase-2 gene expression with 20.5% as compared to control groups (Baharara et al. 2017). This indicated that coating silver nanoparticles with *Salvia officinalis* could have a promising potential as chemotherapeutic agents in future. Similarly silver nanoparticles synthesized from *Viburnum opulus* L. fruits extract and investigated for their anti-inflammatory effect both in vitro on HaCaT cell line and in vivo on Wistar rats were found to have potent anti-inflammatory activity and thus could be used as therapeutic tools for treatment of inflammation (Moldovan et al. 2017). Moreover, silver nanoparticles synthesized using edible mushroom (*Agaricus bisporus*) and forest mushroom (*Ganoderma lucidum*) extract and investigated for their anti-inflammatory effects using the heat-induced hemolysis method were found a protection of between 75 and 84% indicating a good anti-inflammatory activity (Sriramulu and Sumathi 2017). Also, silver nanoparticles synthesized using *Piper nigrum* fruit extract demonstrated inhibitory effect by lipopolysaccharide (LPS)-induced expression of inflammatory cytokines IL-1β and IL-6 in human peripheral blood homonuclear cells by RT-PCR assay (Mani et al. 2015). The anti-inflammatory activity was found to be more active at 10–20 μg/mL concentration which the authors attributed to the synergism between the silver ions with the extract (Mani et al. 2015).

Gold nanoparticles synthesized from the upcycling of jellyfish (*Nemopilema nomurai*) seawater wastes were found to decrease nitric oxide (NO) secretion and inducible nitric oxide synthase (iNOS) expression levels which resulted in anti-inflammatory effects in lipopolysaccharide (LPS)-inflamed macrophages (Ahn et al. 2018). Monodispersed hexagonal gold nanoparticles synthesized by fruit extract of *Prunus serrulata* were demonstrated to reduce expression of inflammatory mediators such as NO, prostaglandin E2 (PGE2), iNOS, and cyclooxygenase-2 (COX-2) and significantly suppressed LPS-induced activation of NF-kB signaling pathway p38 MAPK in RAW 264.7 cells indicating that these biosynthesized gold nanoparticles could be utilized as novel therapeutic agents for the prevention and cure of inflammation (Singh et al. 2018).

4.3.2.7 Anticoagulating Activity

Greener synthesized nanoparticles have also been found to exhibit anticoagulating activity. Coagulation is very complex process which involves fibrin maturation and deposition along with activation, adhesion, and aggregation of platelets (Prentice et al. 1966). Anticoagulation process is required for different diseases such as cardiovascular disorders, allergic responses, and injuries where there are malfunctions in coagulation (Kotcherlakota et al. 2018; Prentice et al. 1966). Silver nanoparticles synthesized from *Petiveria alliacea* as reported by Lateef et al. (2018) were found to inhibit coagulation of human blood. Likewise, silver nanoparticles biosynthesized using cell-free extract of *Bacillus safensis* LAU 13 were also found to inhibit coagulation of human blood and also completely dissolved human blood clots as obtained through light microscopy (Lateef et al. 2016). Also, silver nanoparticles synthesized from the culture supernatant of *Pseudomonas aeruginosa* (Jeyaraj et al. 2013) were found to have a stable anticoagulant effect. Gold and silver nanoparticles biosynthesized using *Brevibacterium casei* (Kalishwaralal et al. 2010), *Panax ginseng* leaves (Singh et al. 2016), and *Aloe vera* (Kamala Priya 2015) were also found to anticoagulating effects.

4.3.2.8 Antioxidant Activity

Antioxidant agents are substances that can regulate the formation of free radicals which are found to cause cellular damage such as brain damage, cancer, and atherosclerosis (Abdel-Aziz et al. 2014; Khandel et al. 2018). The free radicals are generated by reactive oxygen species (ROS) and it is reported that biomolecules such as proteins, lipids, fatty acids and glycoproteins, phenolic, flavonoids, and sugars strongly control the growth formation of the free radicals. The scavenging effect of antioxidants has been found to be useful in the management of several chronic diseases such as diabetes, cancer, AIDS, nephritis, and metabolic disorders (Khandel et al. 2018), and biosynthesized metal nanoparticles contain high phenolic and flavonoids content in the extract which can act as antioxidants. Silver nanoparticle biosynthesized using *Andrographis paniculata* (Suriyakalaa et al. 2013) and *Morinda tinctoria* Roxb leaves (Paramasivam et al. 2017) were found to have very strong antioxidant effect. Gold nanoparticles synthesized from the rind extract of *Citrullus lanatus* were found to have antioxidant activity (Patra and Baek 2015). Silver and gold nanoparticles synthesized from *Bauhinia purpurea* leaf extract displayed a high antioxidant potential with IC_{50} values of 42.37 µg/mL and 27.21 µg/mL, respectively, as measured using DPPH assay (Vijayan et al. 2019).

4.3.2.9 Antidiabetic Activity

Diabetes is a group of metabolic dysfunctions in which a person has high blood sugar levels. Its control is based on the control of certain food, and balanced diet or synthetic insulin drugs which can be a big challenge. Lately, metallic nanoparticles are being used in the management of diabetes mellitus (DM) (Khandel et al. 2018). For example, in a study by Daisy and Saipriya 2012), gold nanoparticles were found to have a good therapeutic activity against diabetes models. In their study they found that the gold nanoparticles synthesized from *Cassia fistula* aqueous extract treated diabetic mice showed a decrease in the HbA (glycosylated hemoglobulin) level in a normal range (Daisy and Saipriya 2012). In another study by Swarnalatha et al. (2012) where they studied silver nanoparticles synthesized from *Sphaeranthus amaranthoides,* they found that the nanoparticles inhibited α-amylase and acarbose sugars in diabetes-induced animal models (Swarnalatha et al. 2012). They also reported that the α-amylase present in the ethanol extract of the *Sphaeranthus amaranthoides* was the main inhibiting compound (BarathManiKanth et al. 2010). Similarly, silver nanoparticles from the aqueous leaf extract of *Pouteria sapota* were found to cause a significant reduction in blood sugar levels in rats which were treated with the leaf extracts indicated their antidiabetic activity and, therefore, could have potential for development of medical applications in the future (Prabhu et al. 2018). In another study by Saratale et al. (2018), silver nanoparticles synthesized using *Punica granatum* leaf extract were found to have antidiabetic potential by exhibiting effective inhibition against α-amylase and α-glucosidase with an IC_{50} of 65.2 and 53.8 μg/mL, respectively. The biosynthesized silver nanoparticles also showed a dose-dependent response against human liver cancer cells (HepG2) with an IC_{50} of 70 μg/mL indicating their efficacy in killing the cancer cells. They also showed free radical scavenging activity, antioxidant activity, and antibacterial activity indicating their potential biomedical applications (Saratale et al. 2018).

4.3.2.10 Antiviral Activity

Viruses are said to enter the host very rapidly and multiply their colonies very often (Khandel et al. 2018). Studies have suggested that biosynthesized nanoparticles using plant extracts can be used as alternative for the treatment and control of the growth of viral pathogens. For example, silver nanoparticles synthesized from plant extracts are said to act as potent antiviral agents for a wide range of viral infections (Khandel et al. 2018). In a study by Sun et al. (2005), they studied the efficiency of biosynthesized silver nanoparticles against HIV pathogens and found that they were effective against HIV action at an early stage of reverse transcription mechanism. They reported that that the nanoparticles can be used as strong antiviral agents since they inhibit the entry of viruses into the host system (Sun et al. 2005). The mechanism for the anti-HIV activity was reported to be due to the multiple binding sites of the nanoparticles and was found to act against cell-free viruses as well as cell-associated viruses (Khandel et al. 2018; Sun et al. 2005).

4.3.2.11 Antiangiogenic Activity

Silver and gold nanoparticles have also been found to have antiangiogenic activity. Angiogenesis is an important phenomenon involved in the normal growth and wound healing processes (Gurunathan et al. 2009). It is the process of development of new blood vessels and has become a major focus of research and is involved in the normal growth and involves endothelial cell growth, differentiation, proliferation, and invasion procedures (Bikfalvi and Bicknell 2002; Kotcherlakota et al. 2018). It is required for physiological procedures and plays a key role in numerous pathological conditions such as tumor growth and metastasis (Baharara et al. 2014a). Hindering angiogenesis process through interfering in its pathway is a favorable methodology to hinder the progression of these diseases (Gurunathan et al. 2009).

Gold nanoparticles have been reported to inhibit the function of pro-angiogenic heparin-binding growth factors (HB-GFs) including vascular endothelial growth factor 165 (VEGF165), basic fibroblast growth factor (bFGF)(Arvizo et al. 2011; Mukherjee et al. 2005). They have also been reported to display excellent biocompatibility, low toxicity, and antiangiogenic effect though the mechanism of the antiangiogenesis is still unknown (Pan et al. 2013). Biosynthesized gold nanoparticles have also been reported to have antiangiogenic activity. For example, gold nanoparticles synthesized using *Memecylon malabaricum* leaf extract by Rekha et al. (2018) were found to have antiangiogenic properties on a mice model.

Silver nanoparticles synthesized from the plant extract of *Salvia officinalis* by Baharara et al. (2014a) were found to exhibit antiangiogenic activity in chick chorioallantoic membrane (CAM) by reducing the hemoglobin content in blood. This prompted the authors to conclude that those biosynthesized silver nanoparticles could be considered as a promising chemotherapeutic agent in cancer treatment (Baharara et al. 2014a). Also, in another study by Baharara et al. (2014b), silver nanoparticles synthesized from *Achillea biebersteinii* flowers extract were found to lead to a 50% decrease in the length and number of vessel-like structures which indicated the antiangiogenic activity of the silver nanoparticles (Baharara et al. 2014b). Similarly, Hullikere et al. (2015) reported silver nanoparticles synthesized from the leaf extract of *Tragia involucrata* which were also found to have antiangiogenic activity. The authors reported that they used the chorioallantoic membrane (CAM) assay and found that the synthesized silver nanoparticles showed significant antiangiogenic effect of the CAM assay. They further reported that though the mechanism of action of the silver nanoparticles in preventing the angiogenesis is unknown, they theorized that the silver nanoparticles may hamper the blood vessel formation either by upregulating the inhibitors or downregulating the stimulators. They however recommended that more studies of silver nanoparticles at the molecular level might help in finding out the mechanism by which the silver nanoparticles act on angiogenesis process (Hullikere et al. 2015). Several other researches have also reported the anti-angiogenesis activity of greener synthesized nanoparticles (He et al. 2016; Paramasivam et al. 2017; Rekha et al. 2018) making these nanoparticles to have potential application in the biomedical field.

4.4 Conclusions

Due to their extreme small sizes, high surface area, and physical and chemical properties, silver and gold nanoparticles have found widespread applications. Owing to their high biocompatibility, chemical stability, convenient surface bioconjugation with molecular probes, and excellent surface plasmon resonance, these nanoparticles have widespread biomedical applications including targeted drug delivery, sensing and imaging, cancer treatment, DNA-RNA analysis, gene therapy, diagnosis, antibacterial agents, and therapeutics among others. Several researchers have reported eco-friendly and cost-effective green methods for the synthesis of these nanoparticles using microorganisms or plant extracts. This book chapter focused on the application of greener synthesized silver and gold nanoparticles in medicine. It has been well established from literature as reported in this chapter that these greener synthesized nanoparticles have potential as anticancer agents, antimicrobial agents, biosensors for disease diagnosis, anticoagulants, and wound healing agents and can also be used in drug formulations for targeted drug delivery. All these applications can transform medicine the way we know, and therefore more research should be done for more practical applications.

References

Abalkhil TA, Alharbi SA, Salmen SH, Wainwright M (2017) Bactericidal activity of biosynthesized silver nanoparticles against human pathogenic bacteria. Biotechnol Biotechnol Equip 31:411–417. https://doi.org/10.1080/13102818.2016.1267594

Abbasi Z, Feizi S, Taghipour E, Ghadam P (2017) Green synthesis of silver nanoparticles using aqueous extract of dried Juglans regia green husk and examination of its biological properties. Green Processes Synth 6:477–485. https://doi.org/10.1515/gps-2016-0108

Abdel-Aziz MS, Shaheen MS, El-Nekeety AA, Abdel-Wahhab MA (2014) Antioxidant and antibacterial activity of silver nanoparticles biosynthesized using Chenopodium murale leaf extract. J Saudi Chem Soc 18:356–363. https://doi.org/10.1016/j.jscs.2013.09.011

Abdel-Raouf N, Al-Enazi NM, Ibraheem IBM (2017) Green biosynthesis of gold nanoparticles using Galaxaura elongata and characterization of their antibacterial activity. Arab J Chem 10:S3029–S3039. https://doi.org/10.1016/j.arabjc.2013.11.044

Abou El-Nour KMM, Aa E, Al-Warthan A, Ammar RAA (2010) Synthesis and applications of silver nanoparticles. Arab J Chem 3:135–140. https://doi.org/10.1016/j.arabjc.2010.04.008

Adebayo-Tayo BC, Inem SA, Olaniyi OA (2019) Rapid synthesis and characterization of gold and silver nanoparticles using exopolysaccharides and metabolites of Wesiella confusa as an antibacterial agent against Esherichia coli. Int J Nano Dimens 10:37–47

Afolabi AS, Abdulkareem AS, Mhlanga SD, Iyuke SE (2011) Synthesis and purification of bimetallic catalysed carbon nanotubes in a horizontal CVD reactor. J Exp Nanosci 6:248–262

Ahmad A, Senapati S, Khan MI, Kumar R, Sastry M (2003) Extracellular biosynthesis of Monodisperse gold nanoparticles by a novel extremophilic actinomycete, Thermomonospora sp. Langmuir 19:3550–3553. https://doi.org/10.1021/la0267721

Ahmad A, Syed F, Imran M, Khan AU, Tahir K, Khan ZUH, Yuan Q (2016a) Phytosynthesis and antileishmanial activity of gold nanoparticles by Maytenus royleanus. J Food Biochem 40:420–427. https://doi.org/10.1111/jfbc.12232

Ahmad A et al (2016b) Isatis tinctoria mediated synthesis of amphotericin B-bound silver nanoparticles with enhanced photoinduced antileishmanial activity: a novel green approach. J Photochem Photobiol B Biol 161:17–24. https://doi.org/10.1016/j.jphotobiol.2016.05.003

Ahmad A et al (2017) Synthesis of phytochemicals-stabilized gold nanoparticles and their biological activities against bacteria and Leishmania. Microb Pathog 110:304–312. https://doi.org/10.1016/j.micpath.2017.07.009

Ahmed S, Ahmad M, Swami BL, Ikram S (2016a) A review on plants extract mediated synthesis of silver nanoparticles for antimicrobial applications: a green expertise. J Adv Res 7:17–28. https://doi.org/10.1016/j.jare.2015.02.007

Ahmed S, Saifullah AM, Swami BL, Ikram S (2016b) Green synthesis of silver nanoparticles using Azadirachta indica aqueous leaf extract. J Radiat Res Appl Sci 9:1–7. https://doi.org/10.1016/j.jrras.2015.06.006

Ahn E-Y, Hwang SJ, Choi M-J, Cho S, Lee H-J, Park Y (2018) Upcycling of jellyfish (Nemopilema nomurai) sea wastes as highly valuable reducing agents for green synthesis of gold nanoparticles and their antitumor and anti-inflammatory activity. Artif Cells Nanomed Biotechnol 46:1–10. https://doi.org/10.1080/21691401.2018.1480490

Akkiraju CP, Tathe SP, Mamillapalli S (2017) Green synthesis of silver nanoparticles from Punica granatum L. and its antimicrobial activity. Adv Appl Sci Res 8:42–49

Alaqad K, Saleh TA (2016) Gold and silver nanoparticles: synthesis methods, characterization routes and applications towards drugs. J Environ Anal Toxicol 6:384. https://doi.org/10.4172/2161-0525.1000384

Ali K, Ahmed B, Dwivedi S, Saquib Q, Al-Khedhairy AA, Musarrat J (2015a) Microwave accelerated green synthesis of stable silver nanoparticles with eucalyptus globulus leaf extract and their antibacterial and antibiofilm activity on clinical isolates. PLoS One 10:e0131178. https://doi.org/10.1371/journal.pone.0131178

Ali ZA, Yahya R, Sekaran SD, Puteh R (2015b) Green synthesis of silver nanoparticles using apple extract and its antibacterial properties. Adv Mater Sci Eng 2016:6

Alivisatos P (2004) The use of nanocrystals in biological detection. Nat Biotechnol 22:47–52. https://doi.org/10.1038/nbt927

Al-Shmgani HSA, Mohammed WH, Sulaiman GM, Saadoon AH (2017) Biosynthesis of silver nanoparticles from Catharanthus roseus leaf extract and assessing their antioxidant, antimicrobial, and wound-healing activities. Artif Cells Nanomed Biotechnol 45:1234–1240. https://doi.org/10.1080/21691401.2016.1220950

Ambrosi A, Airò F, Merkoçi A (2010) Enhanced gold nanoparticle based ELISA for a breast cancer biomarker. Anal Chem 82:1151–1156. https://doi.org/10.1021/ac902492c

Arvizo RR, Rana S, Miranda OR, Bhattacharya R, Rotello VM, Mukherjee P (2011) Mechanism of anti-angiogenic property of gold nanoparticles: role of nanoparticle size and surface charge. Nanomedicine 7:580–587. https://doi.org/10.1016/j.nano.2011.01.011

Ateeq M et al (2015) Green synthesis and molecular recognition ability of patuletin coated gold nanoparticles. Biosens Bioelectron 63:499–505. https://doi.org/10.1016/j.bios.2014.07.076

Azizian Shermeh O, Valizadeh J, Noroozifar M, Ghasemi A, Valizadeh M (2016) Optimization, characterization and anti microbial activity of gold nano particles biosynthesized using aqueous extract of Sambucus ebulus L eco-phytochemical. J Med Plants 4:1–18

Bagherzade G, Tavakoli MM, Namaei MH (2017) Green synthesis of silver nanoparticles using aqueous extract of saffron (Crocus sativus L.) wastages and its antibacterial activity against six bacteria. Asian Pac J Trop Biomed 7:227–233. https://doi.org/10.1016/j.apjtb.2016.12.014

Baharara J, Namvar F, Mousavi M, Ramezani T, Mohamad R (2014a) Anti-angiogenesis effect of biogenic silver nanoparticles synthesized using saliva officinalis on chick chorioalantoic membrane (CAM). Molecules 19:13498

Baharara J, Namvar F, Ramezani T, Hosseini N, Mohamad R (2014b) Green synthesis of silver nanoparticles using Achillea biebersteinii flower extract and its anti-angiogenic properties in the rat aortic ring model. Molecules 19:4624

Baharara J, Ramezani T, Mousavi M, Asadi-Samani M (2017) Antioxidant and anti-inflammatory activity of green synthesized silver nanoparticles using Salvia officinalis extract. Ann Trop Med Public Health 10:1265–1270. https://doi.org/10.4103/atmph.atmph_174_17

Barai AC, Paul K, Dey A, Manna S, Roy S, Bag BG, Mukhopadhyay C (2018) Green synthesis of Nerium oleander-conjugated gold nanoparticles and study of its in vitro anticancer activity on MCF-7 cell lines and catalytic activity. Nano Converg 5:10. https://doi.org/10.1186/s40580-018-0142-5

BarathManiKanth S, Kalishwaralal K, Sriram M, Pandian SRK, H-s Y, Eom S, Gurunathan S (2010) Anti-oxidant effect of gold nanoparticles restrains hyperglycemic conditions in diabetic mice. J Nanobiotechnol 8:16

Bikfalvi A, Bicknell R (2002) Recent advances in angiogenesis, anti-angiogenesis and vascular targeting. Trends Pharmacol Sci 23:576–582. https://doi.org/10.1016/S0165-6147(02)02109-0

Bo R, Ren-Cheng T (2017) Green synthesis of silver nanoparticles with antibacterial activities using aqueous Eriobotrya japonica leaf extract. Adv Nat Sci Nanosci Nanotechnol 8:015014

Cai W, Chen X (2007) Nanoplatforms for targeted molecular imaging in living subjects. Small 3:1840–1854. https://doi.org/10.1002/smll.200700351

Castro L, Blazquez ML, Munoz JA, Gonzalez F, Ballester A (2013) Biological synthesis of metallic nanoparticles using algae. IET Nanobiotechnol 7:109–116. https://doi.org/10.1049/iet-nbt.2012.0041

Che X, Yuan R, Chai Y, Li J, Song Z, Wang J (2010) Amperometric immunosensor for the determination of α-1-fetoprotein based on multiwalled carbon nanotube–silver nanoparticle composite. J Colloid Interface Sci 345:174–180. https://doi.org/10.1016/j.jcis.2010.01.033

Chintamani BR, Salunkhe KS, Chavan MJ (2018) Emerging use of green synthesis silver nanoparticle: an updated review. Int J Pharm Sci Res 9:4029–4055. https://doi.org/10.13040/IJPSR.0975-8232.9(10).4029-55

Cho K, Wang X, Nie S, Chen Z, Shin DM (2008) Therapeutic nanoparticles for drug delivery in cancer. Clin Cancer Res 14:1310–1316. https://doi.org/10.1158/1078-0432.ccr-07-1441

Connor EE, Mwamuka J, Gole A, Murphy CJ, Wyatt MD (2005) Gold nanoparticles are taken up by human cells but do not cause acute cytotoxicity. Small 1:325–327. https://doi.org/10.1002/smll.200400093

Costa P, Amaro A, Botelho A, Inácio J, Baptista PV (2010) Gold nanoprobe assay for the identification of mycobacteria of the Mycobacterium tuberculosis complex. Clin Microbiol Infect 16:1464–1469. https://doi.org/10.1111/j.1469-0691.2010.03120.x

Daisy P, Saipriya K (2012) Biochemical analysis of Cassia fistula aqueous extract and phytochemically synthesized gold nanoparticles as hypoglycemic treatment for diabetes mellitus. Int J Nanomedicine 7:1189

Dang H, Fawcett D, Poinern GEJ (2017) Biogenic synthesis of silver nanoparticles from waste banana plant stems and their antibacterial activity against Escherichia coli and Staphylococcus Epidermis. Int J Res Med Sci 5:3769. https://doi.org/10.18203/2320-6012.ijrms20173947

Das S, Roy P, Mondal S, Bera T, Mukherjee A (2013) One pot synthesis of gold nanoparticles and application in chemotherapy of wild and resistant type visceral leishmaniasis. Colloids Surf B: Biointerfaces 107:27–34. https://doi.org/10.1016/j.colsurfb.2013.01.061

de la Escosura-Muñiz A, Maltez-da Costa M, Sánchez-Espinel C, Díaz-Freitas B, Fernández-Suarez J, González-Fernández Á, Merkoçi A (2010) Gold nanoparticle-based electrochemical magnetoimmunosensor for rapid detection of anti-hepatitis B virus antibodies in human serum. Biosens Bioelectron 26:1710–1714. https://doi.org/10.1016/j.bios.2010.07.069

Dong Y, Hassan WU, Kennedy R, Greiser U, Pandit A, Garcia Y, Wang W (2014) Performance of an in situ formed bioactive hydrogel dressing from a PEG-based hyperbranched multifunctional copolymer. Acta Biomater 10:2076–2085. https://doi.org/10.1016/j.actbio.2013.12.045

Du L, Jiang H, Liu X, Wang E (2007) Biosynthesis of gold nanoparticles assisted by Escherichia coli DH5α and its application on direct electrochemistry of hemoglobin. Electrochem Commun 9:1165–1170. https://doi.org/10.1016/j.elecom.2007.01.007

Duan H, Wang D, Li Y (2015) Green chemistry for nanoparticle synthesis. Chem Soc Rev 44:5778–5792. https://doi.org/10.1039/c4cs00363b

Dumur F, Guerlin A, Dumas E, Bertin D, Gigmes D, Mayer CR (2011) Controlled spontaneous generation of gold nanoparticles assisted by dual reducing and capping agents. Gold Bull 44:119–137. https://doi.org/10.1007/s13404-011-0018-5

Escárcega-González CE et al (2018) In vivo antimicrobial activity of silver nanoparticles produced via a green chemistry synthesis using Acacia rigidula as a reducing and capping agent. Int J Nanomedicine 13:2349–2363. https://doi.org/10.2147/IJN.S160605

Farkhari N, Abbasian S, Moshaii A, Nikkhah M (2016) Mechanism of adsorption of single and double stranded DNA on gold and silver nanoparticles: investigating some important parameters in bio-sensing applications. Colloids Surf B: Biointerfaces 148:657–664. https://doi.org/10.1016/j.colsurfb.2016.09.022

Fayaz AM, Girilal M, Rahman M, Venkatesan R, Kalaichelvan P (2011) Biosynthesis of silver and gold nanoparticles using thermophilic bacterium Geobacillus stearothermophilus. Process Biochem 46:1958–1962

Firdhouse MJ, Lalitha P (2013) Biosynthesis of silver nanoparticles using the extract of Alternanthera sessilis—antiproliferative effect against prostate cancer cells. Cancer Nanotechnol 4:137–143. https://doi.org/10.1007/s12645-013-0045-4

Garg S, Chandra A, Mazumder A, Mazumder R (2014) Green synthesis of silver nanoparticles using Arnebia nobilis root extract and wound healing potential of its hydrogel. Asian J Pharm 8:95. https://doi.org/10.4103/0973-8398.134925

Gatebe E (2012) Nanotechnology: the magic bullet towards attainment of Kenya's vision 2030 on industrialization. J Agric Sci Technol 14:1–2

Geetha R, Ashokkumar T, Tamilselvan S, Govindaraju K, Sadiq M, Singaravelu G (2013) Green synthesis of gold nanoparticles and their anticancer activity. Cancer Nanotechnol 4:91–98. https://doi.org/10.1007/s12645-013-0040-9

Giljohann DA, Seferos DS, Daniel WL, Massich MD, Patel PC, Mirkin CA (2010) Gold nanoparticles for biology and medicine. Angew Chem Int Ed 49:3280–3294. https://doi.org/10.1002/anie.200904359

Gonçalo D et al (2012) Noble metal nanoparticles for biosensing applications. Sensors (Basel, Switzerland) 12:1657–1687. https://doi.org/10.3390/s120201657

Gurunathan S, Lee K-J, Kalishwaralal K, Sheikpranbabu S, Vaidyanathan R, Eom SH (2009) Antiangiogenic properties of silver nanoparticles. Biomaterials 30:6341–6350. https://doi.org/10.1016/j.biomaterials.2009.08.008

Gurunathan S, Han JW, Eppakayala V, Jeyaraj M, Kim JH (2013) Cytotoxicity of biologically synthesized silver nanoparticles in MDA-MB-231 human breast cancer cells. Biomed Res Int 535796:8

Haseeb MT, Hussain MA, Abbas K, Youssif BG, Bashir S, Yuk SH, Bukhari SNA (2017) Linseed hydrogel-mediated green synthesis of silver nanoparticles for antimicrobial and wound-dressing applications. Int J Nanomedicine 12:2845–2855. https://doi.org/10.2147/ijn.s133971

He Y, Du Z, Ma S, Liu Y, Li D, Huang H, Jiang S, Cheng S, Wu W, Zhang K, Zheng X (2016) Effects of green-synthesized silver nanoparticles on lung cancer cells in vitro and grown as xenograft tumors in vivo. Int J Nanomedicine 11:1879–1887. https://doi.org/10.2147/ijn.s103695

Hebeish A, El-Naggar ME, Fouda MM, Ramadan MA, Al-Deyab SS, El-Rafie MH (2011) Highly effective antibacterial textiles containing green synthesized silver nanoparticles. Carbohydr Polym 86:936–940. https://doi.org/10.1016/j.carbpol.2011.05.048

Hu Y, Hua S, Li F, Jiang Y, Bai X, Li D, Niu L (2011) Green-synthesized gold nanoparticles decorated graphene sheets for label-free electrochemical impedance DNA hybridization biosensing. Biosens Bioelectron 26:4355–4361. https://doi.org/10.1016/j.bios.2011.04.037

Huang J, Li Q, Sun D, Lu Y, Yang X, Wang H, Wang Y, Shao W, He N, Hong J, Chen C (2007) Biosynthesis of silver and gold nanoparticles by novel sundried Cinnamomum camphora leaf. Nanotechnology 18:105104

Hullikere MM, Joshi CG, Vijay R, Ananda D, Nivya M (2015) Antiangiogenic, cytotoxic and antimicrobial activity of plant mediated silver nano particle from Tragia involucrate. Res J Nanosci Nanotechnol 5:16–26

Huo Q, Colon J, Cordero A, Bogdanovic J, Baker CH, Goodison S, Pensky MY (2011) A facile nanoparticle immunoassay for cancer biomarker discovery. J Nanobiotechnol 9:20. https://doi.org/10.1186/1477-3155-9-20

Husseiny MI, El-Aziz MA, Badr Y, Mahmoud MA (2007) Biosynthesis of gold nanoparticles using Pseudomonas aeruginosa. Spectrochim Acta A Mol Biomol Spectrosc 67:1003–1006. https://doi.org/10.1016/j.saa.2006.09.028

Ikram S, Ahmed S (2015) Synthesis of gold nanoparticles using plant extract: an overview. Nano Res Appl 1:1

Jason NP, Hitesh KW, Michael GC, William H, Sarah T, Harsh M, Fenil C, Vivek B, Mathew BL, Rajalingam D (2016) Novel synthesis of kanamycin conjugated gold nanoparticles with potent antibacterial activity. Front Microbiol 7:607. https://doi.org/10.3389/fmicb.2016.00607

Jeyaraj M, Varadan S, Anthony KJP, Murugan M, Raja A, Gurunathan S (2013) Antimicrobial and anticoagulation activity of silver nanoparticles synthesized from the culture supernatant of Pseudomonas aeruginosa. J Ind Eng Chem 19:1299–1303. https://doi.org/10.1016/j.jiec.2012.12.031

Justin Packia Jacob S, Finub JS, Narayanan A (2012) Synthesis of silver nanoparticles using Piper longum leaf extracts and its cytotoxic activity against hep-2 cell line. Colloids Surf B: Biointerfaces 91:212–214. https://doi.org/10.1016/j.colsurfb.2011.11.001

Jyh-Gang L et al (2012) The effects of gold nanoparticles in wound healing with antioxidant epigallocatechin gallate and α-lipoic acid. Nanomedicine 8:767–775. https://doi.org/10.1016/j.nano.2011.08.013

Kajani AA, Bordbar A-K, Zarkesh Esfahani SH, Razmjou A (2016) Gold nanoparticles as potent anticancer agent: green synthesis, characterization, and in vitro study. RSC Adv 6:63973–63983. https://doi.org/10.1039/c6ra09050h

Kalangi SK, Dayakar A, Gangappa D, Sathyavathi R, Maurya RS, Narayana Rao D (2016) Biocompatible silver nanoparticles reduced from Anethum graveolens leaf extract augments the antileishmanial efficacy of miltefosine. Exp Parasitol 170:184–192. https://doi.org/10.1016/j.exppara.2016.09.002

Kalishwaralal K, Deepak V, Ram Kumar Pandian S, Kottaisamy M, BarathManiKanth S, Kartikeyan B, Gurunathan S (2010) Biosynthesis of silver and gold nanoparticles using Brevibacterium casei. Colloids Surf B: Biointerfaces 77:257–262. https://doi.org/10.1016/j.colsurfb.2010.02.007

Kamala Priya MR (2015) Applications of the green synthesized gold nanoparticles-antimicrobial activity, water purification system and drug delivery system. Nanosci Technol 2:1–4. https://doi.org/10.15226/2374-8141/2/2/00126

Katas H, Lim CS, Nor Azlan AYH, Buang F, Mh Busra MF (2018a) Antibacterial activity of biosynthesized gold nanoparticles using biomolecules from Lignosus rhinocerotis and chitosan. Saudi Pharm J. https://doi.org/10.1016/j.jsps.2018.11.010

Katas H, Moden NZ, Lim CS, Celesistinus T, Chan JY, Ganasan P, Suleman Ismail Abdalla S (2018b) Biosynthesis and potential applications of silver and gold nanoparticles and their chitosan-based nanocomposites in nanomedicine. J Nanotechnol 2018:1–13. https://doi.org/10.1155/2018/4290705

Kaviya S, Santhanalakshmi J, Viswanathan B, Muthumary J, Srinivasan K (2011) Biosynthesis of silver nanoparticles using citrus sinensis peel extract and its antibacterial activity. Spectrochim Acta A Mol Biomol Spectrosc 79:594–598. https://doi.org/10.1016/j.saa.2011.03.040

Khalil MMH, Sabry DY, Mahdi H (2017) Green synthesis of silver, gold and silver-gold nanoparticles: characterization, antimicrobial activity and cytotoxicity. J Sci Res Sci 34:553–574. https://doi.org/10.21608/jsrs.2018.14711

Khan AKRR, Murtaza G, Zahra A (2014) Gold nanoparticles: synthesis and applications in drug delivery. Trop J Pharm Res 13:1169–1177

Khan I, Saeed K, Khan I (2017) Nanoparticles: properties, applications and toxicities. Arab J Chem. https://doi.org/10.1016/j.arabjc.2017.05.011

Khandel P, Yadaw RK, Soni DK, Kanwar L, Shahi SK (2018) Biogenesis of metal nanoparticles and their pharmacological applications: present status and application prospects. J Nanostruct Chem 8:217–254. https://doi.org/10.1007/s40097-018-0267-4

Khatoon N, Mazumder JA, Sardar M (2017) Biotechnological applications of green synthesized silver nanoparticles. J Nanosci 2:107. https://doi.org/10.4172/2572-0813.1000107

Kim J, Lee J, Kwon S, Jeong S (2009) Preparation of biodegradable polymer/silver nanoparticles composite and its antibacterial efficacy. J Nanosci Nanotechnol 9:1098–1102. https://doi.org/10.1166/jnn.2009.C096

Kotcherlakota R et al (2017) Engineered fusion protein-loaded gold nanocarriers for targeted co-delivery of doxorubicin and erbB2-siRNA in human epidermal growth factor receptor-2+ ovarian cancer. J Mater Chem B 5:7082–7098. https://doi.org/10.1039/c7tb01587a

Kotcherlakota R, Das S, Patra CR (2018) Therapeutic applications of green-synthesized silver nanoparticles, vol 1, 1st edn. Elsevier, Amsterdam

Krishnaraj C, Jagan EG, Rajasekar S, Selvakumar P, Kalaichelvan PT, Mohan N (2010) Synthesis of silver nanoparticles using Acalypha indica leaf extracts and its antibacterial activity against water borne pathogens. Colloids Surf B: Biointerfaces 76:50–56. https://doi.org/10.1016/j.colsurfb.2009.10.008

Kumar V, Yadav SK (2009) Plant-mediated synthesis of silver and gold nanoparticles and their applications. J Chem Technol Biotechnol 84:151–157. https://doi.org/10.1002/jctb.2023

Kundu S (2017) Gold nanoparticles: their application as antimicrobial agents and vehicles of gene delivery. Adv Biotechnol Microbiol 4. https://doi.org/10.19080/AIBM.2017.04.555658.

Kuppusamy P, Mashitah MY, Maniam GP, Govindan N (2014) Biosynthesized gold nanoparticle developed as a tool for detection of HCG hormone in pregnant women urine sample. Asian Pac J Trop Dis 4:237. https://doi.org/10.1016/S2222-1808(14)60538-7

Lateef A, Ojo SA, Oladejo SM (2016) Anti-candida, anti-coagulant and thrombolytic activities of biosynthesized silver nanoparticles using cell-free extract of Bacillus safensis LAU 13. Process Biochem 51:1406–1412. https://doi.org/10.1016/j.procbio.2016.06.027

Lateef A, Bolaji F, Oladejo SM, Oluwadamilare Akinola P, Beukes L, Evariste Bosco GK (2018) Characterization, antimicrobial, antioxidant and anticoagulant activities of silver nanoparticles synthesized from Petiveria alliacea L. leaf extract. Prep Biochem Biotechnol 48:646–652. https://doi.org/10.1080/10826068.2018.1479864

Leena F, Vikram P, Venkataramana RD, Arpit B, Saleem SP, Rahaman IL, Heena S, Subhra D, Rajdeep C, Jitendra P (2017) Biosynthesized protein-capped silver nanoparticles induce ROS-dependent proapoptotic signals and prosurvival autophagy in cancer cells. ACS Omega 2:1489–1504. https://doi.org/10.1021/acsomega.7b00045

Li L, Zhang Z (2016) Biosynthesis of gold nanoparticles using green alga Pithophora oedogonia with their electrochemical performance for determining carbendazim in soil. Int J Electrochem Sci 11:4550–4559

Lim SH, Ahn E-Y, Park Y (2016) Green synthesis and catalytic activity of gold nanoparticles synthesized by Artemisia capillaris water extract. Nanoscale Res Lett 11:474. https://doi.org/10.1186/s11671-016-1694-0

Lima E, Guerra R, Lara V, Guzman A (2013) Gold nanoparticles as efficient antimicrobial agents for Escherichia coli and Salmonella typhi. Chem Cent J 7:7–11

Lin M, Pei H, Yang F, Fan C, Zuo X (2013) Applications of gold nanoparticles in the detection and identification of infectious diseases and biothreats. Adv Mater 25:3490–3496. https://doi.org/10.1002/adma.201301333

Logeswari P, Silambarasan S, Abraham J (2013) Ecofriendly synthesis of silver nanoparticles from commercially available plant powders and their antibacterial properties. Sci Iran 20:1049–1054. https://doi.org/10.1016/j.scient.2013.05.016

Loo C, Lowery A, Halas N, West J, Drezek R (2005) Immunotargeted nanoshells for integrated cancer imaging and therapy. Nano Lett 5:709–711. https://doi.org/10.1021/nl050127s

Luo P, Liu Y, Xia Y, Xu H, Xie G (2014) Aptamer biosensor for sensitive detection of toxin a of Clostridium difficile using gold nanoparticles synthesized by Bacillus stearothermophilus. Biosens Bioelectron 54:217–221. https://doi.org/10.1016/j.bios.2013.11.013

Makarov VV, Love AJ, Sinitsyna OV, Makarova SS, Yaminsky IV, Taliansky ME, Kalinina NO (2014) "Green" nanotechnologies: synthesis of metal nanoparticles using plants. Acta Nat 6:35–44

Mani AK, Seethalakshmi S, Gopal V (2015) Evaluation of in-vitro anti-inflammatory activity of silver nanoparticles synthesised using piper nigrum extract. J Nanomed Nanotechn 6:1

Masurkar SA, Chaudhari PR, Shidore VB, Kamble SP (2011) Rapid biosynthesis of silver nanoparticles using cymbopogan citratus (Lemongrass) and its antimicrobial activity. Nano-Micro Lett 3:189–194. https://doi.org/10.1007/bf03353671

McCall L-I, Zhang W-W, Ranasinghe S, Matlashewski G (2013) Leishmanization revisited: immunization with a naturally attenuated cutaneous Leishmania donovani isolate from Sri Lanka protects against visceral leishmaniasis. Vaccine 31:1420–1425. https://doi.org/10.1016/j.vaccine.2012.11.065

Menon S, Rajeshkumar S, Kumar SV (2017) A review on biogenic synthesis of gold nanoparticles, characterization, and its applications. Resource-Effic Technol 3:516–527. https://doi.org/10.1016/j.reffit.2017.08.002

Mewada A, Oza G, Pandey S, Sharon M (2017) Extracellular synthesis of gold nanoparticles using Pseudomonas denitrificans and comprehending its stability. J Microbiol Biotechnol Res 2:493–499

Mhlanga SD, Coville NJ, Iyuke SE, Afolabi AS, Abdulkareem AS, Kunjuzwa N (2010) Controlled syntheses of carbon spheres in a swirled floating catalytic chemical vapour deposition vertical reactor. J Exp Nanosci 5:40–51

Mhlanga SD, Witcomb MJ, Erasmus RM, Coville NJ (2011) A novel Ca3(PO4)2-CaCO3 support mixture for the CVD synthesis of roughened MWCNT-carbon fibres. J Exp Nanosci 6:49–63

Mishra A, Tripathy SK, Yun S-I (2011) Bio-synthesis of gold and silver nanoparticles from Candida guilliermondii and their antimicrobial effect against pathogenic bacteria. J Nanosci Nanotechnol 11:243–248. https://doi.org/10.1166/jnn.2011.3265

Mittal AK, Chisti Y, Banerjee UC (2013) Synthesis of metallic nanoparticles using plant extracts. Biotechnol Adv 31:346–356. https://doi.org/10.1016/j.biotechadv.2013.01.003

Mocan T, Clichici S, Agoston-Coldea L, Mocan L, Simon Ş, Ilie I, Biris A, Muresan A (2010) Implications of oxidative stress mechanisms in toxicity of nanoparticles (review). Acta Physiol Hung 97:247–255. https://doi.org/10.1556/APhysiol.97.2010.3.1

Moldovan B et al (2017) In vitro and in vivo anti-inflammatory properties of green synthesized silver nanoparticles using Viburnum opulus L. fruits extract. Mater Sci Eng C 79:720–727. https://doi.org/10.1016/j.msec.2017.05.122

Morens DM, Fauci AS (2013) Emerging infectious diseases: threats to human health and global stability. PLoS Pathog 9:e1003467. https://doi.org/10.1371/journal.ppat.1003467

Muhammad G, Hussain MA, Amin M, Hussain SZ, Hussain I, Abbas Bukhari SN, Naeem-ul-Hassan M (2017) Glucuronoxylan-mediated silver nanoparticles: green synthesis, antimicrobial and wound healing applications. RSC Adv 7:42900–42908. https://doi.org/10.1039/c7ra07555c

Mukherjee P et al (2005) Antiangiogenic properties of gold nanoparticles. Clin Cancer Res 11:3530–3534. https://doi.org/10.1158/1078-0432.ccr-04-2482

Mukherjee S, Chowdhury D, Kotcherlakota R, Patra S, Vinothkumar B, Bhadra PM, Sreedhar B, Patra RC (2014) Potential theranostics application of bio-synthesized silver nanoparticles (4-in-1 system). Theranostics 4:316–335. https://doi.org/10.7150/thno.7819

Nabikhan A, Kandasamy K, Raj A, Alikunhi NM (2010) Synthesis of antimicrobial silver nanoparticles by callus and leaf extracts from saltmarsh plant, Sesuvium portulacastrum L. Colloids Surf B: Biointerfaces 79:488–493. https://doi.org/10.1016/j.colsurfb.2010.05.018

Nakkala JR, Mata R, Gupta AK, Sadras SR (2014) Biological activities of green silver nanoparticles synthesized with Acorous calamus rhizome extract. Eur J Med Chem 85:784–794. https://doi.org/10.1016/j.ejmech.2014.08.024

Nanda A, Saravanan M (2009) Biosynthesis of silver nanoparticles from Staphylococcus aureus and its antimicrobial activity against MRSA and MRSE Nanomedicine: nanotechnology. Biol Med 5:452–456. https://doi.org/10.1016/j.nano.2009.01.012

Naraginti S, Tiwari N, Sivakumar A (2017) Green synthesis of silver and gold nanoparticles for enhanced catalytic and bactericidal activity. IOP Conf Ser 263:022009

Narayanan KB, Park HH (2014) Antifungal activity of silver nanoparticles synthesized using turnip leaf extract (Brassica rapa L.) against wood rotting pathogens. Eur J Plant Pathol 140:185–192. https://doi.org/10.1007/s10658-014-0399-4

Natera S, Machuca C, Padrón-Nieves M, Romero A, Díaz E, Ponte-Sucre A (2007) Leishmania spp.: proficiency of drug-resistant parasites. Int J Antimicrob Agents 29:637–642. https://doi.org/10.1016/j.ijantimicag.2007.01.004

Nayak S, Sajankila SP, Rao CV (2018) Green synthesis of gold nanopartciels from banana pith extract and its evaluation of antibacterial activity and catalytic reduction of malachite green dye. J Microbiol Biotechnol Food Sci 7:641–645. https://doi.org/10.15414/jmbfs.2018.7.6.641-645

Ndikau M, Noah NM, Andala DM, Masika E (2017) Green synthesis and characterization of silver nanoparticles using Citrullus lanatus fruit rind extract. Int J Anal Chem 2017:9. https://doi.org/10.1155/2017/8108504

Nguyen DT, Orgill DP, Murphy GF (2009) 4 – the pathophysiologic basis for wound healing and cutaneous regeneration. In: Orgill D, Blanco C (eds) Biomaterials for treating skin loss. Woodhead Publishing, Cambridge, pp 25–57. https://doi.org/10.1533/9781845695545.1.25

Noah N (2018) Green synthesis: characterization and applications of silver and gold nanoparticles, vol 1, 1st edn. Elsevier Publishers, San Diego

Ovais M, Ahmad I, Khalil AT, Mukherjee S, Ayaz M, Raza A, Shinwari ZK (2018) Wound healing applications of biogenic colloidal silver and gold nanoparticles: recent trends and future prospects. Appl Microbiol Biotechnol 102:4305–4318. https://doi.org/10.1007/s00253-018-8939-z

Paciotti GF, Kingston DGI, Tamarkin L (2006) Colloidal gold nanoparticles: a novel nanoparticle platform for developing multifunctional tumor-targeted drug delivery vectors. Drug Dev Res 67:47–54. https://doi.org/10.1002/ddr.20066

Pan Y, Ding H, Qin L, Zhao X, Cai J, Du B (2013) Gold nanoparticles induce nanostructural reorganization of VEGFR2 to repress angiogenesis. J Biomed Nanotechnol 9:1746–1756. https://doi.org/10.1166/jbn.2013.1678

Pandey S, Goswami GK, Nanda KK (2012) Green synthesis of biopolymer–silver nanoparticle nanocomposite: an optical sensor for ammonia detection. Int J Biol Macromol 51:583–589. https://doi.org/10.1016/j.ijbiomac.2012.06.033

Paramasivam G, Kannan R, Paulraj SM, Pandiarajan J (2017) Green synthesis of silver nanoparticles from Morinda tinctoria Roxb and scrutiny of its multi facet on biomedical applications. Pharmaceutical Biol Eval 4:12. https://doi.org/10.26510/2394-0859.pbe.2017.35

Patra JK, Baek KH (2015) Novel green synthesis of gold nanoparticles using Citrullus lanatus rind and investigation of proteasome inhibitory activity, antibacterial, and antioxidant potential. Int J Nanomedicine 10:7253–7264

Patra S, Mukherjee S, Barui AK, Ganguly A, Sreedhar B, Patra CR (2015) Green synthesis, characterization of gold and silver nanoparticles and their potential application for cancer therapeutics. Mater Sci Eng C 53:298–309. https://doi.org/10.1016/j.msec.2015.04.048

Prabhu S, Vinodhini S, Elanchezhiyan C, Rajeswari D (2018) Evaluation of antidiabetic activity of biologically synthesized silver nanoparticles using Pouteria sapota in streptozotocin-induced diabetic rats. J Diabetes 10:28–42. https://doi.org/10.1111/1753-0407.12554

Prasad T, Elumalai EK (2011) Biofabrication of Ag nanoparticles using Moringa oleifera leaf extract and their antimicrobial activity. Asian Pac J Trop Biomed 1:439–442. https://doi.org/10.1016/S2221-1691(11)60096-8

Premkumar T, Lee Y, Geckeler KE (2010) Macrocycles as a tool: a facile and one-pot synthesis of silver nanoparticles using cucurbituril designed for cancer therapeutics. Chem Eur J 16:11563–11566. https://doi.org/10.1002/chem.201001325

Prentice CRM, McNicol GP, Douglas AS (1966) Effects on blood coagulation, fibrinolysis, and platelet aggregation of normal and atheromatous aortic tissue. J Clin Pathol 19:154–158. https://doi.org/10.1136/jcp.19.2.154

Railean-Plugaru V et al (2016) Antimicrobial properties of biosynthesized silver nanoparticles studied by flow cytometry and related techniques. Electrophoresis 37:752–761. https://doi.org/10.1002/elps.201500507

Rajasekharreddy P, Rani PU (2014) Biofabrication of Ag nanoparticles using Sterculia foetida L. seed extract and their toxic potential against mosquito vectors and HeLa cancer cells. Mater Sci Eng C 39:203–212. https://doi.org/10.1016/j.msec.2014.03.003

Ramar M, Beulaja M, Thiagarajan R, Koodalingam A, Marimuthu PN, Muthulakshmi P, Subramanian P, Arumugam M (2015) Biosynthesis of silver nanoparticles using ethanolic petals extract of Rosa indica and characterization of its antibacterial, anticancer and anti-inflammatory activities. Spectrochim Acta A Mol Biomol Spectrosc 138:120–129. https://doi.org/10.1016/j.saa.2014.10.043

Ramasamy M, Lee J-H, Lee J (2016) Potent antimicrobial and antibiofilm activities of bacteriogenically synthesized gold–silver nanoparticles against pathogenic bacteria and their physiochemical characterizations. J Biomater Appl 31:366–378. https://doi.org/10.1177/0885328216646910

Rao Y, Inwati GK, Singh M (2017) Green synthesis of capped gold nanoparticles and their effect on gram-positive and gram-negative bacteria. Future Sci OA 3:FSO239–FSO239. https://doi.org/10.4155/fsoa-2017-0062

Rekha ND, Mallesha L, Vinay G, Santhosh MV (2018) Green synthesis of gold and silver nanoparticles using Memecylon malabaricum leaf extract and their anti-angiogenic activity. Adv Sci Eng Med 10:557–563. https://doi.org/10.1166/asem.2018.2186

Rieger S, Zhao H, Martin P, Abe K, Lisse TS (2015) The role of nuclear hormone receptors in cutaneous wound repair. Cell Biochem Funct 33:1–13. https://doi.org/10.1002/cbf.3086

Rodriguez-Lorenzo L, Alvarez-Puebla RA (2014) 8 – surface-enhanced Raman scattering (SERS) nanoparticle sensors for biochemical and environmental sensing. In: Honeychurch KC (ed) Nanosensors for chemical and biological applications. Woodhead Publishing, Amsterdam, pp 197–230. https://doi.org/10.1533/9780857096722.2.197

Roy S (2017) Green synthesized gold nanoparticles: study of antimicrobial activity. J Bionanosci 11:131–135. https://doi.org/10.1166/jbns.2017.1432

Sadeghi B, Gholamhoseinpoor F (2015) A study on the stability and green synthesis of silver nanoparticles using Ziziphora tenuior (Zt) extract at room temperature. Spectrochim Acta A Mol Biomol Spectrosc 134:310–315. https://doi.org/10.1016/j.saa.2014.06.046

Sadeghi B, Rostami A, Momeni SS (2015) Facile green synthesis of silver nanoparticles using seed aqueous extract of Pistacia atlantica and its antibacterial activity. Spectrochim Acta A Mol Biomol Spectrosc 134:326–332. https://doi.org/10.1016/j.saa.2014.05.078

Sankar R, Karthik A, Prabu A, Karthik S, Shivashangari KS, Ravikumar V (2013) Origanum vulgare mediated biosynthesis of silver nanoparticles for its antibacterial and anticancer activity. Colloids Surf B: Biointerfaces 108:80–84. https://doi.org/10.1016/j.colsurfb.2013.02.033

Santra TS, Tseng F-G, Barik TK (2014) Green biosynthesis of gold nanoparticles and biomedical applications. J Nano Res Appl Spec Issue 2:5–12. https://doi.org/10.11648/j.nano.s.2014020602.12

Saratale RG, Shin HS, Kumar G, Benelli G, Kim D-S, Saratale GD (2018) Exploiting antidiabetic activity of silver nanoparticles synthesized using Punica granatum leaves and anticancer potential against human liver cancer cells (HepG2). Artif Cells Nanomed Biotechnol 46:211–222. https://doi.org/10.1080/21691401.2017.1337031

Sarkar J, Ray S, Chattopadhyay D, Laskar A, Acharya K (2012) Mycogenesis of gold nanoparticles using a phytopathogen Alternaria alternata. Bioprocess Biosyst Eng 35:637–643. https://doi.org/10.1007/s00449-011-0646-4

Satyavani K, Ramanathan T, Gurudeeban S (2011) Green synthesis of silver nanoparticles by using stem derived callus extract of bitter apple (Citrullus colocynthis). Dig J Nanomater Biostruct 6:1019–1024

Scriba MR, Arendse C, Härting M, Britton DT (2008) Hot-wire synthesis of Si nanoparticles. Thin Solid Films 516:844–846. https://doi.org/10.1016/j.tsf.2007.06.191

Shaik M, Albalawi GH, Khan ST, Khan M, Adil SF, Kuniyil M, Al-Warthan A (2016) "Miswak" based green synthesis of silver nanoparticles: evaluation and comparison of their microbicidal activities with the chemical synthesis. Molecules 21:1478

Shankar S, Jaiswal L, Aparna R, Prasad V, Kumar P, Murthy Manohara C (2015) Wound healing potential of green synthesized silver nanoparticles prepared from Lansium domesticum fruit peel extract. Mater Express 5:159–164. https://doi.org/10.1166/mex.2015.1225

Shittu KO, Bankole MT, Abdulkareem AS, Abubakre OK, Ubaka AU (2017) Application of gold nanoparticles for improved drug efficiency. Adv Nat Sci Nanosci Nanotechnol 8(3):035014

Singaravelu G, Arockiamary JS, Kumar VG, Govindaraju K (2007) A novel extracellular synthesis of monodisperse gold nanoparticles using marine alga, Sargassum wightii Greville. Colloids Surf B: Biointerfaces 57:97–101. https://doi.org/10.1016/j.colsurfb.2007.01.010

Singh AK, Srivastava ON (2015) One-step green synthesis of gold nanoparticles using black cardamom and effect of pH on its synthesis. Nanoscale Res Lett 10:353. https://doi.org/10.1186/s11671-015-1055-4

Singh M, Kalaivani R, Manikandan S, Sangeetha N, Kumaraguru AK (2013) Facile green synthesis of variable metallic gold nanoparticle using Padina gymnospora, a brown marine macroalga. Appl Nanosci 3:145–151. https://doi.org/10.1007/s13204-012-0115-7

Singh P, Kim YJ, Yang DC (2016) A strategic approach for rapid synthesis of gold and silver nanoparticles by Panax ginseng leaves. Artif Cells Nanomed Biotechnol 44:1949–1957. https://doi.org/10.3109/21691401.2015.1115410

Singh T, Jyoti K, Patnaik A, Singh A, Chauhan R, Chandel SS (2017) Biosynthesis, characterization and antibacterial activity of silver nanoparticles using an endophytic fungal supernatant of Raphanus sativus. J Genet Eng Biotechnol 15:31–39. https://doi.org/10.1016/j.jgeb.2017.04.005

Singh P, Ahn S, Kang JP, Veronika S, Huo Y, Singh H, Chokkaligam M, El-Agamy FM, Aceituno VC, Kim YJ, Ynag DC (2018) In vitro anti-inflammatory activity of spherical silver nanoparticles and monodisperse hexagonal gold nanoparticles by fruit extract of Prunus serrulata: a green synthetic approach. Artif Cells Nanomed Biotechnol 46:2022–2032

Singhal G, Bhavesh R, Kasariya K, Sharma AR, Singh RP (2011) Biosynthesis of silver nanoparticles using Ocimum sanctum (Tulsi) leaf extract and screening its antimicrobial activity. J Nanopart Res 13:2981–2988. https://doi.org/10.1007/s11051-010-0193-y

Solano-Umaña V, Vega-Baudrit RJ (2015) Gold and silver nanotechology on medicine. J Chem Biochem 3:21–33. https://doi.org/10.15640/jcb.v3n1a2

Soltani Nejad M, Shahidi Bonjar GH, Khaleghi N (2015) Biosynthesis of gold nanoparticles using streptomyces fulvissimus isolate. Nanomed J 2:153–159

Srinath BS, Ravishankar Rai V (2015) Rapid biosynthesis of gold nanoparticles by Staphylococcus epidermidis: its characterisation and catalytic activity. Mater Lett 146:23–25. https://doi.org/10.1016/j.matlet.2015.01.151

Sriramulu M, Sumathi S (2017) Photocatalytic, antioxidant, antibacterial and anti-inflammatory activity of silver nanoparticles synthesised using forest and edible mushroom. Adv Nat Sci Nanosci Nanotechnol 8:045012. https://doi.org/10.1088/2043-6254/aa92b5

Sun RW-Y, Chen R, Chung NPY, Ho C-M, Lin C-LS, Che C-M (2005) Silver nanoparticles fabricated in Hepes buffer exhibit cytoprotective activities toward HIV-1 infected cells. Chem Commun 40:5059–5061. https://doi.org/10.1039/b510984a

Sunderam V, Thiyagarajan D, Lawrence AV, Mohammed SSS, Selvaraj A (2018) In-vitro antimicrobial and anticancer properties of green synthesized gold nanoparticles using Anacardium occidentale leaves extract. Saudi J Biol Sci 26:455–459. https://doi.org/10.1016/j.sjbs.2018.12.001

Sunkari S, Gangapuram BR, Dadigala R, Bandi R, Alle M, Guttena V (2017) Microwave-irradiated green synthesis of gold nanoparticles for catalytic and anti-bacterial activity. J Anal Sci Technol 8:13. https://doi.org/10.1186/s40543-017-0121-1

Suriyakalaa U, Antony JJ, Suganya S, Siva D, Sukirtha R, Kamalakkannan S, Pichiah PBT, Achiraman S (2013) Hepatocurative activity of biosynthesized silver nanoparticles fabricated using Andrographis paniculata. Colloids Surf B: Biointerfaces 102:189–194. https://doi.org/10.1016/j.colsurfb.2012.06.039

Swarnalatha L, Rachela C, Ranjan P, Baradwaj P (2012) Evaluation of in vitro antidiabetic activity of Sphaeranthus amaranthoides silver nanoparticles. Int J Nanomater Biostruct 2:25–29p

Swihart MT (2003) Vapor-phase synthesis of nanoparticles. Curr Opin Colloid Interface Sci 8:27–33

Syed B, Prasad N, Satisha S (2016) Endogenic mediated synthesis of gold nanoparticles bearing bactericidal activity. J Microsc Ultrastruct 4:162–166. https://doi.org/10.1016/j.jmau.2016.01.004

Tang D, Yuan R, Chai Y (2006) Ligand-functionalized core/shell Ag@Au nanoparticles label-free amperometric immun-biosensor. Biotechnol Bioeng 94:996–1004. https://doi.org/10.1002/bit.20922

Tedesco S, Doyle H, Blasco J, Redmond G, Sheehan D (2010) Oxidative stress and toxicity of gold nanoparticles in Mytilus edulis. Aquat Toxicol 100:178–186. https://doi.org/10.1016/j.aquatox.2010.03.001

Thanganadar Appapalam S, Panchamoorthy R (2017) Aerva lanata mediated phytofabrication of silver nanoparticles and evaluation of their antibacterial activity against wound associated bacteria. J Taiwan Inst Chem Eng 78:539–551. https://doi.org/10.1016/j.jtice.2017.06.035

Thirumurugan A, Ramachandran S, Tomy NA, Jiflin GJ, Rajagomathi G (2012) Biological synthesis of gold nanoparticles by Bacillus subtilis and evaluation of increased antimicrobial activity against clinical isolates. Korean J Chem Eng 29:1761–1765. https://doi.org/10.1007/s11814-012-0055-7

Tomar A, Garg G (2013) Short review on application of gold nanoparticles. Global J Pharmacol 7:34–38

Tonnesen MG, Feng X, Clark RAF (2000) Angiogenesis in wound healing. J Invest Dermatol Symp Proc 5:40–46. https://doi.org/10.1046/j.1087-0024.2000.00014.x

Torchilin VP (2010) Passive and active drug targeting: drug delivery to Tumors as an example. In: Schäfer-Korting M (ed) Drug delivery. Springer, Berlin/Heidelberg, pp 3–53. https://doi.org/10.1007/978-3-642-00477-3_1

Usman AI, Aziz AA, Noqta OA (2019) Application of green synthesis of gold nanopartciels: a review. J Teknologi (Sci & Eng) 8:171–182. https://doi.org/10.11113/jt.v81.11409

Usmani A, Dash PP, Mishra A (2018) Metallic nanoformulations: green synthetic approach for advanced drug delivery. Mater Sci 2:1–4. https://doi.org/10.18063/msacm.v2i2.729

Venkatesan J, Manivasagan P, Kim S-K, Kirthi AV, Marimuthu S, Rahuman AA (2014) Marine algae-mediated synthesis of gold nanoparticles using a novel Ecklonia cava. Bioprocess Biosyst Eng 37:1591–1597. https://doi.org/10.1007/s00449-014-1131-7

Verma A, Mehata MS (2016) Controllable synthesis of silver nanoparticles using Neem leaves and their antimicrobial activity. J Radiat Res Appl Sci 9:109–115. https://doi.org/10.1016/j.jrras.2015.11.001

Vijayan R, Joseph S, Mathew B (2019) Anticancer, antimicrobial, antioxidant, and catalytic activities of green-synthesized silver and gold nanoparticles using Bauhinia purpurea leaf extract. Bioprocess Biosyst Eng 42:305–319

Wabuyele MB, Vo-Dinh T (2005) Detection of human immunodeficiency virus type 1 DNA sequence using plasmonics nanoprobes. Anal Chem 77:7810–7815. https://doi.org/10.1021/ac0514671

Wang C, Mathiyalagan R, Kim YJ, Castro-Aceituno V, Singh P, Ahn S, Wang D, Yang DC (2016) Rapid green synthesis of silver and gold nanoparticles using Dendropanax morbifera leaf extract and their anticancer activities. Int J Nanomedicine 11:3691–3701. https://doi.org/10.2147/ijn.s97181

Wen L, Zeng P, Zhang L, Huang W, Wang H, Chen G (2016) Symbiosis theory-directed green synthesis of silver nanoparticles and their application in infected wound healing. Int J Nanomedicine 11:2757–2767. https://doi.org/10.2147/IJN.S106662

Xiang F, Minghui J, Xin D, Yuhong Y, Ren Z, Zhengzhong S, Zhao X, Chen X (2013) Green synthesis of silk fibroin-silver nanoparticle composites with effective antibacterial and biofilm-disrupting properties. Biomacromolecules 14:4483–4488. https://doi.org/10.1021/bm4014149

Xuwang Z, Yuanyuan Q, Wenli S, Jingwei W, Huijie L, Zhaojing Z, Shuzhen L, Jiti Z (2016) Biogenic synthesis of gold nanoparticles by yeast Magnusiomyces ingens LH-F1 for catalytic reduction of nitrophenols. Colloids Surf A Physicochem Eng Asp 497:280–285. https://doi.org/10.1016/j.colsurfa.2016.02.033

Yang N, WeiHong L, Hao L (2014) Biosynthesis of Au nanoparticles using agricultural waste mango peel extract and its in vitro cytotoxic effect on two normal cells. Mater Lett 134:67–70

Yarramala DS, Doshi S, Rao CP (2015) Green synthesis, characterization and anticancer activity of luminescent gold nanoparticles capped with apo-α-lactalbumin. RSC Adv 5:32761–32767. https://doi.org/10.1039/c5ra03857j

Yuan C-G, Huo C, Gui B, Cao W-P (2017) Green synthesis of gold nanoparticles using Citrus maxima peel extract and their catalytic/antibacterial activities. IET Nanobiotechnol 11:523–530

Zhou W, Ma Y, Yang H, Ding Y, Luo X (2011) A label-free biosensor based on silver nanoparticles array for clinical detection of serum p53 in head and neck squamous cell carcinoma. Int J Nanomedicine 6:381–386. https://doi.org/10.2147/ijn.s13249

Nanofinished Medical Textiles and Their Potential Impact to Health and Environment

Eman Osman

Abstract

In spite of the fact that nanofinished medical textiles have gigantic potential for a large group of utilizations, their unfavorable impacts on living cells have raised genuine concerns after utilization in the human services and customer segments; due to that they are too small and can penetrate the living cell walls easily, so they can have a very harmful effect on both the environment and the population. Besides the importance of engineered nanomaterials and all their varieties of applications, yet there is an information hole between the innovative advancement in nanotechnology and nanosafety, which made a few concerning organizations to give the most noteworthy need to researches related to the safety and health in occupational settings.

This chapter introduces an overview of the nanofinished medical textiles both reusable and disposable and their potential effects on both the consumer and the environment. Moreover, the safety and health concerns identified with nanomaterials and the adverse potential of their waste are displayed.

Keywords

Nanofinished textiles · Potential impact · Medical textile · Nanowaste · Life cycle analysis LCA · Risk assessment

E. Osman (✉)
Textile Metrology Laboratory, Chemical Metrology Division, National Institute of Standards, Giza, Egypt

5.1 Smart, Technical, and Medical Textiles

Before we begin our survey, a few articulations or terms must be explained. Hereafter are the contrasts between smart, technical, and medical textiles.

5.1.1 Smart Textiles

Smart textiles can be characterized as the materials and structures which sense or can detect the natural conditions or upgrades (temperature, light, heartbeat, and so on), i.e., environmentally sensitive, whereas intelligent textiles can be defined as material structures that can not only detect but also respond and react to ecological conditions or stimuli. Smart textiles can be made by fusing smart materials, conductive polymers, encapsulated phase change materials, shape-memory polymers and materials, and other electronic sensors and communication equipment. Every single smart material includes an energy transfer from the stimuli to response given out by the material (Syduzzaman et al. 2015; Munima 2019).

The degree of smartness can be partitioned in three subgroups (Amit and Arif n.d.; Oakes et al. 2005; Wen 2013; Vanlangenhove 2013):

- *Passive smart materials:* They can just detect the environment, they are sensors; for instance, a highly insulated coat would remain protecting to a similar degree independent of the outside temperature.
- *Active smart materials:* They can detect the stimuli from the environment and furthermore respond to them. Active smart materials are shape memory, chameleonic, water-resistant and vapor permeable (hydrophilic/nonporous), heat storage, thermo regulated, vapor absorbing, and heat evolving fabric and electrically heated suits.
- *Ultrasmart materials:* They are the third generation of smart materials, which can detect, respond, and embrace themselves to natural conditions or improvements. A smart or intelligent material basically comprises of a unit, which works like the cerebrum, with comprehension, thinking, and actuating limits. The generation of intelligent materials is currently a reality after an effective marriage of customary materials and dressing innovation with different parts of science like material science, basic mechanics, sensor and actuator innovation, advance handling innovation, correspondence, artificial intelligence, science, and so forth.

5.1.2 Technical and Protective Textiles

Technical textiles are characterized as textile materials and items utilized principally for their specialized exhibition and useful properties instead of their stylish or improving qualities. Different terms utilized for characterizing technical textiles incorporate industrial textiles, functional textiles, performance textiles, engineering textiles, invisible textiles, and hi-tech textiles. Presently, technical textile materials are most generally utilized in filter clothing, furniture, hygiene medicals, and construction material. A technical textile can be woven or non-woven and blends of both. It tends to be made up as a solitary layer or numerous layers and can be delivered as a composite or a covered as well as impregnated material. It very well may be produced using any fiber yarn or fiber of purely natural or synthetic or mix of the two kinds (Alhayat and Omprakash 2014).

Protective clothing is a noteworthy piece of technical or industrial textiles. Protective clothing alludes to pieces of clothing designed to shield the wearer from harsh ecological impacts that may result in wounds or even death (Elias 2015).

Technical textiles represent about 31% of the absolute textile production. The worldwide technical textile market is geographically sectioned into five key districts: North America, Latin America, Eastern and Western Europe, Asia Pacific, and Africa and Middle East (Indian Technical 2016; Knowledge Paper 2016; https://technical-textile.net/articles/report-on-medical-textiles-and-sport-outdoor-textiles-5032).

5.1.2.1 Selection of Protective Cloths

The first step in selecting protective clothing is to determine the hazard, evaluate the potential for exposure, and select the degree of protection required. The consequences of direct skin contact can range from minor diseases like dermatitis to systemic poisoning and cancer. Further, we have to consider the environment in/from which we want protection.

There are various sorts of protective clothing and it tends to be isolated into groups dependent on their end uses. The most accepted and generally utilized scheme plan has been given by Techtextil (leading international trade organization for technical textiles). Techtextil characterizes 12 primary application zones of technical textiles (Deepti 2011; Protective Textiles Introduction 2018; https://textilestudycenter.com/protective-textiles-introduction/):

1. Agrotech (Agriculture, horticulture and forestry)
2. Buildtech (building and construction)
3. Clothtech (technical components of shoes and clothing)
4. Geotech (geotextiles, civil engineering)
5. Hometech (components of furniture, household textiles and floor coverings)
6. Indutech (filtration, cleaning and other industrial usage)
7. Medtech (hygiene and medical)
8. Mobiltech (automobiles, shipping, railways, and aerospace)
9. Oekotech (environmental protection)
10. Packtech (packaging)
11. Protech (personal and property protection)
12. Sportech (sport and leisure)

5.1.3 Medical Textiles

Medical textile is a new field which is considered as the mating product between textile and medical technology having applications in the field of medical and clinical considerations. They are otherwise called healthcare textiles and can be characterized according to "Textile Terms & Definitions" as (Mclistyre and Daniels 1995): a general term which depicts a material structure which has been planned and created for use in any of an assortment of medical applications, including implantable applications.

One of the most important issues in emergency clinics and healthcare institutions is microbial sullying of surfaces, including material textures, which can prompt contaminations and, subsequently, lead to cross-diseases. Independent of their applications, internal (surgical threads and various implants) or external (different extracorporeal gadgets, for example, catheters and empty strands for dialyzers, gauzes, bandages, nappies, tampons, etc.), medical textiles, must contain essential bioactive properties, particularly antimicrobial. The correct choice of the used materials, either natural (e.g., cotton and silk) or synthetic (e.g., polyester, polyamide, polyethylene, glass, etc.), should be chosen very carefully depending on the end-use application; all is done for improving the consumer comfort (Madalina and Fulga 2014). Medical textile improvement can be considered as one such advancement, which is truly intended to change difficult patient days into agreeable days (Raaz n.d.).

Materials that are utilized in medicinal applications incorporate fibers, yarns, fabrics, and composites. Contingent on the application, the real prerequisites of medical textiles are absorbency, tenacity, flexibility, softness, and at times biostability or biodegradability.

5.1.3.1 Classification of Medical Textile Products

The significant results of the medical textiles based on their applications are arranged into four divisions, to be specific, implantable materials, non-implantable materials, extracorporeal gadgets, and healthcare items (Medical Textiles 2019; Onar and Sarnsik 2002), as indicated in Table 5.1.

Implantable Materials
Implant items were identified as the largest application segment accounting for 33.20% of global revenue in 2018. Examples of implantable materials include artificial tendons, artificial ligaments, artificial skin, artificial Lumina, eye contact lenses, orthopedic implants, artificial joints, artificial bones, cardiovascular implants, vascular grafts, and heart valves (Farooq et al. 2014).

Non-implantable Materials
Non-implantable materials are utilized on the body; more often they have direct contact with the human skin, accounting for 32.24% of the global market volume in 2018. They include wound dressings, absorbent pads, simple and elastic bandages, plasters, gauze, pressure garments, wadding, and orthopedic belts. Surgical dressings are one of the fundamental kinds of non-implantable medical textile items that

Table 5.1 Classification of medical textiles according to their usage (https://slideplayer.com/slide/6467899/)

Non-Implantables	Healthcare Products	Implantables	Extra Corporeal
Absorbents with and without ex-ray detactable • Abdominal pad with /without x-Ray • Cotton & viscose gauze • Dressing packs • Wadding Gauze Bandages • W/W bandage (open bandages or gauze bandages) • Triangular & POP Bandages • Synthetic cast Extensible Bandages • Cotton crepe bandages • Elastic bandages (with rubber or lycra) • Compression bandages • Elastic adhesive & Cohesive bandages Tubular bandages • Knitted fabric in tubular form • Surgical hosiery Wound dressing and medicated bandages • Chlorhexidine gauze dressing • Elastic Adhesive dressing • Framycetin gauze dressing • Paraffin gauze dressing	• Surgical gowns • Surgical caps • Surgical mask • Surgical drapes • Wipes • Hospital bed sheets, pillows, pillow covers, blankets, mattresses • Patient clothing (summer & winter) • Burns clothing • Operation theatre clothing	• Sutures Biodegradable/ Non-biodegradable • Bifurcated arterial prosthetic graft • Artificial Joints • Dialysers • Artificial Tendon (Mesh) • Artificial Vascular Grafts • Artificial heart valve etc • Art. skin	• Art. Kidney • Art. Liver / Lungs

are utilized to conceal, guard, and hold the harmed body part. These dressings additionally help to absorb liquid coming out of the wound (Shilpi 2014).

Extracorporeal Gadgets

Extracorporeal gadgets are utilized to keep up the function of basic organs and incorporate counterfeit kidneys (dialyzers), artificial livers, and mechanical lungs.

The worldwide medical textiles share as indicated by their applications in 2018 is demonstrated as follows (Fig. 5.1):

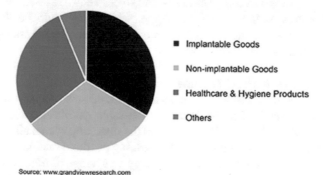

■ Implantable Goods

▦ Non-implantable Goods

■ Healthcare & Hygiene Products

■ Others

Source: www.grandviewresearch.com

Fig. 5.1 The global textile market share % of medical textiles according to their applications, 2018 (Medical Textiles 2019)

5.1.3.2 Prerequisites of Medical Textiles

Huge quantities of healthcare and hygiene products of medical textiles are utilized in the operating theater and in emergency clinic wards. These products are either washable or disposable. Items utilized in a working auditorium incorporate surgical gowns, caps, masks, patient drapes, and cover cloths. Surgical gowns, caps, and masks go about as a barricade to avert the arrival of pollutant particles into the air as they are a conceivable source of infection to patients. All these items are available in reusable or disposable forms. They should have a high level of air permeability and high filter capacity, as well as being lightweight and non-allergenic. Cover cloths and surgical drapes are utilized to cover the patient and the regions around the person in question. They should be totally impervious to microorganisms and porous to substantial sweat and wounds (Chinta and Veena 2013).

These medical textiles should have a couple of specific properties to be profitable as medical material, since their displays rely on interaction with the cells and particular fluids delivered by the body.

Depending on the application, the requirements of material texture for medical applications are (Meena et al. n.d.):

- *In the patient:* Biocompatible, nontoxic, nonallergic, sterility standards for highest level
- *On the patient:* Nontoxic, nonallergic, sterile, biocompatibility
- *Near the patient:* Sterile, lint-free
- *For the patient:* Clean
- *General properties:* Noncarcinogenic; antistatic in nature; optimum fatigue endurance; flame resistance; great protection from dissolvable bases, acids, and microorganisms; great dimensional stability; flexibility

5.1.3.3 Disposable Medical Textiles

Disposable textiles are materials designed to be utilized in or as clothing for functional applications and intended to be discarded after use. They are bound by "use and throw" concept, for example, disposable surgical gowns, surgical drapes, masks, bandage, gauze, shoe covers, aprons, caps, and so on. They can be either non-woven or woven textures as indicated by the type, while they could be differentiated according to their application to medical protection or surgical dressing (Mayekar 2008; Disposable medical textiles 2019).

5.1.3.4 Reusable Medical Textiles

They are items produced to be used several times, i.e., rewashable, such as gowns and drapes (Michael 2012). The American Reusable Textile Association (ARTA) researchers performed comparison project in the operating rooms aimed to replace disposable surgical items by the reusable ones. It was found that the medical waste resulted from the operating rooms is reduced by an average of 65% when using the reusable products. So, it was concluded that reusable surgical gowns are better for the environment than disposable gowns as they decrease the environmental burden (Conrardy et al. 2010; John 2014).

Moreover, an investigative study by the American Reusable Textile Association (ARTA) and the International Association for Healthcare Textile Management (IAHTM) has found that reusable surgical gowns are significantly better for the environment than disposable gowns in areas like energy consumption, water use, greenhouse gas emissions, and waste management. The study looked at the entire life cycle of both disposable and reusable surgical items (Life Cycle 2018; Stacy 2018)

Also, reusable healthcare textiles also improve patient care, as when asked which they prefer, 72% of patients said they liked cloth (reusable) better than paper (disposable). And in a study of operating room employee performance, hospital employees who used reusable textiles had better reaction times and made fewer mistakes (https://www.prudentialuniforms.com/blog/advantages-of-reusable-healthcare-textiles/).

Nanotechnologies
Nanotechnologies can be perceived as the design, characterization, production, and application of structures, devices, and systems by controlling shape and size of material particles on nanometer scale with widespread applications including the production and application of physical, chemical, and biological systems at scales ranging from individual atoms or molecules to around 100 nm (Christoph 2005). One of these applications is the use of nanotechnology in the field of medical textiles (Jitendra et al. 2016; Rothen-Rutishauser et al. 2006). The small size of the particles provides a larger surface area because of which there is a greater bioavailability of the drugs and other substance and also assures sustain release of drugs. The fibers encased with such nanoparticles show greater antimicrobial activity (Sawhney et al. 2008) (Table 5.2).

5.1.4   Nanotechnology Applications

Nanotechnology is a budding field with a wide range of applications into a number of spheres. One of these applications is the use of nanotechnology in the field of medical textiles. The most common use is the use of fibers treated with the

Table 5.2 Definitions of nanoparticles according to various global organizations: ISO, ASTM, NIOSH, SCCP, BAuA (Nowack 2010)

ISO	A particle spinning 1–100 nm (diameter)	
ASTM	An ultrafine particle whose length in 2 or 3 places is 1–100 nm	
NIOSH	A particle with diameter between 1 and 100 nm or a fiber spanning the range of 1–100 nm	
SCCP	At least one side is in the nanoscale range	Materials for which at least one side or internal structure is in the nanoscale
BSI	All the fields or diameters are in the nanoscale range	Materials for which at least one side or internal structure is in the nanoscale
BAuA	All the fields or diameters are in the nanoscale range	Materials consisting or a nanostructure or a nanosubstance

nanosynthesized materials so as to enhance the antimicrobial activity of the fibers. The nanomaterial has both positive and negative impacts on the environment and its organisms, especially when these nanotreated medical items are released into the environment such as the washout water from reusable gowns or the disposal of disposable gowns, caps, drapes, and so on (Teli n.d.).

5.2   Nanofinished Medical Textiles

Nanofibers are an important class of one-dimensional nanomaterials that are being used in many medical applications due to their unique properties, high surface area to volume ratio, film thinness, nanoscale fiber diameter porosity of structure, and lighter weight, so they can be considered as engineered scaffolds with broad application in the field of tissue engineering (Silas et al. 2007). Nanofibers are manufactured using biocompatible or biodegradable materials that have a high potential in the biomedical and healthcare sectors owing to their unique properties and functionalities (Suprava et al. 2015). The most known widespread applications of nanofinished medical textiles are in the form of gowns (disposable and reusable), face masks, caps, aprons, drapes, etc. (Alper and Bekir 2018).

The constitution of the nanofinished fibers of a high surface area and porosity helps them to enhance the adhesion of cells as well as various proteins and drug molecules. These attributes make them superior to their micro and macro counterparts composed of the same materials (Global Medical Textiles Market 2018).

5.3   Potential Impact of Nanotechnology

Some have depicted nanotechnology as a two-edged sword. On one hand, some are worried about that the nanoscale particles may enter and collect in indispensable organs (lungs and minds), possibly making damage or demise people and creatures, and that the dissemination of nanoscale particles may hurt environments (Approaches to Safe Nanotechnology 2009; Musee et al. 2012). Then again, some trust that nanotechnology can possibly convey significant environmental, health and safety (EHS) advantages, for example, decreasing energy consumption, pollution, and greenhouse gas emissions; remediating ecological harm; curing, overseeing, or preventing diseases; and offering new safety enhancing materials that are more stronger, self-fixing, and ready to adapt to give protection (Auffan et al. 2009; Joob and Wiwanitkit 2019).

Before nanomaterials are permitted to be utilized in day-by-day life exercises, it is significant for nanotoxicology research to reveal and see how nanomaterials impact the environment with the goal that their bothersome properties can maintain a strategic distance from both population and environment. To address issues concerning potential impacts of rising nanotechnologies on the living organisms, any survey should examine the ongoing advances on danger and ecological effect of nanomaterials (Mueller et al. 2012; Paresh et al. 2009).

One of the territories of most noteworthy worry about dangers is the investigation, use, disposal, or reusing of engineering nanomaterials. To decide whether the unique chemical and physical properties of new nanoparticles have brought about explicit toxicological properties, the nanotechnology network needs better approaches to assess the risk and to evaluate the hazard factor, and along these lines an endeavor must be made to focus solely on potential health risks. This negative effect is named as nanopollution or contamination (Almeidal and Ramos 2017).

5.4 Nanopollution

Nanopollution is the waste created through assembling procedure of nanogadgets or by utilizing nanoitems if not appropriately discharged into the environment. Ecotoxicological impact of nanoparticles and its potential for bioaccumulation in plants and microorganisms is a flow subject for research, as nanoparticles are considered to exhibit novel ecological effects. The US government has spent US $710 million out of 2002 on nanotechnology inspections; where $500,000 was spent on evaluation of effect of nanoparticles on environment.

In addition, the 2014 version of the ecological label GOTS (Global Organic Textile Standard) completely bans the presence of nanofinishes in textile materials. Additionally, in the ongoing discourse of the new form of the EU Ecomark for materials, there were a few voices to reject nanomaterials (Kamalja and Khatik 2015).

With respect to medical textiles, various vital issues have to be considered carefully because of the mostly used antibacterial agents which are often nanomaterials. A few examinations on antibacterial finished fabrics treated with silver particles either nano or ordinary structures, can possibly discharge significant measures of Ag into washing fluid, with percentages discharged in the first washing of up to 20–30%. The washing of nanosilver-treated materials conceivably discharges both dissolved and particulate silver, with a portion of the Ag discharged including particles bigger than 450 nm (Michael 2012). The presence of nano-Ag particles in the clothes washout water could have ramifications for both the environment and for the living creatures too (Troy et al. 2010; Denise et al. 2014). Besides, utilizing silver nanoparticles in wound dressings advances recuperating while at the same time lessening potential diseases; in any case, these equivalent nanoparticles could likewise annihilate supportive microorganisms in nature. Along these lines, the Environmental Protection Agency (EPA) has put a limitation on any use of silver nanoparticles on account of its apparent capability in devastating great microorganisms whenever discharged into nature (Chandra et al. 2009).

5.5   Nanotoxicity and Risks

Early science on nanotechnology conveys sufficient evidence to demonstrate that nanoparticles may have noxious properties that are unmistakable. The nanoparticles can react with the human body by means of three distinct methods for contact of penetration: inhalation, ingestion, and skin contact. Albeit all the three pathways that can be related to textiles, skin contact is obviously the most relevant (Mueller and Nowack 2008). Global associations are calling for adequate and appropriate oversight, security testing, and valuation of the developing field of nanoinnovation.

The risk is a matter of exposing the body to certain dose of the nanoparticles for a period of time. So, the risk of using nanomaterials depends mainly on how long the body will be in contact with the hazardous material (Kai et al. 2015). Subsequently, the risk for population can be minimized by using the PPE (personal protective equipment) and for the environment; the nanowaste must be managed as explained later (https://mercatormedical.eu/products/Disposable-apparel).

The expanded use and transfer of items containing produced nanomaterials will unavoidably result in their gathering in soil, water, air and organisms by means of direct data sources as well as spillover from contaminated destinations. Evaluation of their exposure and hazards in the genuine condition through the life cycle of nanoitems must depend on standardized testing conventions (counting monitoring tools) and coordinated hazard investigation strategies (Boxall et al. 2007).

5.6   Life Cycle Assessment (LCA) (Scott 2015)

LCA is a method used to assess environmental impacts throughout a product or process'life, which analyzes all stages of the product or process' life, including, raw material extractionand processing, manufacture, distribution, use, maintenance and repair, recycling, and disposal.Established guidelines for performing detailed LCAs are well documented by the EnvironmentalProtection Agency (EPA), Society for Environmental Toxicologists and Chemists (SETAC), andthe International Organization of Standardization (ISO) (Fava et al. 1991; Vigon et al. 1992; ISO 2006a).

According to ISO 14040 standards, an LCA is defined by four steps, as shown in Fig. 5.2.

LCAs start with the goal and scope definition, which expressly sets the setting of the investigation and characterizes the exact amounts of what item to be broke down inside the system boundary and the degree to which an item's life cycle is analyzed. Moreover, designating the resulted impact categories (e.g., global warming, acidification, eutrophication, carcinogenic impacts, respiratory effects, etc.).

The second step of the LCA is to perform inventory analysis in which the precise amounts of emissions, materials, and energy to and from the technosphere (i.e., synthetic materials/items) are incorporated.

Fig. 5.2 Life cycle assessment steps (ISO 2006b)

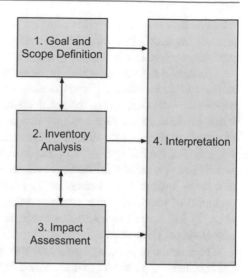

Pursued by the inventory analysis is the impact assessment, which totals the LCI information into environmental impact category. Common impact assessment systems incorporate the Tool for the Reduction and Assessment of Chemical and Other Environmental Impacts (TRACI) (Pre 2013a, b).

The fourth step of an LCA is interpretation, which is performed all through each step of the LCA. LCA interpretation identifies, quantifies, and evaluates information and results from the other three steps. Interpretation will recognize critical issues dependent on the results obtained.

LCA can help in recognizing the most benign technology and on account of single-use healthcare products can help in contrasting the ecological effects of elective items by following their effects all through their assembling, use, and transfer. Through utilization of LCA, it is conceivable to see which procedure (or procedures) drives ecological and human health impacts and may offer knowledge into limiting effects all through an item's life. On account of developments for new single-use and reprocessed human service items, LCA can give knowledge into the effects and trade-offs of choices during the innovation procedure.

5.7   Nanowaste

As of now, there is no official meaning of the expression "nanowaste." One proposed definition is "independently gathered or collectable waste materials which are or contain designed nanomaterials" (Brouwer et al. 2015). Because of the different uses of nanomaterials, nanowaste is definitely not a uniform sort of waste. The discarded nanoparticles are normally as yet having their individual properties; they are simply not being appropriately utilized any longer. More often than not, they are lost because of contact with various conditions. For instance, nano silver particles treated garments as antibacterial agent; those particles are lost after washing. The way that they are as yet working and are so little is the thing that makes nanowaste obscure impacts. Most human-made nanoparticles do not show up in nature, so living beings might not have appropriate means to deal with nanowaste.

The fuse of the nanomaterials in the waste relies upon the underlying item it was utilized in. Nonetheless, the disposal of medical waste should be done such that neither the environment nor the health conditions of individuals are put in danger as those medical nanowastes are very dangerous. It is crucial to guarantee that the disposal of such waste does not cause antagonistic environmental and health impacts. So, the life cycle of an item, from treatment of nanoparticles to handling, production, consumer utilization, and its disposal, is imperative to completely assess the level and seriousness of associated risks. Age of perceptions on hazard appraisal of nanocomposite or nanoparticles is a procedure in progress, and conceivably anticipated to restricting guidelines of using those products by the open associations (Thiry 2007).

The absence of learning about how nanoparticles may influence or meddle with the biochemical pathways and procedures of the human body is especially troublesome. Researchers are essentially worried about toxicity, characterization, and exposure pathways. Other than the undeniable potential dangers to patients, there are other toxicological dangers related with nanodrug. There are likewise legitimate worries over the transfer of nanowaste and ecological pollution from the assembling of nanomedicinal gadgets (Berger 2007).

Nanowaste transfer requires broad research and, furthermore, assessment; it additionally necessitates that extremely exacting and clear standards and strategies be embraced. Researchers, policy makers, and other included partners must cooperate and gain from one another in request to make powerful and suitable standards.

5.7.1 Classification of Hazardous Nanowastes

Presently, our insight on the issues identified with nanotechnology and nanowaste is excessively juvenile; what's more, ineffectively comprehended to choose whether all nanomaterials and nanoparticles can be dealt with the same route as far as disposal and neutralization. The facts demonstrate that numerous materials inside the equivalent material characterization gathering act in the same way and have comparable properties. Notwithstanding, it is likewise certain that there are various classifications of nanomaterials and nanoparticles (e.g., organic and inorganic, natural and manufactured, spheres and clusters, nanofibers, wires and rods, thin films and plates, bulk nanomaterials, and so on). Depending upon the kind of the material, they can have disparate physical and chemical properties, including melting point, size, hardness, and so on. Because of these varying properties, strategies utilized for one sort of material may not really be relevant to other gatherings of nanomaterials (Faunce 2017).

The inborn attributes of the potential introduction exposure and hazard information at the transfer life cycle portrayed above, nanowastes, can extensively be sorted

Table 5.3 Nanowaste classification as a function of constituent NM toxicity and exposure potency as a function of NM loci in the utilized nanotechnology (Musee et al. 2008a)

Classes	Nanotoxicity description	Comments	Examples of waste streams
Class I	Hazard: Nontoxic Exposure: Low to high	Concerns on waste management may only arise if the bulk parent materials (Trojan horse effects) can cause toxicity to humans and the environment through accumulation beyond a certain threshold concentration limit. Otherwise, nanowaste can be handled as benign/safe. No special disposal requirements. **Risk profile**: *None to very low*	Display backplanes of television screens, solar panels, memory chips, polishing agents
Class II	Hazard: Harmful or toxic Exposure: Low to medium	Toxicity of NMs may warrant establishing potential acute or chronic effects to determine the most suitable and optimal management approach during handling, transportation, or disposal processes **Risk profile**: *Low to medium*	Display backplane, memory chips, polishing agents, solar panels, paints, and coatings
Class III	Hazard: Toxic to very toxic Exposure: Low to medium	Protocols appropriate for managing hazardous waste streams in the entire waste management chain are desirable/recommended. Need for research to determine if current waste management infrastructure is adequate to deal with hazardousness of waste streams due to nanoscale materials **Risk profile**: *Medium to high*	Food packaging, food additives, wastewater containing personal care products, polishing agents, pesticides
Class IV	Hazard: Toxic to very toxic Exposure: Medium to high	Waste streams should be disposed only in specialized hazardous waste designated sites **Risk profile**: *High*	Paints and coatings, personal care products, pesticides, etc.
Class V	Hazard: Very toxic to extremely toxic Exposure: Medium to high	Dispose only in specialized hazardous waste streams designated sites. Poor waste management can cause extensive nanopollution to diverse ecological and watersystems, which may prove to be costly, laborious, and time-consuming to remediate. Immobilization and neutralization techniques among the most effective treatment techniques **Risk profile**: *High to very high*	

into five subjective likely classes, listed in Table 5.3, as given by Musee et al. This nanowaste order routine can help in overseeing various sorts of nanowaste and creating precautionary and practical methodologies through (Musee et al. 2008b):

1. Separating waste sorts that could be releaed or generated during operations.
2. Taking care of transportation and storage framework.

3. Permitting successful and proper methods of treating, reusing, and discarding different kinds of nanowastes to moderate against any type of negative effects to the people and the environment dependent on level of hazard (Resent 2016).

5.8   Concept of Storage and Disposal of Nanomaterials

Nanomaterials are significant for both the general population included and their impact on the environment. However, with no knowledge of their fundamental properties, the degree of the precautionary measures required is obscure. Again, to comprehend the potential effect that any nanomaterial will have on the environment, it is important to indorse the lifecycle viewpoint. This includes evaluating impacts that may happen all through an item's life, including the production, dissemination, taking care of utilization, storing, and disposal. As indicated by the guidance created and utilized by the US Department of Energy (DOE 2007) and the UK Environment Agency (EA) Guidance for the transfer of unsafe materials, an arrangement for storage and disposal of nanomaterials or nanomaterials contaminated waste ought to be created, taking in thought the hazardous nature of those nanomaterials and the amounts included and connected to nanomaterial-bearing waste streams (solid and liquid waste), including (Brief for GSDR 2016):

(a) Pure nanomaterials
(b) Items contaminated with nanomaterials, i.e., compartments, wipes, and expendable PPE
(c) Fluid suspensions containing nanomaterials

   We have reviewed in the above mentioned sections the precautionary measures which must be taken when managing the generation of the nanomaterials, discussed below is the manner by which to store the nanomaterials and how to gather and arrange the waste or the contaminated apparatuses which are utilized in dealing with these nanomaterials.

5.8.1   Storage of Nanomaterial Waste Prior to Disposal

Coming up next are proper methodologies for accumulation and storage of nanomaterial wastes preceding transfer:

(a) Storage in waste compartments: those are in good condition to prevent the spillage of the nanomaterials. Mark the waste compartment with a portrayal of the waste and incorporate accessible data describing known and suspected properties.
(b) Storage in plastic sacks: Collect paper, wipes, PPE, and different things with free sullying in a plastic pack or other sealable holder put away in the lab hood. At the point when the pack is full, close it and cautiously place it into a second

plastic sack or other fixing holder, to keep away from outside tainting. Remove it from the hood and name the external sack with a fitting waste name.

5.8.2 Disposal of Nanomaterial Waste

It is a sensible most pessimistic scenario supposition to consider all nanomaterial wastes as possibly risky. It can along these lines be discarded as synthetic waste. The List of Waste Regulations 2005 (LOWR) (The List of Wastes 2005) gives point by point data about how to deal with and discard different kinds of waste. Two classifications are especially important: wastes from organic chemical process and wastes from inorganic chemical process. Most nanomaterial wastes can be arranged under these codes.

LOWR additionally gives a rundown of risky properties, including models, for example,

H5: This means *"Harmful"* substances which on the off chance are breathed in or ingested or enter the skin may include restricted health dangers. This would be a sensible most pessimistic scenario suspicion for some insoluble and solvent nanoparticle types.

H6: This means *"Toxic"* substances which on the off chance are breathed in or ingested or penetrated the skin may include genuine, intense, or constant health dangers and even "death." This would be a sensible most pessimistic scenario presumption for carbon nanotubes or other nanomaterials having a fibrous nature.

The guidelines indicate that the most extreme focus for substances delegated harmful is 3% by mass.

Nanomaterials that are dangerous, lethal, or chemically active ought to be neutralized. Because of the wide scope of existing nanomaterials, a solitary method for transfer won't do the trick for all classes of nanomaterials. Consequently, it is essential to comprehend the properties of specific nanowastes before creating compelling disposal rehearses. The created safety measures and disposal strategies essential for taking care of nanowaste must be founded on current learning and consider existing legislation. The disposal procedures must guarantee that the waste is deactivated of its dangerous properties. So, conceivable deactivation arrangements could be done depending upon the sort of the material of nanotechnology-containing waste, i.e., thermal, chemical or physical preparation (Bartlomiej 2016).

5.9 Conclusion

In this chapter, it was very difficult to gather information about the adverse impacts of specifically the nanofinished medical textiles on both human and environment, because of the lack of knowledge about the reaction of each nanosubstance and its individual nanosize with the environment and/or with the consumer using them. So,

I tried to take this issue from the point of view of the hazardous effect of nanomaterials and their waste trying to focus on the medically treated nanomaterials as possible. It was clear that textile-based nanomaterials for medical uses are supposed to improve people's lives and in some cases transform them, taking into account their impact on both humans and environment.

One of the most important negative impacts of nanomaterials is to get rid of them in the wrong way especially the nanofinished medical textiles. The risk of getting rid of the nanofinished medical textiles is that the reusable ones can elute large portion of the nanomaterials in the washout water leading to the series impact on the ecosystem. On the other hand, in case of using disposable nanofinished medical textiles, the risk is bigger because of the absence of knowledge about the hazardous effect of each individual nanosubstance when released into the environment. When dealing with these materials, precautionary measures must be taken when handling them or with any substances contaminated with nanomaterial throughout their life cycle. Through utilization of LCA, it is conceivable to see which procedure (or procedures) drives ecological and human health impacts and may offer knowledge into limiting effects all through an item's life.

Clear and efficient strategies and procedures are required for disposal and, where possible, recycling of these materials. Efforts should be intensified for all relevant organizations and bodies to conduct further research on new materials being discovered day after day calling for adequate and appropriate oversight, security testing, and valuation of the developing field of nanoinnovation.

References

Advantages of Reusable Healthcare Textiles, Uniforms and Apparel. Available from: https://www.prudentialuniforms.com/blog/advantages-of-reusable-healthcare-textiles/

Alhayat G, Omprakash S (2014) Technical fabric as health care material. Biomed Sci Eng 2(2):35–39

Almeidal L, Ramos D (2017) Health and safety concerns of textiles with nano materials. Mater Sci Eng 254:1–7

Alper K, Bekir K (2018) General evaluations of nanoparticles. El-Cezerî J Sci Eng 5(1):191–236

Amit DK, Arif KN (n.d.) A paper on protective textiles. Available from: www.technicaltextile.net

Approaches to Safe Nanotechnology Managing the Health and Safety Concerns Associated with Engineered Nanomaterials. NOISH Publication (2009). Available form: http://www.cdc.gov/noish/docs/2009-129/

Auffan M, Rose J, Bottero J, Lowry G, Jolivet J, Wiesner M (2009) Towards a definition of inorganic nanoparticles from an environmental, Health and Safety Perspective. Nat Nanotechnol 4(634):1–9

Bartlomiej K (2016) Nanotechnology, nanowaste and their effects on ecosystems: a need for efficient monitoring, disposal and recycling. Available from: https://authorzilla.com/JBwQa/nanotechnology-nanowaste-and-their-effects-on-ecosystems-a.html

Berger M (2007) Ethical aspects of nanotechnology in medicine, Department of Energy Nanoscale Science Research Centers approach to nanomaterial ES&H, revision 2, June. Available from: https://www.nanowerk.com/spotlight/spotid=3938.php

Boxall ABA, Tiede K, Chaudhry Q (2007) Engineered nano-materials in Soilsand water: how do they behave and could they pose a risk to human health. Nanomedicine 2(6):919–927

Brief for GSDR – 2016 Update: nanotechnology, nanowaste and their effects on ecosystems: a need for efficient monitoring, disposal and recycling, BartlomiejKolodziejczyk

Brouwer D, Kuijpers E, Bekker C, Asbach C, Kuhlbusch TAJ (2015) Field and laboratory measurements related to occupational and consumer exposure. In: Safety of nanomaterials along their lifecycle: release, exposure and human hazard. Taylor & Francis, Boca Raton

Chandra PR, Yu H, Fu P (2009) Toxicity and environmental risks of Nanomaterials: challenges and future needs. J Environ Sci Health C 27:1–35

Chinta SK, Veena KV (2013) Impact of textiles in medical field. Int J Latest Trends Eng Technol 2:142–145

Christoph L (2005) Opportunities and risks of nanotechnologies – Allianz report in Cooperation with the OECD international future program, 3–6, 35 (June)

Conrardy J, Hillanbrand M, Myers S, Nussbaum GF (2010) Reducing medical waste. AORNJ 91(6):711–721

Deepti G (2011) Functional clothing – definition and classification. Indian J Fibre Text Res 36:321–326

Denise MM, Elisa R, Adrian W, Rolf E, Murray H, Nowack B (2014) Presence of nanoparticles in wash water from conventional silver and nano-silver textiles. Available from: www.acsnano.org

Department of Energy Nanoscale Science Research Centers Approach to Nanomaterial ES&H, Revision 2, June (2007)

Disposable Medical Textiles Market (2019) Global industry size, share, business growth, revenue, trends, global market demand penetration and forecast to 2024|360 Market updates. Posted: Mar 29, 2019. Available from: http://www.wboc.com/story/40219718/disposable-medical-textiles-market-2019-global-industry-size-share-business-growth-revenue-trends-global-market-demand-penetration-and-forecast-to

Disposal Apparel: Medical protective clothing encompasses: caps, masks, aprons, and shoe covers. Available form: https://mercatormedical.eu/products/Disposable-apparel

Elias K (2015) A technical overview on protective clothing against chemical hazards. AASCIT J Chem 2(3):67–76

Farooq A, Irfan AS, Tanveer H, Iftikhar A, Soniya M, Mariyam Z (2014) Developments in health care and medical textiles – a mini review-1. Pak J Nutr 13(12):780–783

Faunce T (2017) Bartlomiej Kolodziejczyk, Nanowaste: need for disposal and recycling standards, Available from: www.G20insights.org

Fava JA, Denison R, Jones B, Curran MA, Vignon B, Selke S, Barnum J (eds) (1991) A technical framework for life-cycle assessment. SETAC and SETAC Foundation for Environmental Education, Inc., Washington, DC

Global Medical Textiles Market (2018–2022) Available from: https://www.technavio.com/report/global-medical-textiles-market-analysis-share 2018?utm_source=t9&utm_medium=bw_wk49&utm_campaign=businesswire

Indian Technical Textile Industry (2016) Current scenario & opportunities, Techtextil, India symposium 08, September, 2016 by ICRA Management Consulting Services Limited. Available from: http://textilevaluechain.in/2019/03/13/technical-textiles-current-scenario-a-review/

ISO (2006a) Environmental management – life cycle assessment – requirements and guidelines. International Organization for Standardization, Geneva

ISO (2006b) Life cycle assessment (ISO 14040). Building

Jitendra ST, Roger AH, Hanks C, Trybula W, Fazarro D (2016) Addressing ethical and safety issues of nanotechnology in health and medicine in undergraduate engineering and technology curriculum. Glob J Eng Educ 18(1):30–34

John J (2014) Comparative life cycle assessment of reusable vs. Disposable Textiles, Exponent Inc., USA, 2014

Joob B, Wiwanitkit V (2019) Review article: nanotechnology for health: a new useful technology in medicine. Med J Dr. D.Y. Patil Univ:401–405

Journal of Fashion Technology & Textile Engineering, available from: https://www.scitechnol.com/textile-engineering/medical-textile.php

Kai S, Ulrika B, Derk B, Bengt F, Teresa F, Thomas K, Robert L, Iseult L, Lea P (2015) Nanosafety in Europe 2015–2025: towards safe and sustainable nano materials and nanotechnology innovations, Finnish Institute of Occupational Health. Available from: www.ttl.f/en/publications/electronic_publications/pages/default.aspx

Kamalja MD, Khatik SF (2015) Application of nanotechnology in medical textiles & its impact on environment. J Emerg Trends Model Manuf 1(2):35–37

Knowledge Paper on Technical Textiles: Towards a Smart Future Technotex 2016 – Fifth international exhibition & conference on technical textiles, Mumbai, India, 21–23 April 2016

Life Cycle Analysis of Isolation Gowns Gives the Advantage to Reusables, Feb 06, 2018. Available from: https://www.infectioncontroltoday.com/personal-protective-equipment/life-cycle-analysis-isolation-gowns-gives-advantage-reusables

Madalina Z, Fulga T (2014) Antimicrobial reagents as functional finishing for textiles intended for biomedical applications. I. Synthetic organic compounds. Chem J Mold 9(1):14–32

Mayekar A (2008) Disposable textiles – future of India. In: Proceedings of international conference on technical textiles & Nonwovens, ICTN-2008, IIT Delhi, 11–13 November 2008

Mclistyre JE, Daniels PN (1995) Textile terms and definition, 10th edn. The Textile Institute, Manchester. pp 206

Medical Textiles (2019) Market size, share & trends analysis report by fabric (Non-woven, Knitted, Woven), By Application And Segment Forecasts, 2019–2025. https://www.grandviewresearch.com/industry-analysis/medical-textiles-market

Meena CR, Nitin A, Pranayak S (n.d.) Medical textiles. https://www.technicaltextile.net/articles/medical-textiles-2587

Michael O (2012) Review article: a comparison of reusable and disposable perioperative textiles: sustainability state-of-the-art. Int Anesth Res Soc 114(5):1055–1066

Mueller N, Nowack B (2008) Exposure Modeling of engineered nanoparticles in the environment. Environ Sci Technol 42(12):4447–4453

Mueller NC, Jurgen B, Johannes B, Miroslav C, Peter R, David R, Bernd N (2012) Application of nanoscale zero valent iron (NZVI) for groundwater remediation in Europe. Environ Sci Pollut Res 19(2):50–558

Munima H (2019) Nano fabrics in the 21st century: a review, Article 2. Asian J Nanosci Mater 2(2):120–256

Musee N, Aldrich C, Lorenzen L (2008a) New methodology for hazardous waste classification using fuzzy set theory: part II. Intelligent decision support system. J Hazard Mater 157:94–105

Musee N, Lorenzen L, Aldrich C (2008b) New methodology for hazardous waste classification using fuzzy set theory: part I. knowledge acquisition. J Hazard Mater 154:1040–1051

Musee N, Foladori G, Azoulay D (2012) Social and environmental implications of nanotechnology development in Africa. CSIR (Nanotechnology Environmental Impacts Research Group, South Africa). Available from: https://ipen.org/pdfs/nano_booklet_sept_5.pdf

Nowack B (2010) Pollution prevention and treatment using nanotechnology. In: Krug HF (ed) Nanotechnology. Springer. Available from: https://application.wiley-vch.de/books/sample/352731735X_c01.pdf

Oakes J, Batchelor SN, Dixon S (2005) New method for obtaining proper initial clusters to perform FCM algorithm for colour image clustering. Color Technol 12:237–244

Onar N, Sarnsik M (2002) Properties and classification of medical textiles, 1st International Technical Textile Congress, 11–12 October 2002, Izmir

Paresh CR, Hongtao Y, Peter PF (2009) Toxicity and environmental risks of Nanomaterials: challenges and future needs. J Environ Sci Health C Environ Carcinog Ecotoxicol Rev 27(1):1–35

Pre (2013a) Eco indicator 99. http://www.presustainability.com/download/manuals/EI99_Manual.pdf

Pre (2013b) ReCiPe. Available from: http://www.lcia-recipe.net/

Protective Textiles Introduction (2018, March). Available from: https://textilestudycenter.com/protective-textiles-introduction/

Raaz NA (n.d.) The classification of medical textile. Available from: http://textilemerchandising.com/classification-medical-textile/

Raw materials for medical textiles. Available from: https://slideplayer.com/slide/6467899/

Report on Medical Textiles and Sport/Outdoor Textiles. Available from: https://technicaltextile. net/articles/report-on-medical-textiles-and-sport-outdoor-textiles-5032

Resent I (2016) Risk of nanowastes. Inżynieria i OchrŚrodowiska 19:469–478. https://doi. org/10.17512/ios.2016.4.3

Rothen-Rutishauser BM, Schurch S, Haenni B (2006) Interaction of fine particles and nanoparticles with red blood cells visualized with advanced microscopic techniques. Environ Sci Technol 40:4353–4359

Sawhney APS, Condon B, Singh KVS, Pang S, Li G, David H (2008) Modern applications of nanotechnology in textiles. Text Res J 78(8):731–739

Scott U (2015) Sustainable solutions for medical devices and services. A dissertation presented in partial fulfillment of the requirements for the degree doctor of philosophy, Arizona State University, December

Shilpi A (2014) Medical textiles: significance and future prospect in Bangladesh. Eur Sci J 10(12):488–502

Silas J, Hansen J, Lent T (2007) The future of fabric. Health Care. In conjunction with health care without harm's research collaborative. Available from: http://www.noharm.org/us. Accessed 14 Nov 2018

Stacy (2018) Study finds reusable surgical gowns are better for the environment than disposable gowns on Fri, May 25, 2018. Available from: http://www.medicleanse.com/blog/ reusable-surgical-gowns-study/

Suprava P, Rachita N, Sibasish S (2015) Review article: nanotechnology in healthcare: applications and challenges. Med Chem J 5(12):528–533

Syduzzaman MD, SarifUllah P, Kaniz F, Sharif A (2015) Smart textiles and nano-technology: a general overview. J Text Sci Eng 5:181

Teli MD (n.d.) Functional textiles & apparels, written by: new cloth market. Available from: https://www.technicaltextile.net/articles/functional-textiles-and-apparels-3292

The List of Wastes (England) Regulations (LOWR), Great Britain SI, 895 (2005)

Thiry MC AATCC review, (June 2007), 23–27

Troy B, Bridget C, Kiril H, Jonathan DP, Paul W (2010) The release of Nanosilver from consumer products used in the home. J Environ Qual 39(6):1875–1882

Vanlangenhove L (2013) Smart textiles for protection: an overview. Woodhead Publishing Limited, Philadelphia

Vigon BW, Tolle DA, Cornaby BW, Latham HC, Harrison CL, Boguski TL, Sellers JD (1992) Life-cycle assessment: inventory guidelines and principles. US Environmental Protection Agency, Battelle and Franklin Associates Ltd, Cincinnati

Wen Z (2013) Introduction to healthcare and medical textiles. DESTech Publications Inc., Lancaster

What are Protective Textiles? 15 types of protective textiles. Available from:https://www.fiber2apparel.com/2018/07/protective-textiles-definition-types-importance.html

Therapeutic Applications of Graphene Oxides in Angiogenesis and Cancers

6

Ayan Kumar Barui, Arpita Roy, Sourav Das,
Keerti Bhamidipati, and Chitta Ranjan Patra

Abstract

The application of nanotechnology in biology and medicine is generally termed as nanomedicine which reforms the strategic platforms of the modern healthcare system associated with diagnosis and therapy of different diseases. Since past decades, several research groups including ours demonstrated the diverse biomedical applications of different inorganic nanoparticles. Among these nanomaterials, graphene oxide (GO) nanoparticles are of great attraction for their potent applications in angiogenesis as well as cancer therapy due to its unique physicochemical and biological properties such as large surface area, high drug loading efficacy, biocompatibility, biodegradability, etc. This book chapter illustrates the overview of recent applications of GO in angiogenesis including pro-angiogenic activity, anti-angiogenic activity and wound healing potential. Moreover, the therapeutic (anticancer activity, drug/gene delivery, photothermal/immuno therapy) and bio-imaging applications of GO for different cancer diseases are also described in a concise manner. Additionally, in view of future clinical applications, pharmacokinetics, toxicity and clearance studies of GO are briefly demonstrated. Finally, this book chapter provides the global market overview along with challenges and future directions of GO in biomedical applications.

Keywords

Graphene oxide · Nanomedicine · Angiogenesis · Cancer therapy · Biomedical applications

A. K. Barui · A. Roy · S. Das · K. Bhamidipati · C. R. Patra (✉)
Academy of Scientific and Innovative Research (AcSIR), Ghaziabad, Uttar Pradesh, India

Department of Applied Biology, CSIR-Indian Institute of Chemical Technology, Hyderabad, India
e-mail: crpatra@iict.res.in

© Springer Nature Singapore Pte Ltd. 2020
A. K. Shukla (ed.), *Nanoparticles and their Biomedical Applications*,
https://doi.org/10.1007/978-981-15-0391-7_6

Abbreviations

A549	Human alveolar adenocarcinoma epithelial cell line
Ab	Antibody
ADM	Acellular dermal matrix
AIE	Aggregation-induced emission
APCs	Antigen-presenting cells
B16	Murine melanoma cells
Bcl-2	B-cell lymphoma-2
bFGF	Basic fibroblast growth factor
BSA	Bovine serum albumin
CAD	Cis-aconitic anhydride-modified doxorubicin
CAM	Chick chorioallantoic membrane
CEA	Chick embryo angiogenesis
CMC	Carboxymethylcellulose
CpG	Cytosine-phosphate-guanine
CS-PVA	Chitosan-polyvinyl alcohol
CT	Computed tomography
Cx43	Connexin43
DCs	Dendritic cells
DDS	Drug delivery system
DIM	Diindolylmethane
DNA	Deoxyribonucleic acid
DOX	Doxorubicin
E. coli	*Escherichia coli*
ECM	Extracellular matrix
EGFP	Enhanced green fluorescence protein
eNOS	Endothelial nitric oxide synthase
EPR	Enhanced permeability and retention
FA	Folic acid
FDA	Food and drug administration
fGO	Functionalized graphene oxide
FITC	Fluorescein isothiocyanate
FSHR	Follicle-stimulating hormone receptor
GelMA	Methacrylated gelatin
GF	Graphene foams
GIC	Graphene intercalation compounds
GO	Graphene oxide
GPD	GO-PEG-PAMAM
H_2SO_4	Sulphuric acid
HBD	Heparin-binding domain
HDAC	Histone deacetylases
HeLa	Human cervical cancer cells
HGF	Hepatocyte growth factor
HNO_3	Nitric acid

HUVECs	Human umbilical vein endothelial cells
IDO	Immune checkpoint overexpressed in tumours
IL1β	Interleukin-1 beta
IL6	Interleukin 6
IR800	Infrared 800
IUPAC	International union of pure and applied chemistry
KClO$_3$	Potassium chlorate
LDI	Laser Doppler imaging
LHT7	Low molecular weight heparin
LSECs	Liver sinusoidal endothelial cells
MBA-MB-231	Human breast cancer cell line
MCF-7	Human breast cancer cell line
MCP-1	Monocyte chemotactic protein 1
MDR	Multidrug-resistant
MIA PaCA-2	Human pancreatic carcinoma
MMP-9	Matrix metallopeptidase 9
MRI	Magnetic resonance imaging
MSCs	Mesenchymal stem cells
NIR	Near-infrared
NO	Nitric oxide
OVA	Ovalbumin
PAACA	Poly(acryloyl-6-aminocaproic acid)
PAH	Polyallylamine hydrochloride
PAMAM	Polyamidoamine dendrimer
PDDA	Poly(diallyldimethylammonium chloride)
PDGF	Platelet-derived growth factor
pDNA	Plasmid DNA
PEG	Polyethylene glycol
PEI	Polyethyleneimine
PET	Positron emission tomography
PGO	Porphyrin graphene oxide
PI	Propidium iodide
PMAA	Poly(methacrylic acid)
PPa	Pyropheophorbide-a
PTX	Paclitaxel
PVP	Poly N-vinylpyrrolidone
RAW 264.7	Murine macrophage cell line
RES	Reticuloendothelial system
rGO	Reduced graphene oxide
RNA	Ribonucleic acid
ROS	Reactive oxygen species
S. aureus	*Staphylococcus aureus*
SCC-7	Mouse head and neck carcinoma cell line
shRNA	Short hairpin RNA
SiHa	Cervical squamous cancer cells

siRNA   Short interfering RNA
SPION   Superparamagnetic iron oxide nanoparticles
SRGO   Sorafenib reduced graphene oxide
TiO_2 Titanium dioxide
TLR   Toll-like receptor
TNFα   Tumour necrosis factor alpha
U118   Human brain glioma cells
U87   Human primary glioblastoma cell line
UCNPs   Upconversion nanoparticles
Ure B   Urease B
VAR   Peptide probe
VEGF   Vascular endothelial growth factor
ZnO   Zinc oxide
ZnPc   Zn(II)-phthalocyanine

6.1   Introduction

Nanotechnology is an interdisciplinary field which involves physical, chemical, environmental and biological sciences (Dai 2006; Schaefer 2010; Teli et al. 2010). It deals with different kinds of nanomaterials which possess unique physicochemical properties (e.g. optical, mechanical, electrical, biological, etc.) compared to the corresponding bulk substances owing to the high surface-area-to-volume ratio. Owing to its distinctive characteristics, nanoparticles exhibited numerous applications in different fields including agriculture (Khot et al. 2012), catalysis (Enterkin et al. 2011), cosmetics (Raj et al. 2012), electronics (Millstone et al. 2010), energy (Lohse and Murphy 2012), etc. Besides these applications, nanotechnology especially alters the paradigm of modern healthcare research, offering an alternative diagnostic and therapeutic strategies for different diseases (Caruso et al. 2012). Because of the size similarity, nanomaterials can easily interact with the cellular membrane, proteins, DNA/RNA, etc. which might be the plausible reason for their prevalent biomedical applications throughout the world (Winter 2007). Since the past decades, several research groups designed and developed various inorganic nanomaterials such as noble metal nanoparticles (Leteba and Lang 2013; Ouay and Stellacci 2015), lanthanide nanoparticles (Ahmad et al. 2015; Das et al. 2012; Patra et al. 2011; Zhao et al. 2016), transition metal oxide nanoparticles (Barui et al. 2012; Meghana et al. 2015; You et al. 2016), quantum dots (Wang et al. 2013), etc. for diverse biomedical applications. Besides these nanomaterials, carbon-based GO has emerged as one of the most attractive materials for modern research in healthcare sector due to its several advantages such as water dispensability, larger surface area, easier functionalization, high drug loading efficiency, biocompatibility, biodegradability, etc. (Muazim and Hussain 2017). GO exhibited promising therapeutic applications for several diseases including cancers (Akhavan et al. 2012; Robinson et al. 2011; Shim et al. 2014; Zhao et al. 2015), cardiovascular diseases/ischaemic diseases (Paul et al. 2014; Sun et al. 2013), wounds (Fan et al. 2014; Zhou et al. 2016), microbial diseases (Nanda et al. 2016; Perreault et al. 2015), etc. Additionally, GO could also be employed for diagnosing disease conditions through bio-imaging (Hong et al. 2012; Li et al. 2017; Shi et al. 2013) and biosensing (Li et al. 2012; Liu et al. 2010; Lu et al. 2016). Recently, few research groups including ours found pro- as well as anti-angiogenic properties of different GO nanomaterials, offering their therapeutic potential for angiogenesis-related diseases (Mukherjee et al. 2015; Wierzbicki et al. 2013). This book chapter will specifically discuss only the recent advances of GO nanomaterials for diverse therapeutic applications in angiogenesis and cancers. Additionally, toxicity study, pharmacokinetics, global market and future perspective of GO in biomedical applications are concisely described.

6.2   History of Graphene and Graphene Oxides

Carbon, one of the most abundant elements on Earth's crust, is highly essential for any biological system (Cotton and Wilkinsion 1972). It is well-known that the human body system consists of 18% carbon element. All the biomacromolecules such as carbohydrates, proteins, lipids, nucleic acids, etc. contain carbon in their structure (Falkowski et al. 2000). Carbon has several allotropes such as graphite, diamond, amorphous carbon, etc. Interestingly, graphite has a long history of use since the Neolithic era of Southeastern Europe especially for painting purposes (Nicol 2015). The term 'graphite' was basically originated from the Greek word 'graphein' which refers 'to write' (Dreyer et al. 2010). Graphite oxide and graphite intercalation compounds (GIC) were first time reported by a German researcher Schafhaeutl in 1840 (Dreyer et al. 2010; Schafhaeutl 1840). Later, in 1859, this work was followed by a British scientist Brodie who made the interaction of potassium chlorate ($KClO_3$) as well as strong acids such as sulphuric acid (H_2SO_4) and nitric acid (HNO_3) to graphite leading to the formation of graphite oxide. The surface modification of this graphite oxide could further produce GO, rGO and GIC (Brodie 1859; Dreyer et al. 2010).

Graphite is generally used as the precursor for the synthesis of graphene which has different forms such as graphene sheets, GO, reduced graphene oxide (rGO), etc. (Muazim and Hussain 2017). Graphene has a single layer of honeycomb-like hexagonal carbon lattice (Byun 2015). In 1986, the term 'graphene' was first time recommended by Boehm and co-workers who suggested that –ene suffix in 'graphene' refers to polycyclic aromatic hydrocarbons (Boehm et al. 1994; Dreyer et al. 2010). However, after 11 years, in 1997, IUPAC adopted the term 'graphene' in Compendium of Chemical Technology as 'The term graphene should be used only when the reactions, structural relations or other properties of individual layers are discussed' (Dreyer et al. 2010; McNaught and Wilkinson 1997; Tan and Lee 2013). Recently, Novoselov, Geim and co-workers isolated graphene through a simple mechanical exfoliation in 2004, and thereafter the massive applications of graphene, GO and rGO were explored in various fields throughout the world (Dreyer et al. 2010; Tan and Lee 2013). These scientists were awarded the prestigious Nobel Prize in Physics in 2010 for their innovative research work (Jaleel et al. 2017). Presently, graphene-based materials including GO and rGO are considered as one of the most explored topics for material science and biomedical research.

6.3   Biomedical Applications of Graphene Oxides

Due to the possession of a large aromatic surface, graphene is hydrophobic in nature. However, GO (oxidized form of graphene) contains different functional groups (e.g. hydroxyl, carboxylic acid and epoxide) on its surface, enhancing its hydrophilicity and water dispensability (Nejabat et al. 2017). Several physicochemical and biological properties which make GO an excellent material for diverse biomedical applications include large surface area, high thermal conductivity, high drug loading efficiency, water dispensability, biocompatibility, biodegradability, cost-effectiveness, etc. The functional groups present on the surface of GO facilitate the easier conjugation of therapeutic molecules as well as fluorophore, leading to augment its various biomedical applications including drug delivery (Shim et al. 2014; Zhao et al. 2015), gene delivery (Yin et al. 2017; Zhang et al. 2011), bioimaging (Hong et al. 2012; Li et al. 2017; Shi et al. 2013), etc. GO could also be employed as a photothermal agent for cancer therapy (Akhavan et al. 2012; Robinson et al. 2011) owing to its high near-infrared (NIR) absorbance. GO is also employed for biosensing application for the detection of different biomolecules such as glucose (Liu et al. 2010), amino acid (Li et al. 2012), DNA (Lu et al. 2016), etc. Scientists also found that GO could be useful for tissue engineering (Nie et al. 2017; Shin et al. 2016a; Shin et al. 2016b) as well as antimicrobial (Nanda et al. 2016; Perreault et al. 2015) applications. Very recently, scientists observed the applications of GO in angiogenesis which provides the basis for its future therapeutic potential for ischaemic diseases, wound repairment, cancers etc. where angiogenesis is the major target (Fan et al. 2014; Lai et al. 2016; Mukherjee et al. 2015; Sun et al. 2013). Although GO exhibits several biomedical applications (Table 6.1), considering the limited scope, this book chapter particularly focuses on recent advances of GO for its therapeutic applications in angiogenesis and cancers.

6.4   Therapeutic Applications of Graphene Oxides

The following section briefly illustrates the various biomedical applications of GO in angiogenesis and cancers.

6.4.1   Angiogenesis

Angiogenesis is a critical process for the establishment of new blood vessels from pre-existing vasculature (Bikfalvi and Bicknell 2002; Folkman 1995). It is highly essential for several physiological (e.g. embryonic development, wound healing,

Table 6.1 List of recent research on graphene oxide-related biomedical applications

S. No.	Target cell/tissue/ material	Biomedical application	References
1.	Ischaemic muscle	Therapeutic angiogenesis	Sun et al. (2013)
2.	HUVECs	Pro-angiogenic activity	Mukherjee et al. (2015)
3.	Skin tissue (SD rat)	Wound healing	Fan et al. (2014)
4.	HUVECs	Anti-angiogenic activity	Lai et al. (2016)
5.	U87 and U118 cells	Anticancer activity	Jaworski et al. (2015)
6.	MCF-7 cells	Cancer therapy: photothermal agent	Robinson et al. (2011)
7.	SiHa cells	Drug delivery	Zhao et al. (2015)
8.	KB cells	Drug delivery	Shim et al. (2014)
9.	HeLa cells	Gene delivery	Zhang et al. (2011)
10.	MIA PaCa-2	Gene delivery	Yin et al. (2017)
11.	Ischaemic muscle	Bio-imaging	Sun et al. (2013)
12.	4 T1 tumour vasculature	Bio-imaging	Hong et al. (2012)
13.	HeLa cells	Bio-imaging	Li et al. (2017)
14.	Glucose	Biosensing	Liu et al. 2010)
15.	Amino acid	Biosensing	Li et al. (2012)
16.	Cardiac tissue	Tissue engineering	Shin et al. (2016b)
17.	Bone tissue	Tissue engineering	Nie et al. (2017)
18.	*E. coli* and *E. faecalis*	Antibacterial activity	Nanda et al. (2016)
19.	*E. coli*	Antibacterial activity	Perreault et al. (2015)

hair growth, menstrual cycle, etc.) as well as pathological (atherosclerosis, diabetic retinopathy, cancer, ischaemic heart disease, ischaemic limb disease, etc.) courses. Both the pro-angiogenic and anti-angiogenic molecules/agents control the process of angiogenesis by promoting and inhibiting the formation of new vasculature, respectively. While several pro-angiogenic cytokines (e.g. vascular endothelial growth factor, VEGF; basic fibroblast growth factor, bFGF; hepatocyte growth factor, HGF; platelet-derived growth factor, PDGF; etc.) have conventionally been employed for the treatment of ischaemic heart disease, ischaemic limb disease, wounds, etc. (Barui et al. 2012; Patra et al. 2008), several anti-angiogenic drugs (e.g. cannabinoids, thalidomide, bevacizumab, etc.) have been used for the treatment of cancers and retinal disorders (Bergers and Hanahan 2008). However, the therapeutic use of these pro- as well as anti-angiogenic molecules/drugs is often associated with several limitations including non-specificity, less bioavailability, side effects, prolonged treatment as well as high cost. Therefore, to overcome these aforesaid challenges, some alternative therapeutic treatment strategies have been developed in the form of nanomaterials by various research groups all over the world. In this context, several groups including ours developed different pro-angiogenic nanomaterials such as europium hydroxide (Patra et al. 2008), zinc oxide (Barui et al. 2012), ceria (Das et al. 2012), carbon nanotube (Chaudhuri et al. 2010), gold (Bartczak et al. 2013), silver (Kang et al. 2011), etc. as well as anti-angiogenic nanomaterials such as gold (Arvizo et al. 2011), silver (Gurunathan et al. 2009), ceria (Hijaz et al. 2016), silicate (Hu et al. 2015), etc. that could be useful for the treatment of different angiogenesis-related diseases. Besides these nanoparticles, recent reports demonstrated the applications of GO nanomaterials in angiogenesis, which are concisely described in the following sections.

6.4.1.1 Pro-angiogenic Property/Therapeutic Angiogenesis

Some research groups including ours illustrated the pro-angiogenic properties of GO nanomaterials. Additionally, the therapeutic angiogenesis based on these materials through cytokine delivery, angiogenic gene delivery, etc. is also explored. These reports are briefly discussed in the following sections.

Recently, our group demonstrated the synthesis of GO as well as rGO and investigated their angiogenic properties employing different in vitro and in vivo assays (Mukherjee et al. 2015; Patra 2015). Cell viability assay in HUVECs (human umbilical vein endothelial cells) and ECV-304 cells (presumptive endothelial cells) exhibited that both GO and rGO promoted endothelial cell proliferation (key step of angiogenesis) at lower doses (1–100 ng/mL), suggesting their pro-angiogenic properties. However, GO and rGO treatments beyond 100 ng/mL induced inhibition of endothelial cell proliferation, indicating their anti-angiogenic nature at higher doses. The in vivo chick embryo angiogenesis (CEA) assay further showed that lower doses (5–10 ng/mL) of GO and rGO promoted mature blood vessel formation in terms of blood vessel size, length and junction compared to control experiment, confirming their pro-angiogenic properties (Fig. 6.1). The mechanistic study revealed that controlled formation of reactive oxygen species (ROS: H_2O_2, O_2^-) and nitric oxide (NO), leading to Akt-mediated activation of eNOS signaling pathway, might play a crucial role behind the pro-angiogenic properties of GO and rGO. In another study, Park and co-workers developed MSC-rGO spheroid employing as synthesized rGO flakes (~2–5 μm) and MSC (mesenchymal stem cell) spheroids via a modified hanging-drop method and investigated the therapeutic potential of MSCs for ischaemic heart disease in presence of rGO (Park et al. 2015). The results exhibited that the incorporation of rGO into MSC spheroids enhanced cell-ECM interactions leading to higher expression of a paracrine factor, having reparative actions during myocardial injury. Additionally, the angiogenic factor Cx43 (inducing cardiac repair) was also upregulated because of the electrical conductance of rGO and activation of the paracrine factor. Further, the administration of MSC-rGO spheroid to infarcted hearts of BALB/c nude (nu/nu) mice augmented the cardiac repair compared to alone rGO and MSC spheroids, suggesting the therapeutic efficacy of MSC-rGO spheroid system for ischaemic heart disease. Similarly, Sun et al. developed VEGF decorated IR800 (near-infrared fluorescent dye) conjugated functionalized GO (GO-IR800-VEGF) and studied its potent efficiency for image-guided targeted therapeutic angiogenesis employing mice hind limb ischaemia model (Sun et al. 2013). The administration of GO-IR800-VEGF to mice with hind limb ischaemia exhibited a higher rate of blood perfusion recovery compared to other control groups as observed from laser Doppler imaging (LDI), indicating the therapeutic efficiency of the delivery system (Fig. 6.2). Additionally, photoacoustic imaging revealed that GO-IR800-VEGF treatment to ischaemic mice enhanced the oxygen

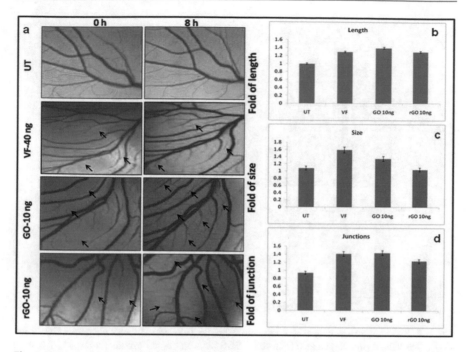

Fig. 6.1 In vivo CEA assay in presence of GO and rGO. (**a**) Increase of matured blood vessel formation (marked by black arrows) was observed in embryo treated with GO and rGO at low dose (10 ng mL^{-1}) in a time-dependent manner. VEGF (40 ng mL^{-1}) was used as positive control experiment. (**b–d**) Several angiogenic parameters such as blood vessel length, size and junction were quantified and presented as histogram. Statistical significance was calculated by using t -test. All data are statistically significant (p < 0.05). (Figure reproduced with permission from Ref. (Mukherjee et al. 2015). Copyright © 2015 WILEY-VCH Verlag GmbH & Co. KGaA, Weinheim)

supply to ischaemic tissues 14 days post-surgery in comparison with other groups, which might be attributed to the improved capillary formation and sprouting of small arteries in ischaemic tissues. Moreover, positron emission tomography (PET) imaging showed the uptake of GO-IR800-VEGF was significantly higher in the ischaemic hind limb of mouse 1 week post administration compared to other control groups, confirming the potent therapeutic angiogenesis of the delivery system. The research group headed by Dr. Khademhosscini also developed an injectable and biocompatible methacrylated gelatin hydrogel (GelMA), containing functionalized GO (fGO) nanomaterials loaded with DNA$_{VEGF}$ to form fGO$_{VEGF}$/GelMA gene delivery system for cardiac repairment (Paul et al. 2014). The authors exhibited that intramyocardial administration of the nanocomposite system to Lewis rat model of myocardial infarction leads to exert therapeutic effects, as observed from the significant enhancement in myocardial capillary density as well as a decrease in scar area of the infarcted hearts compared to sham and other control groups. Additionally, 2 weeks post administration of fGO VEGF/GelMA showed better cardiac activity in echocardiography in comparison with other control groups. Altogether, this study

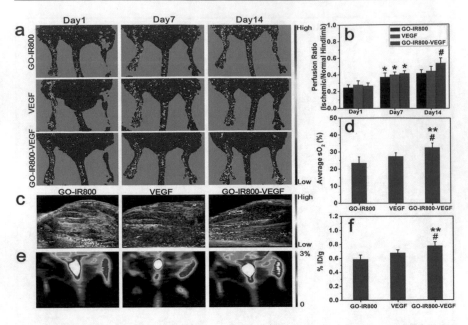

Fig. 6.2 Therapeutic angiogenesis of ischaemic muscle. (**a**) Laser Doppler images of GO-IR800 (top), free VEGF (middle)- and GO-IR800-VEGF (bottom-treated groups on days 1, 7 and 14 post-surgery. (**b**) Quantitative analysis of tissue blood perfusion in ischaemic limbs and control non-ischaemic limbs (∗, $P < 0.05$, as compared to day 1; #, $P < 0.05$, as compared with the VEGF-treated group on day 14). (**c**) Photoacoustic (PA) images for tissue oxygen saturation (sO_2) detection of GO-IR800 (left)-, free VEGF (middle)- and GO-IR800-VEGF (right)-treated groups on day 14 post-surgery, respectively. (**d**) Quantitative analysis of tissue oxygen saturation (%) within ischaemic limbs of the above three groups (∗∗, $P < 0.01$, as compared with the GO-IR800 group; #, $P < 0.05$, as compared with the free VEGF group). (**e**) Representative positron emission tomography (PET) images at 1 h after intravenous injection of ^{18}F-AlF-NOTA-PRGD2 (^{18}F-Alfatide) in the above three groups. (**f**) Quantitative analysis of tracer uptake in ischaemic tissue three groups above (∗∗, $P < 0.01$, as compared with the GO-IR800 group; #, $P < 0.05$, as compared with the free VEGF group). (Figure reproduced with permission from Ref. (Sun et al. 2013). Copyright © 2013 Royal Society of Chemistry)

provides the platform for the advancement of hydrogel-mediated gene delivery system for myocardial infarction using functionalized GO nanomaterials.

6.4.1.2 Anti-angiogenic Property

In the early 1970s, Dr. Judah Folkman first time proposed the concept of anti-angiogenic therapy especially for cancer diseases (Folkman 1995). Anti-angiogenic drugs/materials could attenuate endothelial cell proliferation, migration as well as tube formation, thereby reducing tumour growths. These drugs/materials basically downregulate the expression of different cytokines as well as suppress the phosphorylation of angiogenesis-regulating proteins. Besides different nanomaterials (e.g. gold (Arvizo et al. 2011), silver (Gurunathan et al. 2009), ceria (Hijaz et al. 2016), etc.), researchers have recently explored the anti-angiogenic properties of GO-based nanomaterials which are briefly described in the following section.

Lai et al. developed bovine serum albumin decorated GO (BSA-GO) that was highly stable in physiological solution (Lai et al. 2016). The binding affinity of functionalized GO to VEGF-A$_{165}$ was found to be five times stronger than highly abundant plasma proteins (e.g. fibrinogen, transferrin, immunoglobulin and human serum albumin G). Considering this strong binding affinity, the authors studied the anti-angiogenic efficacy of BSA-GO. The results revealed that BSA-GO attenuated the human umbilical vein endothelial cell (HUVEC) proliferation and migration, suggesting its anti-angiogenic properties. Besides endothelial cell proliferation and migration, tube formation is another fundamental criterion for angiogenesis. The authors observed that the BSA-GO inhibited the VEGF-A$_{165}$ which induced tube formation significantly in HUVECs, in a dose-dependent manner compared to the control experiment (Fig. 6.3). Additionally, the nanocomposite system inhibited the vascular growth as observed in chick chorioallantoic membrane (CAM) assay and attenuated VEGF-A$_{165}$-mediated generation of blood vessels in a rabbit model of corneal neovascularization. The authors claimed that their findings could be useful for the treatment of retinal disorder and cancer by suppressing angiogenesis. As described in the earlier section (*5.1.1. Pro-angiogenic property/therapeutic angiogenesis*), our group illustrated that GO and rGO could exhibit anti-angiogenic properties at a higher dose (> 100 ng/mL) as observed from the inhibition of human umbilical vein endothelial cell (HUVEC) proliferation. Additionally, a higher dose

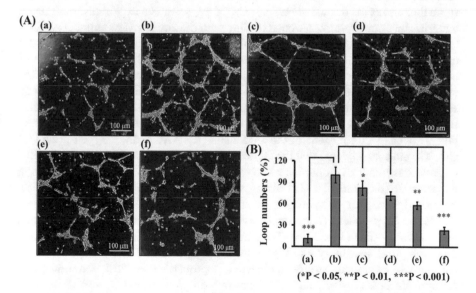

Fig. 6.3 Effect of BSA-GO on the VEGF-A$_{165}$-induced HUVEC tube formation. (**A**) Representative microscopic images of HUVECs inoculated on Matrigel and treated (a) without and (b–f) with VEGF-A$_{165}$ (1.0 nM) in the (b) absence or presence of (c) BSA-GO ([GO] = 5 µg mL^{-1}), (d) BSA-GO ([GO] = 10 mg mL^{-1}), (e) BSA-GO ([GO] = 15 mg mL^{-1}) or (f) BSA-GO ([GO] = 30 mg mL^{-1}) for 18 h. (**B**) The number of loops formed by the corresponding HUVECs is plotted. Error bars in (**B**) represent the standard deviations from four replicate experiments. (Figure reproduced with permission from Ref. (Lai et al. 2016). Copyright © 2016 Elsevier B.V)

of GO and rGO attenuated the growth of vasculature in CEA assay, confirming their anti-angiogenic nature (Mukherjee et al. 2015). The excessive generation of intracellular ROS (H_2O_2 and O_2.-) was found to be the plausible mechanism behind the anti-angiogenic nature of GO and rGO. Interestingly, Wierzbicki and co-workers demonstrated the angiogenesis study of different carbon-based nanomaterials including diamond nanoparticles, graphite nanoparticles, graphene nanosheets, multiwall nanotubes and C60 fullerenes (Wierzbicki et al. 2013). CAM assay revealed that diamond nanoparticles and multiwall nanotubes had maximum anti-angiogenic properties, while graphene nanosheets had no significant effect, and fullerene exhibited pro-angiogenic activity. In summary, the authors provided insight on the utility of various carbon-based nanomaterials for anti-angiogenic therapy.

6.4.1.3 Wound Healing

While any damage or cut or blow to our body part/living tissue can be considered as a wound, wound healing can be referred to as a natural restorative process of tissue injuries (Kumar and Chatterjee 2016; Nguyen et al. 2009; Rieger et al. 2015). This complex course involves many cell types such as endothelial, fibroblast, epidermal, etc. It is well-known that angiogenesis plays a pivotal role during wound healing. Earlier reports demonstrated that different cytokines/growth factors augment this process (Velnar et al. 2009). However, the use of these cytokines is often related to some limitations (as discussed earlier). Therefore, scientists explored various nanomaterials that could promote wound healing with the aim of overcoming those limitations. Considering the aim of this book chapter, the following section illustrates only the recent advancement of GO nanomaterials in wound healing application.

Cong et al. developed GO-based poly(acryloyl-6-aminocaproic acid) (PAACA) polymer composite hydrogel, having pH-stimulated self-healing nature (Cong et al. 2013). The double networks of the hydrogel are activated by GO and Ca^{2+} ions. Basically, Ca^{2+} ions activated the generation of 3D cross-linked network via interaction of polar group present in the side chain of PAACA and oxygen groups of GO. The authors depicted that the GO-based hydrogel composite system exhibited improved mechanical properties while retaining the self-healing ability, suggesting its potential for biological scaffolding as well as drug delivery. Also, Fan et al. developed a number of hydrogels by cross-linking of N, N'-methylene bisacrylamide and Ag/graphene (Ag: graphene = 5:1) composites and studied their wound healing ability (Fan et al. 2014). The nanocomposites were found to be biocompatible with good sensitivity and high swelling property. Simultaneously, the nanocomposites exhibited strong antibacterial properties against Gram-negative *Escherichia coli* (*E. coli*) and Gram-positive *Staphylococcus aureus* (*S. aureus*) bacteria and showed excellent wound healing ability to the artificial wound created on a Sprague Dawley rat model within 15 days of treatment. Altogether, this study shed lights for the development of GO-based hydrogel materials for wound dressing application. Similarly, Lu and co-workers fabricated chitosan-PVA (CS-PVA) decorated graphene-based nanocomposite materials and demonstrated their wound healing properties, employing in vivo wound healing models of C57/BL6 mice and van

Beveren rabbit (Lu et al. 2012). The application of the nanocomposite to the wound region of mice showed faster recovery of wounds compared to only CS-PVA nanofibers and control groups, suggesting the wound healing activities of the nanocomposite (Fig. 6.4). Similar results were also observed in rabbit wound model co-relating with the mice model data. In another study, Zhou et al. prepared silver/silver chloride (Ag/AgCl)-decorated rGO (Ag/AgCl/rGO) and investigated their efficacy in burn wound healing model of ICR mice (Zhou et al. 2016). The topical administration of the nanocomposite to wound area exhibited significantly faster wound healing compared to the positive control group. By day 14, the wounds of mice were fully closed for Ag/AgCl/rGO-administered group. The wound healing properties of the nanocomposite might be attributed to their antibacterial efficacy (as observed against *E. coli* and *S. aureus*) that would encounter infection during

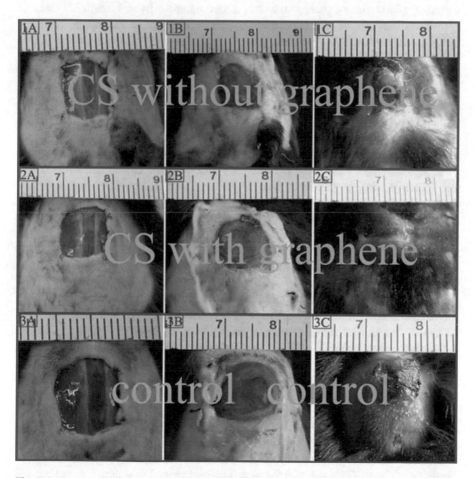

Fig. 6.4 Images of mice wound healing under: (1) pure chitosan-PVA nanofibers, (2) chitosan-PVA nanofibers containing graphene and (3) control. (Figure reproduced with permission from Ref. (Lu et al. 2012). Copyright © 2012 Royal Society of Chemistry)

burn conditions. Li et al. also developed a 3D graphene foam (GF) scaffold, conjugated with mesenchymal stem cells (MSCs) for wound healing purposes (Li et al. 2015b). The scaffold was found to be biocompatible, and it could induce the proliferation of MSCs. The authors further exhibited that the application of MSCs loaded GF scaffold to a rat wound model lead to a faster rate of healing compared to control and only GF groups, suggesting the potent wound healing efficacy of the composite system. Additionally, reduced scarring was observed for MSC-loaded GF scaffold group. The wound healing potential of the composite system might be explained by the upregulation of VEGF and bFGF due to MSC transplantation. Altogether this study depicted that the 3D-GF scaffold with MSC could be beneficial for wound healing process through neovascularization. Further, Chu et al. synthesized a collagen-based artificial acellular dermal matrix (ADM) scaffold conjugated with PEGylated GO and quercetin for diabetic wound healing (Chu et al. 2018). The research group demonstrated that this biocompatible hybrid scaffold material exerted various advantages: (1) escalated the attachment of mesenchymal stem cell (MSC), (2) induced MSC proliferation, (3) facilitated differentiation of MSCs into adipocytes as well as osteoblasts due to highly stable and adjustable conduction potential of quercetin and (4) induced collagen deposition and angiogenesis thereby accelerating diabetic wound healing in mice model. The affectivity was as evidenced by the protein expression of Col I, Col III and α-SMA by western blot analysis. Finally, the authors concluded that this GO-based hybrid biodegradable scaffold material could be useful for multiple applications such as tissue engineering, regenerative medicines, drug delivery as well as stem cell-based therapies. Thangavel et al. also designed a nanocomposite scaffold made of rGO conjugated with isabgol for wound healing in diabetic as well as normal rats (Thangavel et al. 2018). The authors first synthesized rGO using solar radiation and dispersed it into isabgol solution for making the nanocomposite scaffold isab+rGO. In vitro results in mouse fibroblast (NIH-3T3) cells revealed the biocompatible nature of the scaffold. On the other hand, in vivo studies in diabetic and normal Wistar rats exhibited that the scaffold-based dressing induced the shrinking of the wound area, collagen synthesis and shortening the time of re-epithelialization significantly in comparison with the control group. Further studies demonstrated that the scaffold dressing promoted angiogenesis and collagen deposition in the wounds through the reduction of inflammation phase and engaging macrophages. Further, Lu and co-workers developed polydopamine-reduced GO (pGO)-coated chitosan and silk fibroin-based scaffold (pGO-CS/SF) and demonstrated its efficacy as a wound dressing to heal wounds (Tang et al. 2019). The authors showed that the incorporation of pGO on the scaffold not only improved its mechanical properties but also increased its electroactivity for regulating cellular activities as well as induced antioxidant properties to scavenge excessive ROS that could be essential for wound healing purpose. The administration of the pGO-CS/SF scaffold in SD rats of wound regeneration model exhibited efficient wound healing as compared to the control group, indicating the potential of the pGO-based scaffold as an efficient wound dressing.

6.4.2   Cancer

Cancer can be referred to as a complex disease associated with uncontrolled growth of cells (Ma et al. 2015). It involves different mutated genes/proteins in malignant cells, with the possibility to spread to different body parts. It is well-known that cancer is one of the deadliest diseases all over the world in a modern scenario (Mulcahy 2008). The number of cancer incidence is growing day by day which could majorly be attributed to environmental pollution and urban lifestyle. The conventional therapy for cancer includes chemotherapy, radiation therapy, hormonal therapy, etc. all of which are associated with several limitations such as adverse side effects, poor bioavailability, high treatment cost, etc. In this context, nanotechnology plays an important role to overcome the aforesaid challenges. Since the past decades, several research groups including ours developed different nanomaterials (e.g. gold, silver, silica, ZnO, TiO_2, iron oxide, lanthanide, GO, etc.) for the treatment of cancers (Yaacoub et al. 2016). In view of the limited scope of this book chapter, the following section describes only the recent advances of GO for their diverse applications in cancer therapy.

6.4.2.1 Anticancer Activity

Recent literature demonstrates that GO-based nanomaterials themselves could exhibit toxicity to cancer cells, suggesting their anticancer properties. For example, Chang et al. synthesized GO nanomaterials of three different sizes (s-GO, 160 ± 90 nm; m-GO, 430 ± 300 nm; and l-GO, 780 ± 410 nm) using modified Hummers' method (Chang et al. 2011). The cell viability assay in A549 cells revealed that at higher concentration, all of the GO nanoparticles inhibited cancer cell proliferation, indicating their anticancer potential. However, the inhibitory effect was more pronounced for s-GO treatment which might be due to more production of intracellular ROS as compared to that of m-GO and l-GO. In another study, Jaworski et al. investigated the toxic effect of GO and rGO against different glioma cells (U87 and U118) (Jaworski et al. 2015). Both GO and rGO exhibited inhibition of cancer cell proliferation in a dose-dependent manner, suggesting their anticancer activity. However, the authors illustrated that rGO possessed better anticancer efficacy compared to GO. The mechanistic study further showed that the nanomaterials lead to toxic effects on glioma cells through an apoptotic pathway. Luo et al. also prepared GO employing a modified Hummers' method and reduced it to form rGO using pyrogallol which acted as reducing as well as a stabilizing agent during synthesis (Luo et al. 2017). The cell viability assay in HeLa cells exhibited the attenuation of cell proliferation at higher doses, suggesting their anticancer properties. However, the toxic response to cancer cells was better for rGO treatment compared to that of GO. Further, Mallick et al. developed polyethyleneimine-decorated self-assembled GO-based nanocomposite system (PEI-GTC-NPs) containing topoisomerase I inhibitor topotecan as well as chemotherapeutic drug cisplatin (Mallick et al. 2019). The GO-based nanoparticles (diameter around 170 nm) exhibited high positive zeta potential which is essential for targeting mitochondria. The confocal microscopy demonstrated the successful accumulation of PEI-GTC-NPs in the mitochondria of HeLa cells within 6 h of treatment, leading to damage of mitochondrial membrane and production of ROS, thereby ultimately killing the cancer cells. Altogether, the authors speculated that the GO-based nanosystem could be employed for advanced cancer therapy, targeting mitochondria in the near future.

6.4.2.2 Drug Delivery

The failure of the chemotherapeutic treatment in cancer is a serious concern (Chidambaram et al. 2011; Nolan and DeAngelis 2015). The plausible reasons may be due to several issues such as lack of specific target, non-specific cytotoxicity, resistance mechanisms (Chen et al. 2014), adverse side effects, poor solubility, improper biodistribution, failure in the clinical trials and also the high cost of the drugs (Chen et al. 2016). To overcome these obstacles, since the past few decades, nanotechnology is being used to deliver potent cancer therapeutics successfully. The fundamental criteria for an effective drug delivery system include the slow and sustained release of drugs as well as their delivery to the desired area of body system (Cho et al. 2008). To achieve the site-specific delivery of drugs, two strategies are generally employed, namely, passive targeting and active targeting. While the passive targeting mainly relies on enhanced permeability and retention (EPR) effect to reach tumour site, active targeting strategy requires targeting ligands such as antibodies, recombinant proteins, peptides, small molecules, etc. (Bae and Park 2011; Danhier et al. 2010; Torchilin 2010). Different types of nanoparticles such as metallic (e.g. gold, silver, zinc oxide, iron oxide, titanium dioxide, etc.), quantum dots, liposomes, dendrimers, polymeric nanoparticles, carbon-based nanomaterials, etc. are widely used for the delivery of different anticancer drugs (Jong and Borm 2008). Among various nanoparticles, GO is presently emerged as an efficient drug delivery vehicle due to its unique physicochemical properties such as high stability (Novoselov et al. 2004), water dispensability, large surface area, high drug loading efficiency (Yang et al. 2008), biocompatibility, biodegradability, etc. (Kiew et al. 2016). Recently, several research groups demonstrated the effective anticancer drug delivery systems based on GO nanomaterials which are concisely described in the next section.

Sun et al. demonstrated the fabrication of PEGylated nanosized sheet-like GO (PEG-NGO) following modified Hummers' method, followed by conjugation with a targeting antibody Rituxan (CD20+) as well as an FDA-approved anticancer drug doxorubicin (DOX) to form NGO-PEG-Ab/DOX drug delivery system (DDS) (Sun et al. 2008). The administration of this DDS to Raji cells exhibited better antiproliferative activity compared to NGO-PEG/DOX or free DOX in a dose-dependent manner, suggesting the efficacy of the DDS. Similarly, Zhao et al. illustrated the synthesis of poly(methacrylic acid)(PMAA) functionalized PEGylated GO nanoparticles (PMAA-GON-PEG) followed by conjugation with DOX to design an effective DDS (Zhao et al. 2015). The authors showed that the application of this DDS to SiHa cells (cervical squamous cancer cells) significantly inhibited cell proliferation compared to that of LSECs (liver sinusoidal endothelial cells), indicating the cancer cell-specific delivery of DOX employing the DDS. Likewise, GO was first time used as a nanocarrier for dual-drug delivery by Tran et al. with the aim of overcoming drug resistance via chemo-phototherapy (Tran et al. 2015). The authors developed a DDS (GO-DI) based on poloxamer 188 functionalized GO, loaded with dual anticancer drugs DOX and irinotecan via π–π stacking for photothermal cancer therapy using near-infrared (NIR) laser. To identify the photothermal efficacy of GO-DI, the researchers employed live (green)/dead (red) assay using calcein-AM (green) and propidium iodide (PI: red) dyes in SCC-7, MCF-7 and drug-resistant MBA-MB-231 cells (Fig. 6.5). The results showed that GO-DI treatment along with NIR leads to maximum dead cells through apoptosis compared to control or GO or GO-DI treatments without NIR laser, in both dose and time-dependent manner. This study suggested the effective application of the dual-drug chemo-phototherapy employing GO-based DDS. Western blot analysis further revealed that GO-DI treatment could upregulate different apoptotic proteins (p53, p27 and p21) in cancer cells, which might be the plausible reasons behind the anti-proliferative activity of the DDS. In another approach, Shim et al. developed taurocholate derivative of low molecular weight heparin (LHT7) (Kim et al. 2012)-coated rGO nanosheets and conjugated them with DOX to design a DDS (LHT-rGO/DOX) (Shim et al. 2014). The cell viability assay in KB cells showed that LHT-rGO/DOX treatment attenuated the cell proliferation effectively compared to control experiments, indicating the efficacy of the DDS. Additionally, administration of the DDS to KB tumour containing athymic nude mice leads to significant regression of tumour volume compared to control groups, suggesting that the DDS could efficiently be used for cancer therapy. Tian et al. also designed a DDS (FA/CPT/Pep/GO) based on PEGylated folate and peptide functionalized GO, loaded with camptothecin (CPT; potent inhibitor of topoisomerase I) (Tian et al. 2016). The author's idea of the work is to target folate receptors that are overexpressed in cancer cells and release the drug without harming the normal cells. The results showed that the DDS treatment to HeLa cells inhibited their proliferation significantly compared to free CPT, indicating its therapeutic potential. Moreover, the administration of FA/CPT/Pep/GO to HeLa tumour containing BALB/c nude mice leads to better targeting compared to other control groups as observed from the in vivo fluorescence imaging (Fig. 6.6). Altogether, this study demonstrated the efficacy of GO-based targeted DDS that could be useful for in vivo theranostic applications of cancer.

In another study, Sahne et al. designed a targeted DDS based on folic acid (FA) antibody-loaded GO, coated with cross-linking polymer of carboxymethylcellulose (CMC) and poly N-vinylpyrrolidone (PVP), containing anticancer drug curcumin with very high encapsulation efficiency (94%) (Sahne et al. 2019). The results showed 87% curcumin release from the DDS in the tumour microenvironment, leading to significant inhibition of proliferation of cancerous cells (Saos2 and MCF-7). The authors also demonstrated that the GO-based multifunctional DDS exerted effective tumour growth inhibition in 4 T1 cells bearing breast cancer mice model without any considerable toxic side effects. Further, Zare-Zardini et al. synthesized a new class of DDSs (GR-Arg-Rh2 and GR-Lys-Rh2) by conjugating a new natural anticancer agent ginsenoside Rh2 (Rh2) with arginine- or lysine-treated GO (Zare-Zardini et al. 2018). The authors performed in vitro experiments involving the DDSs

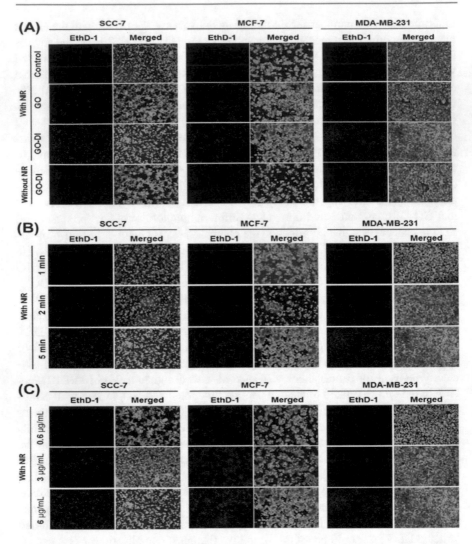

Fig. 6.5 In vitro photothermal ablation of tumour cells using a live/dead staining method. (**a**) Effect of various samples with or without NIR laser exposure, (**b**) effect of exposure time, (**c**) effect of graphene oxide (GO) concentration. Live cells stained green by calcein-AM and dead cells stained red by ethidium homodimer. (Figure reproduced with permission from Ref. (Tran et al. 2015). Copyright © 2015, American Chemical Society)

towards various cell lines (human breast cancer cells, MDA-MB; human ovarian cancer cells, OVCAR-3; human melanoma, A375; and human mesenchymal cells, MSCs) using MTT reagents. The cytotoxicity and TUNEL assays exhibited that the DDSs induced more inhibition of cell proliferation in cancerous cells as compared to non-cancerous MSCs. The authors suggested that it could be a potential strategy to modify amino acids with GO to increase the therapeutic index of any anticancer drug including Rh2, due to the lowering of side effects in normal cells and

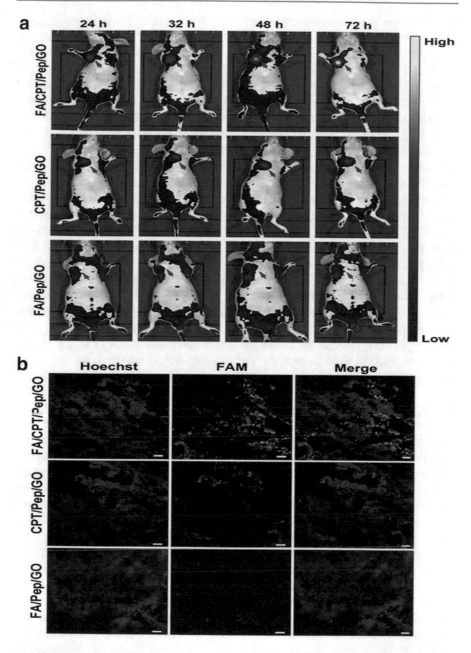

Fig. 6.6 (**a**) Time-dependent in vivo fluorescence imaging on subcutaneous HeLa tumour-bearing mice after intravenous injection of FA/CPT/Pep/GO, CPT/Pep/GO or FA/Pep/GO. The arrows show the tumour sites. (**b**) Fluorescence images of tumour slice for apoptotic assays after 72-h injection. Scale bars: 50 μm. (Figure reproduced with permission from Ref. (Tian et al. 2016). Copyright © 2016 Elsevier B.V)

enhancing the cytotoxicity in cancerous cells. Further, Afarideh et al. developed a DDS (GO/5-FU) based on GO loaded with anticancer drug 5-FU (5- fluorouracil) for effective treatment of cancer (Afarideh et al. 2018). The authors revealed that pristine 5-FU alone has low toxicity towards tumour cells, whereas the DDS GO/5-FU displayed effective anticancer activity towards adenocarcinoma cell line (CT26 dsRED). The IC_{50} value for GO/5-FU (5.2 μg/mL) was found to be lower than that of free 5-FU (8.1 μg/mL), suggesting the potential efficacy of the DDS. Deb et al. also developed chitosan (natural polymer) functionalized folic acid decorated GO-based DDS containing two anticancer drugs camptothecin (CPT) and 3,3′ diin-dolylmethane (DIM) and demonstrated the synergistic anticancer effect of the DDS based on dual drugs against MCF-7 breast cancer cells (Deb et al. 2018). Further, in vivo studies revealed that DIM could effectively mask the adverse toxicity exerted by CPT drug. In another study, Luan and co-workers conjugated mPEG-PLGA (PP) with DOX through a disulphide bond to make a prodrug PP-SS-DOX which was further loaded with PEG-FA-modified GO to form a novel targeted DDS GO/PP-SS-DOX/PEG-FA (Huang et al. 2018). The cell viability assay exhibited that GO/PP-SS-DOX/PEG-FA induced more inhibition of proliferation of different FR-positive cancer cells (MCF-7 and B16) in a dose-dependent manner as compared to nontargeting nanohybrid. The in vivo studies also showed that the administration of GO/PP-SS-DOX/PEG-FA to B16 tumour containing Kunming mice could significantly inhibit the tumour growth as compared to free DOX and nontargeting nanohybrid without exerting any considerable adverse toxicity, suggesting the potential efficacy of the targeted DDS.

6.4.2.3 Gene Delivery

In recent years, gene delivery has evolved as a powerful technology for the treatment of different diseases at a molecular level (Draz et al. 2014). However, there are various challenges associated with gene delivery in terms of targeting efficiency, less bioavailability, cleavage of nucleotide, etc. that have to be faced for targeted and systemic gene delivery (Lu et al. 2010). In this circumstance, considering the unique features of nanomaterials, scientists employed different kinds of nanoparticles for effective delivery of a therapeutic gene to the disease site (Draz et al. 2014). Among such nanoparticles, GO has emerged as one of the excellent gene delivery vehicles as demonstrated by earlier literature (Nurunnabi et al. 2015). For example, Zhang et al. demonstrated a sequential delivery of Bcl-2-targeted siRNA and DOX loaded onto polyethyleneimine (PEI) functionalized GO (PEI-GO) (Zhang et al. 2011). The researchers functionalized GO with PEI through covalent linking of an amide bond. The positive charge of the PEI-GO facilitated the attachment of siRNA via electrostatic interactions. The siRNA-tagged PEI-GO was found to be biocompatible to HeLa cells. Further, knockdown efficiency of the Bcl-2 protein by PEI-GO/Bcl-2 targeted siRNA was observed along with effective co-delivery of DOX for enhanced anticancer effect. The authors claimed this as the first report for sequential delivery of DOX and siRNA using PEI-GO nanocarriers. Yin et al. also fabricated a gene delivery system (FA/GO/(H + K) siRNA) based on PEGylated folic acid (FA)-decorated GO loaded with HDAC1 and K-Ras siRNAs (Yin et al. 2017). The administration of FA/GO/(H + K) siRNA to MIA PaCa-2 bearing athymic nude mice (BALB/cASlac-nu) of xenograft model illustrated the reduction of tumour volume significantly compared to other control groups in presence or absence of NIR, suggesting the cancer therapeutic potential of the delivery system (Fig. 6.7). Likewise, Feng et al. designed a novel GO nano vector for efficient gene transfection (Feng et al. 2011). The researchers synthesized GO using Hummers' method and functionalized it with PEI of two different molecular weights of 1.2 kDa and 10 kDa to form GO-PEI-1.2 k and GO-PEI-10 k, respectively. The in vitro cytotoxicity assay in HeLa cells depicted that GO-PEI-10 k complex showed reduced toxicity compared to only PEI 10 kDa polymer. Further, results demonstrated that GO-PEI-10 k complexes could bind with plasmid DNA for successfully transfecting enhanced green fluorescence protein (EGFP) gene in HeLa cells effectively compared to PEI-10 k. On the other hand, EGFP transfection was not effective for GO-PEI-1.2 k. In another study, Gu et al. designed a dual gene and drug delivery system (GO-PAMAM/DOX/MMP-9) based on polyamidoamine (PAMAM) dendrimer functionalized GO, conjugated with DOX and MMP-9 shRNA plasmid (Gu et al. 2017). The effective co-delivery of DOX and MMP-9 shRNA was observed in MCF-7 cells treated with GO-PAMAM/DOX/MMP-9, suggesting the efficacy of the gene delivery system. This study provides an idea for the promising prospect of GO-based nanomaterials for synergistic delivery of anticancer drug with the therapeutic gene for better efficacy. Ren et al. also developed poly-L-lysine- and Arg-Gly-Asp-Ser-decorated GO containing VEGF-siRNA to form GO-PLL-SDGR/ VEGF-siRNA gene delivery system, which could target the tumours actively with slow and sustained release of VEGF siRNA (Ren et al. 2017). GO-PLL-SDGR exerted low cytotoxicity in HeLa cells, as observed by MTT assay. Further, results showed the downregulation of the expression of VEGF-mRNA as well as VEGF protein in HeLa cells treated with the gene delivery system. Additionally, GO-PLL-SDGR/VEGF-siRNA was found to inhibit tumour growth effectively (51.74%) in S-180 tumour-bearing mice model. Altogether, the authors concluded that GO-PLL-SDGR could be employed as an effective siRNA delivery vehicle for tumour targeting. Further, Liu et al. designed a GO-based dual-delivery system containing anticancer drug cisplatin and antisense microRNA-21 (anti-miR-21) (Liu et al. 2018). The authors demonstrated that the platinated GO could be a potent gene delivery system due to its ability to encapsulate anti-miR-21 with better capacity and improved stability. The system exhibited prominent cytotoxic response to cancer cells upon treatment, due to the synergistic effect of cisplatin and anti-miR-21. The underlying mechanism of enhanced cytotoxicity could be attributed to the apoptosis in cancer cells, augmented by anti-miR-21-mediated gene silencing. Altogether, the authors showed that this GO-based gene-chemo combination therapy could be useful for cancer therapy. Di Santo et al. also synthesized cationic lipid (DOTAP: 1,2-dioleoyl-3-trimethylammonium-propane)-coated GO nanoflakes (GOCL) for gene delivery purpose (Di Santo et al. 2019). GOCL showed significant positive surface charge (ξ = +15 mV) with size <150 nm for gene delivery application. The authors employed the complex of GOCL and plasmid DNA for

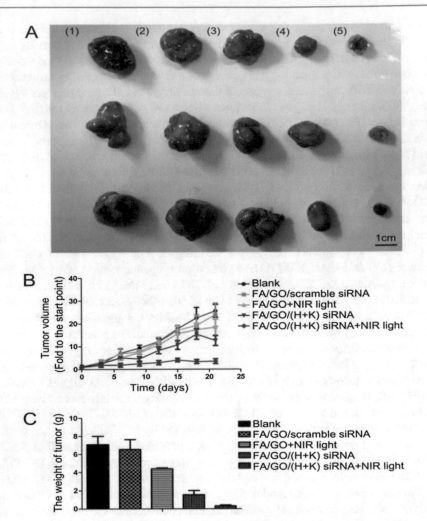

Fig. 6.7 Antitumour activities of GO-based nanoformulations in a MIA PaCa-2 xenograft animal model. (**a**) Representative tumour tissue images of mice treated with (1) PBS, (2) FA/GO/scramble siRNA, (3) FA/GO with NIR light, (4) FA/GO/(H + K) siRNA or (5) FA/GO/(H + K) siRNA with NIR light. Mice treated with FA/GO/(H + K) siRNA with NIR light in the last group exhibited the smallest tumour size. (**b**) Relative changes in tumour volume over time and (**c**) tumour weights of mice treated with the same nanoformulations as in (**a**), respectively. Relative tumour volume was defined as $(V - V_0)/V_0$, where V and V_0 indicate the tumour volume on a particular day and day 0, respectively. Error bars represent SEMs for triplicate data. Mean tumour volumes were analysed using one-way ANOVA. Values represent the means ± SEM, $n = 4$–6 tumours. (Figure reproduced with permission from Ref. (Yin et al. 2017). Copyright © 2017 Ivyspring International Publisher)

transfecting HeLa and HEK-293 cells, while bare GO and bare cationic lipid were used as control experiments. Although GOCL showed similar transfection efficiency, it induced significantly enhanced cell viability as compared to DOTAP cationic lipid. The better performance of GOCL-based DNA complexes could be attributed to their more number, regular size as well as homogeneous distribution than that of DOTAP-based DNA complexes. In another study, Yadav et al. developed a hybrid nanocomposite system (GPD) based on GO functionalized with poly(amidoamine) (PAMAM) and PEG for delivery of siRNA (Yadav et al. 2018). The GPD nanocomposite exhibited high stability at physiological pH and effective binding with EPAC1 siRNA and lower cytotoxicity. The authors showed the efficiency of GPD as a vector for the delivery of EPAC1 siRNA in HUVECs as well as MDA-MB-231 breast cancer cells as evidenced by the very high transfection efficacy of GPD/siRNA complexes, followed by the efficient release of siRNA. Xing and co-workers also designed PEG and PEI functionalized GO-based nanocarrier to deliver Cas9/single-guide RNA (sgRNA) complexes (Yue et al. 2018). The authors demonstrated the efficiency (~39%) of the nanocarrier for editing of the gene in AGS cells. The protection of sgRNA from enzyme-based degradation was also augmented by the nanocarrier, thereby enhancing its stability which is vital for in vivo studies. Altogether, the authors suggested that the GO-based delivery system could be an ideal candidate for genetic engineering.

6.4.2.4 Photothermal Therapy

Nowadays, photothermal therapy is emerging as a newly developed strategy to thermally ablate cancer cells by generating heat, using electromagnetic/NIR (Zou et al. 2016). Several recent reports demonstrated the applications of different nanoparticles based photothermal therapy for cancer treatment, where nanoparticles act as photothermal agents. Among various nanomaterials, GO has been widely reported for cancer treatment through photothermal therapy due to its several advantages such as intrinsic NIR absorption, low toxicity, easy functionalization and high biocompatibility (Bansal and Zhang 2014; Li et al. 2015a). The following section illustrates the recent development of GO-based photothermal therapy in a concise manner.

Robinson et al. demonstrated the application of PEGylated rGO nanomaterials as a potent photothermal agent for in vivo studies (Robinson et al. 2011). rGO nanoparticles were found to have higher NIR absorbance with comparatively less toxicity to cancer cells. For selective targeting, the nano-rGO was loaded with an Arg-Gly-Asp (RGD) peptide (targeting ligand), and the administration of this nanocomposite to glioma cells U87MG showed their effective photoablation in presence of NIR compared to control treatments. This work depicted the efficacy of nano-rGO as a potent photothermal agent. Further, Yang et al. investigated the effect of size and surface chemistry of rGO nanomaterials on photothermal therapy for cancers using PEG coating (Yang et al. 2012). The intravenous administration of rGO-PEG nanocomposite to 4T1 tumour containing BALB/c mice exhibited complete elimination of tumour under NIR, suggesting the efficacy of the nanocomposite for photothermal therapy for cancer. The therapeutic efficacy of the nanocomposite might be explained by the improved NIR absorbance and passive tumour targeting efficacy of rGO-PEG. A similar study reported by Wei et al. depicts the efficiency of a GO-based DDS for photodynamic therapy of tumour (Wei et al. 2016). The authors developed the DDS (PPa-NGO-mAb) using PEGylated pyropheophorbide-a (PPa)-decorated GO nanomaterials, conjugated with $\alpha_v\beta_3$ monoclonal antibody (mAb; targeting antibody). The results revealed that the phototoxicity of PPa on GO nanoparticles could be altered (switched on/off) depending on organic and aqueous media. This switch system facilitated the crossing of a cellular phospholipid membrane, which was observed to potentially active against the tumour through mitochondrial derived apoptosis. Due to the conjugation of mAb, the system was found to have antigen-antibody interactions, and also the targeting efficiency of the system was found to be higher in U87-MG cells compared to the MCF-7 cells, clearly demonstrating the high specificity of the system to $\alpha_v\beta_3$ receptors. While a wide range of investigations focused on the applications of PEGylated GO, a study conducted by Akhavan et al. showed the therapeutic efficacy of green-reduced and -functionalized GO against LNCaP prostate cancer cells (Akhavan et al. 2012). In this study, GO was reduced by glucose and synthesized by chemical exfoliation method with Fe catalyst. A complete lyses of cancer cells were observed with minimum concentration of glucose-reduced GO under near-IR irradiation, indicating the efficacy of GO nanomaterials for photothermal therapy of cancers. Wu et al. proposed a polylysine-based functionalized GO for both photo- and chemotherapy (Wu et al. 2014). Along with the potential activity of GO and its derived complexes, an enhanced therapeutic effect was observed by the graphene-based nanohybrid in combination with DOX and photo sensitizer Zn(II)-phthalocyanine (ZnPc) against HeLa, MCF-7 and B16 cancer cell lines. In another study, Cheon et al. developed a DDS (DOX-BSA-rGO) based on BSA protein-functionalized rGO, conjugated with DOX (Cheon et al. 2016). Under NIR laser, the release of DOX from DOX-BSA-rGO was significantly higher enabling the improved therapeutic activity of the DDS to U87MG brain tumour cells compared to control experiments. Similarly, Su et al. designed a porphyrin-functionalized GO (PGO) that possessed high absorbance at 808 nm (Su et al. 2015). PGO was identified to be more stable in aqueous environment compared to rGO. Similar to stability, the photothermal effect of PGO was effective against glioblastoma cells compared to GO and rGO. The authors also suggested the high efficacy of PGO along with its deep penetrating property that could be useful for treating deep-rooted glioblastoma.

Gulzar et al. also developed GO- and upconversion nanoparticle (UCNP)-based core-shell nanocomposite system containing photosensitizer chlorin e6 (Ce6) and demonstrated its potential application for imaging-guided PDT as well as photothermal therapy (Gulzar et al. 2018). The nanocomposite system exhibited efficient bio-imaging properties due to the presence of UCNPs. Moreover, it could produce ROS under light irradiation of 808 nm, facilitating PDT, as well as translate the photon of the light into thermal energy leading to augmenting photothermal therapy. Altogether, the authors suggested that the multifunctional nanocomposite system could be a potential candidate to be employed for the imaging-guided cancer therapy.

6.4.2.5 Immunotherapy

A better understanding of molecular and cellular processes controlling the immune system leads to a new cancer therapy, generally termed as immunotherapy for averting the drug resistance of tumours in patients (Borghaei et al. 2009). Recently, nanomedicine is found to offer promising utility of various nanoparticles such as liposomes, magnetite nanoparticles, gold nanoparticles, etc. for the selective, controlled and targeted delivery of immunotherapeutic agents against tumours (Krishnamachari et al. 2011; Mattos et al. 2014; Orecchioni et al. 2016b; Steichen et al. 2013). Besides these nanoparticles, scientists also explored the application of GO-based nanomaterials for immunotherapy of cancers. For instance, Tao et al. first time demonstrated photothermally enhanced immunogenicity for immunotherapy of cancer employing GO nanomaterial-mediated delivery of a therapeutic nucleic acid CpG (unmethylated cytosine-phosphate-guanine) (Tao et al. 2014). The authors fabricated polyethylene glycol (PEG) and PEI-functionalized GO, followed by conjugation with CpG that possess immunostimulatory properties. It was observed that the nanocomposite augmented the formation of pro-inflammatory cytokines (TNF-α and IL-6) and improved the immunostimulatory effect of CpG in RAW264.7 cells. The administration of the nanocomposite to CT26 tumour (colon cancer) bearing BALB/c mice under laser irradiation showed the synergistic immunological and photothermal effect leading to significant reduction of tumour volume. The results suggested the efficacy of the CpG-loaded functionalized GO as an efficient immunostimulatory agent for cancer immunotherapy. Yue et al. also developed OVA (ovalbumin, a well-known antigen model)-loaded GO (GO-OVA) and investigated its potential in cancer immunotherapy (Yue et al. 2015). The authors first confirmed the in vivo adjuvant efficacy of this system using transgenic mice (OT-1; specific to OVA). Based on this result, the authors then checked the therapeutic efficacy of the vaccine system in E.G7 tumour-containing C57BL/6 mice. The administration of GO-OVA in tumour-bearing mice exhibited the significant reduction of tumour volume with extended survival time compared to other control groups (Fig. 6.8). The histology study also revealed the lysis cavities of tumour for GO-OVA-administered groups in comparison to control groups. Moreover, results showed the significant infiltration of CD8 T cells (OVA-specific) into tumour for GO-OVA treatment compared to control groups. This study provides an insight for the development of GO-based materials for cancer immunotherapy. Similarly, Xu et al. depicted the synthesis of GO employing Hummers' method and functionalized it with either PEG or PEI or both PEG and PEI to form GO-PEG, GO-PEI and GO-PEG-PEI, respectively, followed by conjugation with *Helicobacter pylori* (carcinogen for gastric cancer)-specific antigen Ure B and investigated the efficacy of the nanocomposites in cancer immunotherapy (Xu et al. 2016). GO-PEG-PEI was found to augment the dendritic cell (DC) maturation and release of cytokines via stimulating multiple Toll-like receptor (TLR) compared to GO-PEG and GO-PEI. Results also revealed that GO-PEG-PEI could successfully deliver Ure B into DCs in an effective manner as compared to GO-PEG and GO-PEI, suggesting the potential of GO-PEG-PEI as vaccine adjuvant. The in vivo studies employing BALB/c mice showed that Ure B-loaded GO-PEG-PEI treatment stimulated the cellular immunity, indicating the

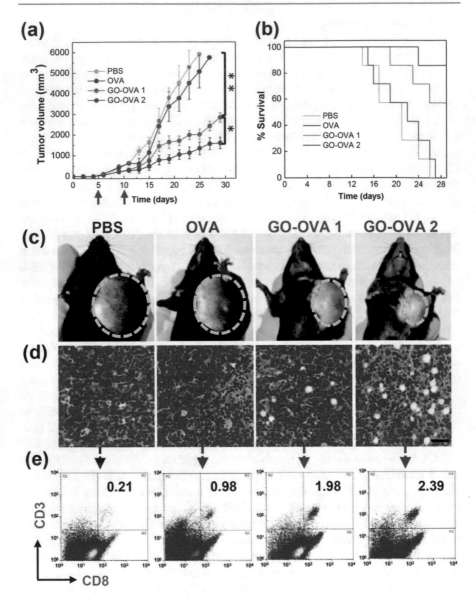

Fig. 6.8 Tumour therapy effect of GO adjuvanted vaccine. (**a**) Tumour growth volumes and survival rate (**b**) of mice bearing E.G7 tumours after different vaccinations. (**c**) Representative photos of tumour-bearing mice after different treatments. (**d**) H&E stained images of tumour sites from vaccine-treated mice. Mononuclear cells transferred into the tumour sites, and obvious cavities were formed during the tumour regression. Scale bar 50 μm. (**e**) FACS plots showing the infiltration of CD8T at the tumour sites. (n = 7, ∗p < 0.05, ∗∗p < 0.01). (Figure reproduced with permission from Ref. (Yue et al. 2015). Copyright © 2015 Royal Society of Chemistry)

potential of GO-based nanocomposite in cancer immunotherapy. In another study, Orecchioni et al. fabricated two GO sheets, namely, small GO (GO-S: <1 μm) and large GO (GO-L: 1–10 μm) following modified Hummers' method (Orecchioni et al. 2016a). The results demonstrated that GO-S exposure to immune cells upregulated several genes associated with immune response and released different cytokines (IL1β and TNFα) compared to that of GO-L, indicating its major impact to immune cells. The whole genome expression analysis of T cells and monocytes treated with GO-S further supported the aforesaid results. The authors claimed that their findings could offer a platform to design novel graphene-based materials that might be employed for immune modulation, especially for cancer immunotherapy.

Further, Zhang et al. developed a nanocomposite system based on yeast ß-glucan (possessing immunostimulatory as well as antitumour activity)-decorated GO loaded with CpG- oligodeoxynucleotides (CpG-ODNs: immunotherapeutic agent with immunostimulatory activity) for cancer immunotherapy applications (Zhang et al. 2018). Results depicted that the ß-glucan functionalization on GO reduced the non-specific protein adsorption, facilitated the biocompatibility as well as enhanced the macrophage targeting ability of the nanocomposite system. Further studies showed that system could efficiently deliver CpG-ODNs to RAW264.7 cells leading to synergistically enhanced secretion of cytokines as well as inhibit the growth of tumour cells. Altogether, the authors concluded that the GO-based nanocomposite systems could be employed for effective cancer immunotherapy applications. In another study, Yan et al. designed a combination of cancer immunotherapy strategy involving phothermal therapy, inhibition of indoleamine-2,3-dioxygenase (IDO: immune checkpoint overexpressed in tumours) and blockade of programmed cell death-ligand 1 (PD-L1) which regulates T cell negatively (Yan et al. 2019). The authors basically developed FA-modified IDOi (inhibitor of IDO: epacadostat)-conjugated multifunctional rGO, having the ability to effectively kill the tumour cells upon exposure of laser radiation, via the synergistic effect of NIR-based photothermal effect and IDOi-mediated immunotherapy. Further, in vivo studies showed that the induced immune response (increase of tumour-infiltrating T cells and NK cells as well as generation of INF-γ) could be enhanced by inhibition of IDO and blockade of PD-L1. The authors suggested that the study could give an insight for targeting various antitumour immune pathways leading to promote synergistic effect to treat cancers.

6.4.2.6 Biosynthesized Graphene Oxides

Biosynthesized nanomaterials are generally fabricated using various biological sources including plant extracts, bacterial cultures, fungus, etc. (Mukherjee et al. 2012). The basic advantage of biosynthesized nanoparticles over chemically synthesized nanomaterials is that the bioactive components present in the bio-resources could be amalgamated on the surface of nanoparticles leading to enhance their biocompatibility as well as therapeutic efficacy. Additionally, biosynthesized nanoparticles are more cost-effective and cheaper compared to chemically prepared nanoparticles. There is a plenty of literature demonstrating the biomedical applications of biosynthesized nanoparticles such as gold, silver, ZnO, etc. Recently, few researchers also illustrated the applications of biosynthesized GO-based nanomaterials for cancer therapy which are briefly discussed in the following section.

Gurunathan et al. fabricated GO using a modified Hummers' and Offeman's method and rGO through biosynthesis employing *Bacillus marisflavi* biomass (Gurunathan et al. 2013). Here the bacterial biomass acted as both reducing and stabilizing agent for biosynthesized rGO. The results showed that both GO and rGO inhibited proliferation of MCF-7 cells in a dose-dependent manner. However, the inhibitory response was better in case of rGO treatment in comparison with GO. This could be attributed to the more production of intracellular ROS as well as lactate dehydrogenase release in the presence of rGO compared to that of GO, leading to facilitate the apoptosis process in cancer cells. In another study, Zhu et al. demonstrated the synthesis of GO following modified Hummers' method and rGO through biosynthesis using aqueous leaf extract of *C. colocynthis* (Zhu et al. 2017). The authors exhibited that both GO and rGO treatments to DU145 prostate cancer cells lead to inhibition of cell proliferation in a dose-dependent manner, indicating their anticancer potential.

On the other hand, Xu et al. also used bio-green approach for fabricating rGO in presence of tyrosine kinase inhibitor anticancer drug sorafenib, employing ascorbic acid as green-reducing agent to treat gastric cancers (Xu et al. 2019). The authors demonstrated that sorafenib-reduced graphene oxide (SRGO) exhibited more inhibition of proliferation of gastric cancer cells (SGC7901) as compared to free drug, suggesting the anticancer potential of the nanoparticulate system. Additionally, the research group observed that SRGO-treated cells displayed transformative nuclei (apoptotic), whereas untreated cells showed round-shaped nuclei as confirmed by Hoechst 33382 staining. Altogether, the authors concluded that SRGO could be used as potential candidate for the treatment of gastric cancers. Further, Lin et al. developed biosynthesized rGO employing the leaf extract of *Euphorbia milii* (Lin et al. 2019). The biosynthesized rGO was conjugated with anticancer drug paclitaxel (PTX) to form the nanocomposite system rGO/PTX which exhibited inhibition of proliferation of A549 lung cancer cells in a dose-dependent manner, suggesting its anticancer potential.

6.5 Bio-imaging Applications of Graphene Oxides

Bio-imaging is basically a complex process to visualize the structural as well as functional alteration of living systems for diagnosis of various diseases (Erathodiyil and Ying 2011). The conventional diagnosis techniques include biopsy, magnetic resonance imaging (MRI), computed tomography (CT), PET, X-ray, etc. Some of these techniques are invasive and painful with adverse side effects because of the complex operating method (Erathodiyil and Ying 2011). In this context, nanomedicine offers suitable bio-imaging process through acquiring data in a non-invasive manner for diagnosis of different diseases especially cancer. Scientists observed that besides enormous therapeutic applications, GO nanomaterials could also be used for bio-imaging applications, related to diagnosis/detection of cancer cells/tumours as discussed in the following section. Sun et al. demonstrated the synthesis of PEGylated GO (PEG-NGO) which was conjugated with Rituxan (CD20:

targeting antibody) and demonstrated the bio-imaging properties of the nanocomposite (Sun et al. 2008). The administration of the nanocomposite to Raji B cells (CD20 positive) exhibited near-infrared (NIR) fluorescence which could be attributed to the inherent photoluminescent property of GO nanomaterials in NIR. In another study, Hong et al. synthesized S-2-(4-isothiocyanatobenzyl)-1,4,7-triazacyclononane-1,4,7-triacetic acid (p-SCN-Bn-NOTA)-modified PEGylated GO and conjugated the nanocomposite with TRC105 antibody (targeting ligand for tumour angiogenesis marker CD 105), followed by labeling with radioactive ^{66}Ga to design ^{66}Ga-NOTA-GO-TRC105 (Hong et al. 2012). The authors successfully demonstrated the higher targeting efficiency of ^{66}Ga-NOTA-GO-TRC105 to 4 T1 breast tumour through PET/CT imaging compared to other control nanocomposite systems, employing an in vivo mice model (Fig. 6.9). Similarly, Yang et al. illustrated the fabrication of p-SCN-Bn-NOTA-decorated PEGylated GO, attached with a monoclonal antibody (mAb) FSHR (follicle-stimulating hormone receptor: marker for tumour vasculature), followed by subsequent labeling with radioactive ^{64}Cu to form ^{64}Cu-NOTA-GO-FSHR-mAb nanocomposite (Yang et al. 2016). The results exhibited that the administration of ^{64}Cu-NOTA-GO-FSHR-mAb to breast cancer (MDA-MB-231) lung metastasis model of nude mice lead to rapid uptake of nanocomposite into tumour modules in lung, even at very early time points through PET scan, suggesting its tumour-targeted bio-imaging properties. Li and co-workers further demonstrated that GO could improve the cytoskeleton imaging efficiency of a cell membrane impermeable fluorophore (5-carboxytetramethylrhodamine: 5-TAMRA; λ_{Ex}: 520–550 nm and λ_{Em}: 560–630 nm) tagged peptide probe VAR, which is selective for microtubules (basic component of cytoskeleton) (Li et al. 2017). The results exhibited that EGFP-α tubulin expressing untreated control HeLa cells or cells treated with only VAR did not fluoresce. On the other hand, cells treated with both VAR and GO showed intense red fluorescence (emission for TAMRA), and the intensity of fluorescence increased both dose- and time-dependant manner, indicating the enhancement of cytoskeleton imaging ability of VAR in the presence of GO (Fig. 6.10). Very recently, Zang et al. fabricated AgInZnS-GO (AIZS-GO) nanocomposites and depicted their bio-imaging property in SK-BR-3 (breast cancer cells) tumour containing nude mice employing their photoluminescence properties (Zang et al. 2017). The results exhibited that intravenous injection of the AIZS-GO nanocomposites to mice leads to broad distribution in tumour tissues, indicating the efficacy of the nanocomposites for in vivo bio-imaging of tumour.

In another study, Sun et al. functionalized GO with PEG to form PEGylated GO that could enhance the stability of GO encapsulated aggregation-induced emission (AIE) nanoparticles in phosphate buffer (Sun et al. 2018). The authors encapsulated highly stable PEGylated GO with a dual-functional molecule TPE-red that possesses both AIE and photosensitizing properties. The results demonstrated that the TPE-red-loaded nanocomposite system exhibited its potential applications for fluorescence bio-imaging in UMUC3 cells as well as mouse ear blood vessels. Further, the nanocomposite system also exerted its efficacy for photodynamic therapy (PDT) both in vitro (UMUC3 cells) and in vivo (UMUC3 xenograft tumour mice model). Finally,

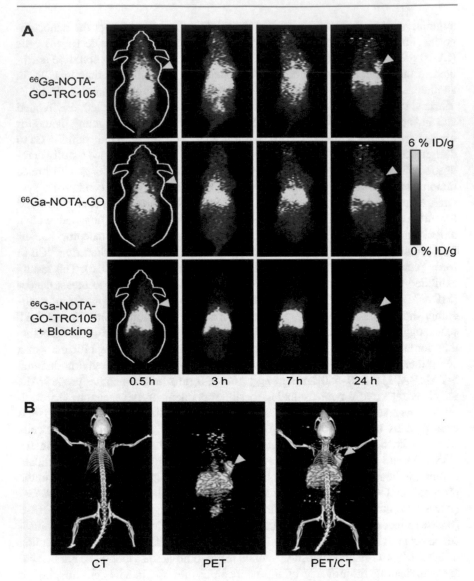

Fig. 6.9 In vivo PET/CT imaging of ^{66}Ga-labeled GO conjugates in 4 T1 tumour-bearing mice. (**a**) Serial coronal PET images of 4 T1 tumour-bearing mice at different time points post-injection of ^{66}Ga-NOTA-GO-TRC105, ^{66}Ga-NOTA-GO or ^{66}Ga-NOTA-GO-TRC105 at 2 h after a blocking dose of TRC105 (denoted as "blocking"). (**b**) Representative PET/CT images of 66Ga-NOTA-GO-TRC105 in 4 T1 tumour-bearing mice at 3 h post-injection. Tumours are indicated by arrowheads. (Figure reproduced with permission from Ref. (Hong et al. 2012). Copyright © 2012 Elsevier B.V)

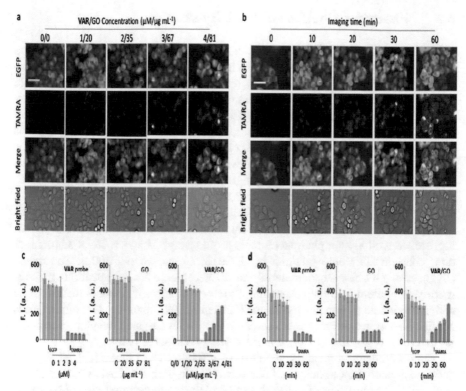

Fig. 6.10 Dose-dependent fluorescence imaging (**a**) and quantification (**c**) of VAR probe, GO and VAR/GO for HeLa cells stably expressing EGFP-α tubulin. Time-dependent fluorescence imaging (**a**) and quantification (**c**) of VAR probe (4 μM), GO (81 μg mL^{-1}) and VAR/GO (4 μM/81 μg mL^{-1}) for HeLa cells stably expressing EGFP-α tubulin. Excitation channels for EGFP and TAMRA are 460–490 and 520–550 nm, and emission channels for EGFP and TAMRA are 500–540 and 560–630 nm, respectively. Scale bar = 50 μm (applicable to all images). (Figure reproduced with permission from Ref. (Li et al. 2017). Copyright © 2017 Royal Society of Chemistry)

the authors concluded that the GO-based system could be used as a potential tool for bio-imaging as well as PDT in the near future. Further, Luo et al. developed superparamagnetic iron oxide nanoparticles (SPION)-decorated GO nanosheets containing cis-aconitic anhydride-modified doxorubicin (CAD) (Luo et al. 2019). The nanocomposite system exhibited pH-responsive release of DOX, leading to more inhibition of proliferation of 4T1 cancer cells as well as inhibition of tumour growth in 4T1 tumour-containing mice as compared to free DOX, suggesting its therapeutic potential. Moreover, the nanocomposite system illustrated its bio-imaging properties (T1-weighted MR imaging) in the same tumour-containing mice model, indicating its potent theranostic efficacy. Song and co-workers also synthesized GO quantum dots (GOQDs) using a cheaper precursor coal and demonstrated their efficient bio-imaging properties (Kang et al. 2019). The in vitro studies exhibited that GOQDs are highly bio-compatible in PANC-1 cells (human pancreatic cancer cells) even at very high dose (5 mg/mL). On the other hand, PANC-1 cells treated with GOQDs exhibited intense green fluorescence, indicating their bio-imaging properties.

6.6   Pharmacokinetics and Toxicity Studies

The pharmacokinetics of any compound inside biological system is one of the prime factors to be considered as ideal drugs. The fate of the administered drug or therapeutic agent is decided by four major factors, namely, absorption, distribution, metabolism and excretion (ADME) which are the fundamental properties for determining the pharmacokinetics (Cho et al. 2013; Yoshioka et al. 2014). To understand the amount of dosing, number of doses, exposure time and bioavailability, it is highly essential to perform the pharmacokinetics study of any material. Since the past decades, various nanoparticles are widely employed for versatile biomedical applications. However, scientists are highly concerned about their toxicity issues prior to clinical applications. The higher exposure of nanoparticles often leads to toxicity causing multiple side effects (Holgate 2010). The toxicity of nanomaterials mainly depends on their physicochemical features (e.g. size, charge, surface coating, etc.) as well as their pharmacokinetic profiles (e.g. dosage, route of administration, concentration, etc.) (Hamidi et al. 2013; Li and Huang 2008). The recent advances of GO-based nanomaterials in diverse biomedical applications make it worthy to contemplate their systemic toxicity including pharmacokinetic profile and metabolic long-term fate (Sahu and Casciano 2009). In the context of numerous applications in healthcare, several research groups investigated the pharmacokinetics and toxicity profiles of GO, which are briefly discussed below.

Yang et al. demonstrated the synthesis of PEGylated GO sheets labeled with ^{125}I and investigated their toxicity and pharmacokinetic profile in BALB/c mice (Yang et al. 2011). The outcome of the work exhibited the major accumulation of the nanocomposite in reticuloendothelial system (RES: liver, spleen, etc.) and their clearance through urine and faeces. The pharmacokinetics of the nanocomposite was carried out by analysing the radioactivity levels in blood of mice, and the result showed two-compartment model in which half-lives were perceived to increase in the second phase compared to first phase of blood circulation. Similarly, the biodistribution of radiolabeled small-sized and large-sized GO were studied in male ICR mice by Liu et al. (Liu et al. 2012). The radiolabeled GO was injected through tail vein, followed by the sacrifice of the mice after the irradiation. The blood and vital organs were collected and analysed for the fate of the system. The authors found that the small-sized GO was having good pharmacological properties compared to the large-sized GO. The small-sized GO entered to blood vessels and dispensed into other organs in more considerable amount compared to large-sized GO. Moreover, the elimination of GO did not depend on the size of the particle. In another study, Yang et al. described that the route of administration played a major role for pharmacokinetic profiles of the nanoparticles (Yang et al. 2013). On the other hand, Jasim et al. for the first time reported tissue distribution and detection of intact functionalized GO sheets in urine sample of the nanomaterial-administered mice (Jasim et al. 2015). The study indicated the focus on design of the graphene-based nanomaterials for theranostic purposes.

Though GO has been explored immensely for its therapeutic applications as a drug carrier, photothermal agent and various other diagnostics purposes, the toxicity of the complexes is one of the major issues for reassigning it from bench to clinical usage. Among the other nanodrugs, GO has already attracted extensive interest due to its comparable less toxicity. However, still the observed side effects caused a serious concern about the safety of using GO for the treatments of different diseases. Furthermore, the investigations confirmed that the dispersion of GO nanoparticles through airways caused various deleterious effects including release to other organs. It was also confirmed that GO could cross the blood-brain and physical barriers (Ou et al. 2016). In a study conducted by Li et al., pulmonary toxicity level of nano-GO (NGO) was elucidated in a detailed manner in C57BL/6 mice (Li et al. 2013). The biodistribution was analysed with the existence of iodine radioactivity in organs after the treatment period. The results showed that the radioactive isotope in NGO (^{125}I-NGO) was found to retain consistently in the lung over the time of exposure and also in low level in other organs including liver and thyroid gland. The comparison of biodistribution between ^{125}I-NGO and Na^{125}I showed better in vivo stability of the former. Later, the confirmation of the clearance of the ^{125}I-NGO was identified from SPECT imaging, as the intensity increased in the bladder within short period of time. Further, the long-term exposure of ^{125}I-NGO in the lung showed black coloration indicating deposition of the material in the lung (Fig. 6.11). The authors also revealed that the NGO were infiltrated to the alveolar-capillary barrier and cleared immediately through the renal system. The effect of pulmonary toxicity was highly dependent on the dosage of nanoparticles used. The authors also suggested the importance of size modulation and differential coating to reduce the toxicity. Similarly, Liao et al. investigated the toxicity profile of GO-derived complexes, with the focus of altering various physical parameters of the complexes (Liao et al. 2011). The results demonstrated that the haemolytic activity of GO was observed to be less in aggregated form than the reduced sized GO. Also, the loosely packed GO caused less toxicity to the fibroblast cells compared to the tightly packed graphene sheets. Similarly, the green rGO (reduced with *Platanus orientalis* leaf extract) was showing relatively lesser toxicity to cardiac cell lines of *Catla catla* (SICH cell lines) compared to the GO at identical concentration. This study revealed that the reduction pattern in biosynthesis also played a role in inducing toxicity (Xing et al. 2016). The importance of physicochemical parameters in toxicity mechanism was also confirmed in another study performed by Mittal et al., wherein the cytotoxicity of rGO by thermal and chemical methods was studied in detail (Mittal et al. 2016). The chemically reduced GO was relatively less toxic to cells compared to the thermally reduced GO. The high toxicity level of thermally reduced GO suggested their inefficiency for further applications. Yang et al. also reported the nontoxic behaviour of PEGylated nanographene sheets as observed by in vivo studies (Yang et al. 2011). The elimination of accumulated particles through faecal renal route was also demonstrated. This study sheds lights for the possible therapeutic implications of GO in the forthcoming future.

In another study, Syama et al. used the Raman spectroscopy to detect the distribution pattern and clearance of PEGylated reduced graphene oxide (PrGO) in Swiss Albino mice, followed by other toxicological consequences after its administration through both the i.p and i.v routes (Syama et al. 2017). Confocal Raman mapping

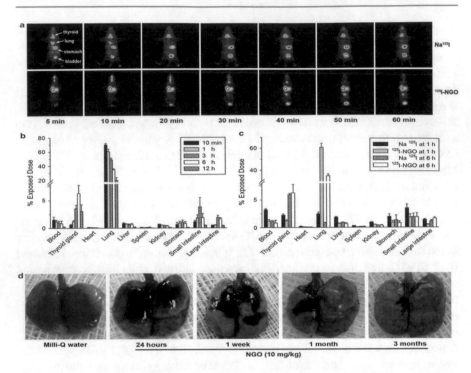

Fig. 6.11 Biodistribution of NGO after intratracheal instillation. (**a**) SPECT images of mice at several time points after intratracheal instillation with ^{125}I-NGO or Na^{125}I. (**b**) Distribution of ^{125}I-NGO in the blood and major organs of mice at five different time points. $N = 5$ in each group. Values are presented as the mean ± s.e.m. (**c**) Comparison of Na^{125}I and ^{125}I-NGO distribution in mice at 1 and 6 h after intratracheal instillation. $N = 5$ in each group. Values are presented as the mean ± s.e.m. (**d**) The morphological observation of the lungs from mice instilled with Milli-Q water or 10 mg kg^{-1} NGO. The dorsal view shows the distribution of NGO (black region). (Figure reproduced with permission from Ref. (Li et al. 2013). Copyright © 2013 Macmillan Publishers Limited, part of Springer Nature)

revealed that PrGO was broadly distributed in vital organs including the brain, kidney, liver, bone marrow and spleen. The authors suggested that the presence of PrGO in the brain depicted its potential to cross the blood-brain barrier. Further studies exhibited the presence of little amount of PrGO in urine, indicating its clearance. The repetitive dosing of PrGO could promote liver injury, blockage of kidney and enhanced proliferation of splenocytes, which limited their applications. Finally, the authors recommended that rigorous safety assessment should be carried out in order to validate its future clinical application as nanomedicine.

6.7   Global Market

Research and development sector for graphene-related materials (e.g. graphene sheets, GO, rGO) is expanding gradually throughout the world, leading to make an immense effect on their global market value (Geim and Novoselov 2007). Recently, several government and industrial agencies have invested a vast amount of money for the development of graphene-based technologies. For example, ~$1.3 billion was invested by the European Union in 'The Graphene Flagship' an association of academic and commercial scientists (Report 2016). Likewise, the UK government has funded $353 million to a graphene-based research organization (Report 2014). Several tech companies such as Samsung, IBM and Nokia are also investing a huge sum of money for grasping the global market of graphene-related products. In this scenario those companies have already applied for several patents related to graphene-based electronic products (Zurutuza and Marinelli 2014). The other graphene-based companies scattered all over the world include Thomas Swan & Co., Graphenea SA, Angstron Materials, Inc., etc. According to Grandviewresearch. com report, the global market value for graphene was $23.7 million in 2015, and it was expected to rise to a CAGR of 36.7% by 2025 (Report 2017). On the other hand, Marketsandmarkets.com report predicted that the global market value for graphene would reach $278.47 million by the year 2020 associated with enormous growth rate (42.8%) in between 2015 and 2020 (Marketsandmarkets.com Report 2017). It is to be mentioned that the major market of graphene-related products is based on their application in different sectors such as energy, electronics, sensors, catalysts, coatings, etc. However, considering the massive growth of graphene and GO-based materials in biomedical research, it could be speculated that some novel nanomedicine products (graphene based) would appear in the market in near future.

6.8   Future Directions of Graphene Oxides in Biomedical Applications

Since the past decades, scientists developed different nanomaterials for versatile applications exploiting their unique physicochemical properties. Several types of nanoparticles such as metal, non-metal, polymeric, liposomes, etc. exhibiting a variety of biomedical applications have evolved during this period. Among these nanomaterials, GO has currently attracted great attention to the researchers for their different applications in healthcare, especially theranostic applications in angiogenesis and cancers. However, the major challenge for clinical translation of GO includes its long-term toxicity concern (Feng and Liu 2011; Rahman et al. 2015). There are conflicting data regarding the in vitro and in vivo toxicity of GO as per recent literature (Muazim and Hussain 2017; Wu et al. 2015). It is often observed that surface functionalization of GO (e.g. PEGylation) might reduce its toxicity both in vitro and in vivo. The size and surface modification of GO could play an effective role for its in vivo biodistribution (Feng and Liu 2011). However, it should be investigated how the size and surface chemistry of GO nanomaterials affect their in vivo fate, especially uptake in RES, targeting efficacy to tumours, excretion through urine/faeces, etc. Moreover, pharmacokinetics and pharmacodynamics studies of these nanomaterials are to be performed to comprehend their toxicity profiles. Therefore, more in vivo toxicological studies of GO-based nanomaterials/ nanocomposites should be carried out in a systematic manner in different animal models prior to their clinical applications. Other than toxicity issues, the commercial applications of GO are also associated with the challenge of reproducibility of functionalized GO nanomaterials (Muazim and Hussain 2017; Orecchioni et al. 2015). However, considering the growth of biomedical research on GO, it is expected that some related nanomedicines would be approved by FDA or other concerned authorities in the near future, leading to expand the market value of GO worldwide (Josefsen and Boyle 2012).

Acknowledgement CRP is grateful to DST-Nanomission, New Delhi, (SR/NM/NS-1252/2013; GAP 570) for financial support. This book chapter is partially supported by 'CSIR-Mayo Clinic Collaboration for Innovation and Translational Research' (CKM/CMPP-09; MLP0020) fund from CSIR, New Delhi and 12th Five Year Plan (FYP) projects (ADD: CSC0302) CSIR, New Delhi, to CRP. A.K.B. and S.D. are thankful to UGC, New Delhi while A.R. and K.B. are thankful to ICMR, New Delhi, for their fellowships. The authors are thankful to the Director, CSIR-IICT for his support and encouragement and for his keen interest in this work. IICT manuscript communication number IICT/Pubs./2019/147 dated April 15, 2019 for this manuscript is duly acknowledged.

References

Afarideh B, Rajabibazl M, Omidi M et al (2018) Anticancer activity of graphene oxide/5-FU on CT26 Ds-Red adenocarcinoma cell line. Orient J Chem 34:2002

Ahmad MW, Xu W, Kim SJ et al (2015) Potential dual imaging nanoparticle: Gd_2O_3 nanoparticle. Sci Rep 5:8549

Akhavan O, Ghaderi E, Aghayee S et al (2012) The use of a glucose-reduced graphene oxide suspension for photothermal cancer therapy. J Mater Chem 22:13773–13781

Arvizo RR, Rana S, Miranda OR et al (2011) Mechanism of anti-angiogenic property of gold nanoparticles: role of nanoparticle size and surface charge. Nanomedicine 7:580–587

Bae YH, Park K (2011) Targeted drug delivery to tumors: myths, reality and possibility. J Control Release 153:198–205

Bansal A, Zhang Y (2014) Photocontrolled nanoparticle delivery systems for biomedical applications. Acc Chem Res 47:3052–3060

Bartczak D, Muskens OL, Sanchez-Elsner T et al (2013) Manipulation of in vitro angiogenesis using peptide-coated gold nanoparticles. ACS Nano 7:5628–5636

Barui AK, Veeriah V, Mukherjee S et al (2012) Zinc oxide nanoflowers make new blood vessels. Nanoscale 4:7861–7869

Bergers G, Hanahan D (2008) Modes of resistance to anti-angiogenic therapy. Nat Rev Cancer 8:592–603

Bikfalvi A, Bicknell R (2002) Recent advances in angiogenesis, anti-angiogenesis and vascular targeting. Trends Pharmacol Sci 23:576–582

Boehm H-P, Setton R, Stumpp E (1994) Nomenclature and terminology of graphite intercalation compounds. Pure Appl Chem 66:1893–1901

Borghaei H, Smith MR, Campbell KS (2009) Immunotherapy of cancer. Eur J Pharmacol 625:41–54

Brodie BC (1859) On the atomic weight of graphite. Philos Trans R Soc London 149:249–259

Byun J (2015) Emerging frontiers of graphene in biomedicine. J Microbiol Biotechnol 25:145–151

Caruso F, Hyeon T, Rotello VM (2012) Nanomedicine. Chem Soc Rev 41:2537–2538

Chang Y, Yang ST, Liu JH et al (2011) In vitro toxicity evaluation of graphene oxide on A549 cells. Toxicol Lett 200:201–210

Chaudhuri P, Harfouche R, Soni S et al (2010) Shape effect of carbon nanovectors on angiogenesis. ACS Nano 4:574–582

Chen Y, Chen HR, Shi JL (2014) Inorganic nanoparticle-based drug codelivery nanosystems to overcome the multidrug resistance of cancer cells. Mol Pharm 11:2495–2510

Chen X, Liu L, Jiang C (2016) Charge-reversal nanoparticles: novel targeted drug delivery carriers. Acta Pharm Sin B 6:261–267

Cheon YA, Bae JH, Chung BG (2016) Reduced graphene oxide nanosheet for chemo-photothermal therapy. Langmuir 32:2731–2736

Chidambaram M, Manavalan R, Kathiresan K (2011) Nanotherapeutics to overcome conventional cancer chemotherapy limitations. J Pharm Pharm Sci 14:67–77

Cho K, Wang X, Nie S et al (2008) Therapeutic nanoparticles for drug delivery in cancer. Clin Cancer Res 14:1310–1316

Cho WS, Kang BC, Lee JK et al (2013) Comparative absorption, distribution, and excretion of titanium dioxide and zinc oxide nanoparticles after repeated oral administration. Part Fibre Toxicol 10:9

Chu J, Shi P, Yan W et al (2018) PEGylated graphene oxide-mediated quercetin-modified collagen hybrid scaffold for enhancement of MSCs differentiation potential and diabetic wound healing. Nanoscale 10:9547–9560

Cong HP, Wang P, Yu SH (2013) Stretchable and self-healing graphene oxide-polymer composite hydrogels: a dual-network design. Chem Mater 25:3357–3362

Cotton FA, Wilkinsion G (1972) Advanced inorganc chemistry, 3rd edn. Wiley, Chichester, ISBN: 0-471-17560-9

Dai L (2006) Carbon nanotechnology recent developments in chemistry, physics, materials science and device applications. Elsevier, Amsterdam, ISBN-10: 044451855X

Danhier F, Feron O, Preat V (2010) To exploit the tumor microenvironment: passive and active tumor targeting of nanocarriers for anti-cancer drug delivery. J Control Release 148:135–146

Das S, Singh S, Dowding JM et al (2012) The induction of angiogenesis by cerium oxide nanoparticles through the modulation of oxygen in intracellular environments. Biomaterials 33:7746–7755

Deb A, Andrews NG, Raghavan V (2018) Natural polymer functionalized graphene oxide for co-delivery of anticancer drugs: in-vitro and in-vivo. Int J Biol Macromol 113:515–525

Di Santo R, Digiacomo L, Palchetti S et al (2019) Microfluidic manufacturing of surface-functionalized graphene oxide nanoflakes for gene delivery. Nanoscale 11:2733–2741

Draz MS, Fang BA, Zhang P et al (2014) Nanoparticle-mediated systemic delivery of siRNA for treatment of cancers and viral infections. Theranostics 4:872–892

Dreyer DR, Ruoff RS, Bielawski CW (2010) From conception to realization: an historial account of graphene and some perspectives for its future. Angew Chem 49:9336–9344

Enterkin JA, Poeppelmeier KR, Marks LD (2011) Oriented catalytic platinum nanoparticles on high surface area strontium titanate nanocuboids. Nano Lett 11:993–997

Erathodiyil N, Ying JY (2011) Functionalization of inorganic nanoparticles for bioimaging applications. Acc Chem Res 44:925–935

Falkowski P, Scholes RJ, Boyle EE et al (2000) The global carbon cycle: a test of our knowledge of earth as a system. Science 290:291–296

Fan ZJ, Liu B, Wang J et al (2014) A novel wound dressing based on Ag/Graphene polymer hydrogel: effectively kill bacteria and accelerate wound healing. Adv Funct Mater 24:3933–3943

Feng L, Liu Z (2011) Graphene in biomedicine: opportunities and challenges. Nanomedicine (Lond) 6:317–324

Feng L, Zhang S, Liu Z (2011) Graphene based gene transfection. Nanoscale 3:1252–1257

Folkman J (1995) Angiogenesis in cancer, vascular, rheumatoid and other disease. Nat Med 1:27–31

Geim AK, Novoselov KS (2007) The rise of graphene. Nat Mater 6:183–191

Gu Y, Guo Y, Wang C et al (2017) A polyamidoamne dendrimer functionalized graphene oxide for DOX and MMP-9 shRNA plasmid co-delivery. Mater Sci Eng C Mater Biol Appl 70:572–585

Gulzar A, Xu J, Yang D et al (2018) Nano-graphene oxide-UCNP-Ce6 covalently constructed nanocomposites for NIR-mediated bioimaging and PTT/PDT combinatorial therapy. Dalton Trans (Cambridge, England) 2003(47):3931–3939

Gurunathan S, Lee KJ, Kalishwaralal K et al (2009) Antiangiogenic properties of silver nanoparticles. Biomaterials 30:6341–6350

Gurunathan S, Han JW, Eppakayala V et al (2013) Green synthesis of graphene and its cytotoxic effects in human breast cancer cells. Int J Nanomedicine 8:1015–1027

Hamidi M, Azadi A, Rafiei P et al (2013) A pharmacokinetic overview of nanotechnology-based drug delivery systems: an ADME-oriented approach. Crit Rev Ther Drug Carrier Syst 30:435–467

Hijaz M et al (2016) Folic acid tagged nanoceria as a novel therapeutic agent in ovarian cancer. BMC Cancer 16:220

Holgate ST (2010) Exposure, uptake, distribution and toxicity of nanomaterials in humans. J Biomed Nanotechnol 6:1–19

Hong H, Zhang Y, Engle JW et al (2012) In vivo targeting and positron emission tomography imaging of tumor vasculature with Ga-66-labeled nano-graphene. Biomaterials 33:4147–4156

Hu H, You YY, He LZ et al (2015) The rational design of NAMI-A-loaded mesoporous silica nanoparticles as antiangiogenic nanosystems. J Mater Chem B 3:6338–6346

Huang C, Wu J, Jiang W et al (2018) Amphiphilic prodrug-decorated graphene oxide as a multifunctional drug delivery system for efficient cancer therapy. Mater Sci Eng C Mater Biol Appl 89:15–24

Jaleel JA, Sruthi S, Pramod K (2017) Reinforcing nanomedicine using graphene family nanomaterials. J Control Release 255:218–230

Jasim DA, Menard-Moyon C, Begin D et al (2015) Tissue distribution and urinary excretion of intravenously administered chemically functionalized graphene oxide sheets. Chem Sci 6:3952–3964

Jaworski S, Sawosz E, Kutwin M et al (2015) In vitro and in vivo effects of graphene oxide and reduced graphene oxide on glioblastoma. Int J Nanomedicine 10:1585–1596

Jong WHD, Borm PJ (2008) Drug delivery and nanoparticles:applications and hazards. Int J Nanomedicine 3:133–149

Josefsen LB, Boyle RW (2012) Unique diagnostic and therapeutic roles of porphyrins and phthalocyanines in photodynamic therapy, imaging and theranostics. Theranostics 2:916–966

Kang K, Lim DH, Choi IH et al (2011) Vascular tube formation and angiogenesis induced by polyvinylpyrrolidone-coated silver nanoparticles. Toxicol Lett 205:227–234

Kang S, Kim KM, Son Y et al (2019) Graphene oxide quantum dots derived from coal for bioimaging: facile and green approach. Sci Rep 9:4101

Khot LR, Sankaran S, Maja JM et al (2012) Applications of nanomaterials in agricultural production and crop protection: a review. Crop Prot 35:64–70

Kiew SF, Kiew LV, Lee HB et al (2016) Assessing biocompatibility of graphene oxide-based nanocarriers: a review. J Control Release 226:217–228

Kim JY, Shim G, Choi HW et al (2012) Tumor vasculature targeting following co-delivery of heparin-taurocholate conjugate and suberoylanilide hydroxamic acid using cationic nanolipoplex. Biomaterials 33:4424–4430

Krishnamachari Y, Geary SM, Lemke CD et al (2011) Nanoparticle delivery systems in cancer vaccines. Pharm Res 28:215–236

Kumar S, Chatterjee K (2016) Comprehensive review on the use of graphene-based substrates for regenerative medicine and biomedical devices. ACS Appl Mater Interfaces 8:26431–26457

Lai PX, Chen CW, Wei SC et al (2016) Ultrastrong trapping of VEGF by graphene oxide: anti-angiogenesis application. Biomaterials 109:12–22

Leteba GM, Lang CI (2013) Synthesis of bimetallic platinum nanoparticles for biosensors. Sensors (Basel) 13:10358–10369

Li SD, Huang L (2008) Pharmacokinetics and biodistribution of nanoparticles. Mol Pharm 5:496–504

Li SH, Aphale AN, Macwan IG et al (2012) Graphene oxide as a quencher for fluorescent assay of amino acids, peptides, and proteins. ACS Appl Mater Interfaces 4:7068–7074

Li B, Yang J, Huang Q et al (2013) Biodistribution and pulmonary toxicity of intratracheally instilled graphene oxide in mice. NPG Asia Materials 5:e44

Li Y, Dong H, Li Y et al (2015a) Graphene-based nanovehicles for photodynamic medical therapy. Int J Nanomedicine 10:2451–2459

Li Z, Wang H, Yang B et al (2015b) Three-dimensional graphene foams loaded with bone marrow derived mesenchymal stem cells promote skin wound healing with reduced scarring. Mater Sci Eng C Mater Biol Appl 57:181–188

Li QR, Jiao JB, Li LL et al (2017) Graphene oxide-enhanced cytoskeleton imaging and mitosis tracking. Chem Commun (Camb) 53:3373–3376

Liao KH, Lin YS, Macosko CW et al (2011) Cytotoxicity of graphene oxide and graphene in human erythrocytes and skin fibroblasts. ACS Appl Mater Interfaces 3:2607–2615

Lin S, Ruan J, Wang S (2019) Biosynthesized of reduced graphene oxide nanosheets and its loading with paclitaxel for their anti cancer effect for treatment of lung cancer. J Photochem Photobiol B 191:13–17

Liu Y, Yu D, Zeng C et al (2010) Biocompatible graphene oxide-based glucose biosensors. Langmuir 26:6158–6160

Liu JH, Yang ST, Wang HF et al (2012) Effect of size and dose on the biodistribution of graphene oxide in mice. Nanomedicine 7:1801–1812

Liu P, Wang S, Liu X et al (2018) Platinated graphene oxide: a nanoplatform for efficient gene-chemo combination cancer therapy. Eur J Pharm Sci 21:319–329

Lohse SE, Murphy CJ (2012) Applications of colloidal inorganic nanoparticles: from medicine to energy. J Am Chem Soc 134:15607–15620

Lu CH, Zhu CL, Li J et al (2010) Using graphene to protect DNA from cleavage during cellular delivery. Chem Commun 46:3116–3118

Lu B, Li T, Zhao H et al (2012) Graphene-based composite materials beneficial to wound healing. Nanoscale 4:2978–2982

Lu C, Huang PJ, Liu B et al (2016) Comparison of graphene oxide and reduced graphene oxide for dna adsorption and sensing. Langmuir 32:10776–10783

Luo L, Xu L, Zhao H (2017) Biosynthesis of reduced graphene oxide and its in-vitro cytotoxicity against cervical cancer (HeLa) cell lines. Mater Sci Eng C Mater Biol Appl 78:198–202

Luo Y, Tang Y, Liu T et al (2019) Engineering graphene oxide with ultrasmall SPIONs and smart drug release for cancer theranostics. Chem Commun 55:1963–1966

Ma H, Liu J, Ali MM et al (2015) Nucleic acid aptamers in cancer research, diagnosis and therapy. Chem Soc Rev 44:1240–1256

Mallick A, Nandi A, Basu S (2019) Polyethylenimine coated graphene oxide nanoparticles for targeting mitochondria in cancer cells. ACS Appl Bio Mater 2:14–19

Marketsandmarkets.com Report (2017) Graphene market worth 278.47 Million USD by 2020. http://www.marketsandmarketscom/PressReleases/grapheneasp

Mattos AJP, Raquel FE, Anna DR (2014) Gold nanoparticle mediated cancer immunotherapy. Nanomedicine 10:503–514

McNaught AD, Wilkinson A (1997) IUPAC. Compendium of chemical terminology, 2nd edn. Blackwell Scientific Publications, Oxford

Meghana S, Kabra P, Chakraborty S et al (2015) Understanding the pathway of antibacterial activity of copper oxide nanoparticles. RSC Adv 5:12293–12299

Millstone JE, Kavulak DF, Woo CH et al (2010) Synthesis, properties, and electronic applications of size-controlled poly(3-hexylthiophene) nanoparticles. Langmuir 26:13056–13061

Mittal S, Kumar V, Dhiman N et al (2016) Physico-chemical properties based differential toxicity of graphene oxide/reduced graphene oxide in human lung cells mediated through oxidative stress. Sci Rep 6:15860

Muazim K, Hussain Z (2017) Graphene oxide - a platform towards theranostics. Mater Sci Eng C Mater Biol Appl 76:1274–1288

Mukherjee S, Sushma V, Patra S et al (2012) Green chemistry approach for the synthesis and stabilization of biocompatible gold nanoparticles and their potential applications in cancer therapy. Nanotechnology 23:455103

Mukherjee S, Sriram P, Barui AK et al (2015) Graphene oxides show angiogenic properties. Adv Healthc Mater 4:1722–1732

Mulcahy N (2008) Cancer to become leading cause of death worldwide. Medscape

Nanda SS, Yi DK, Kim K (2016) Study of antibacterial mechanism of graphene oxide using Raman spectroscopy. Sci Rep 6:28443

Nejabat M, Charbgoo F, Ramezani M (2017) Graphene as multifunctional delivery platform in cancer therapy. J Biomed Mater Res A 105:2355–2367

Nguyen DT, Orgill DP, Murphy GF (2009) The pathophysiologic basis for wound healing and cutaneous regeneration, Biomaterials for treating skin loss. Woodhead Publishing, Boca Raton, pp 25–57

Nicol W (2015) A material supreme: how graphene will shape the world of tomorrow digital trends

Nie W, Peng C, Zhou X et al (2017) Three-dimensional porous scaffold by self-assembly of reduced graphene oxide and nano-hydroxyapatite composites for bone tissue engineering. Carbon 116:325–337

Nolan CP, DeAngelis LM (2015) Neurologic complications of chemotherapy and radiation therapy. Continuum (Minneap Minn) 21:429–451

Novoselov KS, Geim AK, Morozov SV et al (2004) Electric field effect in atomically thin carbon films. Science 306:666–669

Nurunnabi M, Parvez K, Nafiujjaman M et al (2015) Bioapplication of graphene oxide derivatives: drug/gene delivery, imaging, polymeric modification, toxicology, therapeutics and challenges. RSC Adv 5:42141–42161

Orecchioni M, Cabizza R, Bianco A et al (2015) Graphene as cancer theranostic tool: progress and future challenges. Theranostics 5:710–723

Orecchioni M, Jasim DA, Pescatori M et al (2016a) Molecular and genomic impact of large and small lateral dimension graphene oxide sheets on human immune cells from healthy donors. Adv Healthc Mater 5:276–287

Orecchioni M, Menard-Moyon C, Delogu LG et al (2016b) Graphene and the immune system: challenges and potentiality. Adv Drug Deliv Rev 105:163–175

Ou LL, Song B, Liang H et al (2016) Toxicity of graphene-family nanoparticles: a general review of the origins and mechanisms. Part Fibre Toxicol 13:57

Ouay LB, Stellacci F (2015) Antibacterial activity of silver nanoparticles: a surface science insight. Nano Today 10:339–354

Park J, Kim YS, Ryu S et al (2015) Graphene potentiates the myocardial repair efficacy of mesenchymal stem cells by stimulating the expression of angiogenic growth factors and gap junction protein. Adv Func Mater 25:2590–2600

Patra CR (2015) Graphene oxides and the angiogenic process. Nanomedicine (Lond) 10:2959–2962

Patra CR, Bhattacharya R, Patra S et al (2008) Pro-angiogenic properties of europium(III) hydroxide nanorods. Adv Mater 20:753–756

Patra CR, Kim JH, Pramanik K et al (2011) Reactive oxygen species driven angiogenesis by inorganic nanorods. Nano Lett 11:4932–4938

Paul A, Hasan A, Kindi HA et al (2014) Injectable graphene oxide/hydrogel-based angiogenic gene delivery system for vasculogenesis and cardiac repair. ACS Nano 8:8050–8062

Perreault F, de Faria AF, Nejati S et al (2015) Antimicrobial properties of graphene oxide nanosheets: why size matters. ACS Nano 9:7226–7236

Rahman M, Akhter S, Ahmad MZ et al (2015) Emerging advances in cancer nanotheranostics with graphene nanocomposites: opportunities and challenges. Nanomedicine 10:2405–2422

Raj S, Jose S, Sumod US et al (2012) Nanotechnology in cosmetics: opportunities and challenges. J Pharm Bioallied Sci 4:186–193

Ren L, Zhang Y, Cui C et al (2017) Functionalized graphene oxide for anti-VEGF siRNA delivery: preparation, characterization and evaluation in vitro and in vivo. RSC Adv 7:20553–20566

Report B (2014) Autumn statement 2014: Manchester to get £235m science research centre. http://www.bbccom/news/uk-england-30309451

Report D (2016) Graphene: research now, reap next decade. https://www2.deloitte.com/global/en/pages/technology-media-and-telecommunications/articles/tmt-pred16-tech-graphene-research-now-reap-next-decade.html

Report Gc (2017) Graphene market Size and trend analysis by product (Nanoplatelets, Oxide), by application (Electronics, Composites, Energy), by region (North America, Europe, Asia Pacific, Rest of the World), and Segment forecasts, 2014–2025. http://www.grandviewresearchcom/industry-analysis/graphene-industry

Rieger S, Zhao H, Martin P et al (2015) The role of nuclear hormone receptors in cutaneous wound repair. Cell Biochem Funct 33:1–13

Robinson JT, Tabakman SM, Liang YY et al (2011) Ultrasmall reduced graphene oxide with high near-infrared absorbance for photothermal therapy. J Am Chem Soc 133:6825–6831

Sahne F, Mohammadi M, Najafpour GD (2019) Single-layer assembly of multifunctional carboxymethylcellulose on graphene oxide nanoparticles for improving in vivo curcumin delivery into tumor cells. ACS Biomater Sci Eng 5(5):2595–2609

Sahu SC, Casciano DA (eds) (2009) Nanotoxicity: from in vivo and in vitro models to health risks. Wiley, Chichester

Schaefer H-E (2010) Nanoscience. The science of the small in physics, engineering, chemistry, biology and medicine. Springer, New York

Schafhaeutl C (1840) LXXXVI. On the combinations of carbon with silicon and iron, and other metals, forming the different species of cast iron, steel, and malleable iron. London, Edinburgh, Dublin Philos Mag J Sci 16:570–590

Shi S, Yang K, Hong H et al (2013) Tumor vasculature targeting and PET imaging in living mice with reduced graphene oxide. Eur J Nucl Med Mol Imaging 40:S153–S153

Shim G, Kim JY, Han J et al (2014) Reduced graphene oxide nanosheets coated with an anti-angiogenic anticancer low-molecular-weight heparin derivative for delivery of anticancer drugs. J Control Release 189:80–89

Shin SR, Li YC, Jang HL et al (2016a) Graphene-based materials for tissue engineering. Adv Drug Deliv Rev 105:255–274

Shin SR, Zihlmann C, Akbari M et al (2016b) Reduced graphene oxide-GelMA hybrid hydrogels as scaffolds for cardiac tissue engineering. Small 12:3677–3689

Steichen SD, Caldorera-Moore M, Peppas NA (2013) A review of current nanoparticle and targeting moieties for the delivery of cancer therapeutics. Eur J Pharm Sci 48:416–427

Su SH, Wang JL, Wei JH et al (2015) Efficient photothermal therapy of brain cancer through porphyrin functionalized graphene oxide. New J Chem 39:5743–5749

Sun X, Liu Z, Welsher K et al (2008) Nano-graphene oxide for cellular imaging and drug delivery. Nano Res 1:203–212

Sun ZC, Huang P, Tong G et al (2013) VEGF-loaded graphene oxide as theranostics for multi-modality imaging-monitored targeting therapeutic angiogenesis of ischemic muscle. Nanoscale 5:6857–6866

Sun X, Zebibula A, Dong X et al (2018) Aggregation-induced emission nanoparticles encapsulated with PEGylated nano graphene oxide and their applications in two-photon fluorescence bioimaging and photodynamic therapy in vitro and in vivo. ACS Appl Mater Interfaces 10:25037–25046

Syama S, Paul W, Sabareeswaran A et al (2017) Raman spectroscopy for the detection of organ distribution and clearance of PEGylated reduced graphene oxide and biological consequences. Biomaterials 131:121–130

Tan YB, Lee JM (2013) Graphene for supercapacitor applications. J Mater Chem A 1:14814–14843

Tang P, Han L, Li P et al (2019) Mussel-inspired electroactive and antioxidative scaffolds with incorporation of polydopamine-reduced graphene oxide for enhancing skin wound healing. ACS Appl Mater Interfaces 11:7703–7714

Tao Y, Ju EG, Ren JS, Qu XG (2014) Immunostimulatory oligonucleotides-loaded cationic graphene oxide with photothermally enhanced immunogenicity for photothermal/immune cancer therapy. Biomaterials 35:9963–9971

Teli MK, Mutalik S, Rajanikant GK (2010) Nanotechnology and nanomedicine: going small means aiming big. Curr Pharm Des 16:1882–1892

Thangavel P, Kannan R, Ramachandran B et al (2018) Development of reduced graphene oxide (rGO)-isabgol nanocomposite dressings for enhanced vascularization and accelerated wound healing in normal and diabetic rats. J Colloid Interface Sci 517:251–264

Tian J, Luo Y, Huang L et al (2016) Pegylated folate and peptide-decorated graphene oxide nano-vehicle for in vivo targeted delivery of anticancer drugs and therapeutic self-monitoring. Biosens Bioelectron 80:519–524

Torchilin VP (2010) Passive and active drug targeting: drug delivery to tumors as an example. Handb Exp Pharmacol 197:50

Tran TH, Nguyen HT, Pham TT et al (2015) Development of a graphene oxide nanocarrier for dual-drug chemo-phototherapy to overcome drug resistance in cancer. ACS Appl Mater Interfaces 7:28647–28655

Velnar T, Bailey T, Smrkolj V (2009) The wound healing process: an overview of the cellular and molecular mechanisms. J Int Med Res 37:1528–1542

Wang Y, Hu R, Lin G et al (2013) Functionalized quantum dots for biosensing and bioimaging and concerns on toxicity. ACS Appl Mater Interfaces 5:2786–2799

Wei Y, Zhou F, Zhang D et al (2016) A graphene oxide based smart drug delivery system for tumor mitochondria-targeting photodynamic therapy. Nanoscale 8:3530–3538

Wierzbicki M, Sawosz E, Grodzik M et al (2013) Comparison of anti-angiogenic properties of pristine carbon nanoparticles. Nanoscale Res Lett 8:195

Winter JO (2007) Nanoparticles and nanowires for cellular engineering. Nanotechnologies for the life sciences. Wiley, New York

Wu CH, He QM, Zhu AN et al (2014) Synergistic anticancer activity of photo- and chemorespon-sive nanoformulation based on polylysine-functionalized graphene. ACS Appl Mater Interfaces 6:21615–21623

Wu SY, An SS, Hulme J (2015) Current applications of graphene oxide in nanomedicine. Int J Nanomedicine 10(Spec Iss):9–24

Xing FY, Guan LL, Li YL et al (2016) Biosynthesis of reduced graphene oxide nanosheets and their in vitro cytotoxicity against cardiac cell lines of Catla catla. Environ Toxicol Pharmacol 48:110–115

Xu LG, Xiang J, Liu Y et al (2016) Functionalized graphene oxide serves as a novel vaccine nano-adjuvant for robust stimulation of cellular immunity. Nanoscale 8:3785–3795

Xu X, Tang X, Wu X et al (2019) Biosynthesis of sorafenib coated graphene nanosheets for the treatment of gastric cancer in patients in nursing care. J Photochem Photobiol B 191:1–5

Yaacoub K, Pedeux R, Tarte K et al (2016) Role of the tumor microenvironment in regulating apoptosis and cancer progression. Cancer Lett 378:150–159

Yadav N, Kumar N, Prasad P et al (2018) Stable dispersions of covalently tethered polymer improved graphene oxide nanoconjugates as an effective vector for siRNA delivery. ACS Appl Mater Interfaces 10:14577–14593

Yan M, Liu Y, Zhu X et al (2019) Nanoscale reduced graphene oxide-mediated photothermal therapy together with IDO inhibition and PD-L1 blockade synergistically promote antitumor immunity. ACS Appl Mater Interfaces 11:1876–1885

Yang XY, Zhang XY, Liu ZF et al (2008) High-efficiency loading and controlled release of doxo-rubicin hydrochloride on graphene oxide. J Phys Chem C 112:17554–17558

Yang K, Wan JM, Zhang SA et al (2011) In vivo pharmacokinetics, long-term biodistribution, and toxicology of PEGylated graphene in mice. ACS Nano 5:516–522

Yang K, Wan JM, Zhang S et al (2012) The influence of surface chemistry and size of nanoscale graphene oxide on photothermal therapy of cancer using ultra-low laser power. Biomaterials 33:2206–2214

Yang K, Gong H, Shi XZ et al (2013) In vivo biodistribution and toxicology of functionalized nano-graphene oxide in mice after oral and intraperitoneal administration. Biomaterials 34:2787–2795

Yang DZ, Feng L, Dougherty CA et al (2016) In vivo targeting of metastatic breast cancer via tumor vasculature-specific nano-graphene oxide. Biomaterials 104:361–371

Yin F, Hu K, Chen Y et al (2017) SiRNA delivery with PEGylated graphene oxide nanosheets for combined photothermal and genetherapy for pancreatic cancer. Theranostics 7:1133–1148

Yoshioka Y, Higashisaka K, Tsunoda S, Tsutsumi Y (2014) The absorption, distribution, metabolism, and excretion profile of nanoparticles. In: Akashi M, Akagi T, Matsusaki M (eds) Engineered cell manipulation for biomedical application. Nanomedicine and nanotoxicology. Springer, Tokyo, pp 259–271

You DG, Deepagan VG, Um W et al (2016) ROS-generating TiO_2 nanoparticles for non-invasive sonodynamic therapy of cancer. Sci Rep 6:23200

Yue H, Wei W, Gu Z et al (2015) Exploration of graphene oxide as an intelligent platform for cancer vaccines. Nanoscale 7:19949–19957

Yue H, Zhou X, Cheng M et al (2018) Graphene oxide-mediated Cas9/sgRNA delivery for efficient genome editing. Nanoscale 10:1063–1071

Zang Z, Zeng X, Wang M et al (2017) Tunable photoluminescence of water-soluble AgInZnS-graphene oxide (GO) nanocompositesand their application in-vivo bioimaging. Sens Actuators B Chem 252:1179–1186

Zare-Zardini H, Taheri-Kafrani A, Amiri A et al (2018) New generation of drug delivery systems based on ginsenoside Rh2-, Lysine- and Arginine-treated highly porous graphene for improving anticancer activity. Sci Rep 8:586

Zhang LM, Lu ZX, Zhao QH et al (2011) Enhanced chemotherapy efficacy by sequential delivery of siRNA and anticancer drugs using PEI-grafted graphene oxide. Small 7:460–464

Zhang M, Kim JA, Huang AYC (2018) Optimizing tumor microenvironment for cancer immunotherapy: β-Glucan-based nanoparticles. Front Immunol 9:341–341

Zhao X, Yang L, Li X et al (2015) Functionalized graphene oxide nanoparticles for cancer cell-specific delivery of antitumor drug. Bioconjug Chem 26:128–136

Zhao H, Osborne OJ, Lin S et al (2016) Lanthanide hydroxide nanoparticles induce angiogenesis via ROS-sensitive signaling. Small 12:4404–4411

Zhou Y, Chen R, He T et al (2016) Biomedical potential of ultrafine Ag/AgCl nanoparticles coated on graphene with special reference to antimicrobial performances and burn wound healing. ACS Appl Mater Interfaces 8:15067–15075

Zhu X, Xu X, Liu F et al (2017) Green synthesis of graphene nanosheets and their in vitro cytotoxicity against human prostate cancer (DU 145) cell lines. Nanomater Nanotechnol 7. https://doi.org/10.1177/1847980417702794

Zou L, Wang H, He B et al (2016) Current approaches of photothermal therapy in treating cancer metastasis with nanotherapeutics. Theranostics 6:762–772

Zurutuza A, Marinelli C (2014) Challenges and opportunities in graphene commercialization. Nat Nanotechnol 9:730–734

Use of Nanoparticles to Manage *Candida* Biofilms

7

Douglas Roberto Monteiro, Laís Salomão Arias,
Heitor Ceolin Araujo, Anne Caroline Morais Caldeirão,
Bianca Fiorese Gulart, Joseane de Oliveira,
Marilene Batista dos Santos, Gordon Ramage,
and Juliano Pelim Pessan

Abstract

Candida species constitute an important part of the human oral microbiome and may be found as commensal colonizers in the oral cavity, as well as in the digestive and vaginal tracts. However, disturbances in host homeostasis may cause an overgrowth of these species, resulting in various types of candidiasis. This aspect, in conjunction with its ability to form organized and resistant structures, namely biofilms, has stimulated interest in nanotechnology-based therapies to fight biofilms and improve individuals' health. This chapter approaches some of the clinical implications of *Candida* biofilms and their mechanisms of resistance to conventional antimicrobials, as well as the main types of nanoparticles used in controlling and preventing biofilms formed by different *Candida* species.

Keywords

Antimicrobials · Biofilms · *Candida* · Nanomaterials · Nanoparticles

7.1 Introduction

Humans can be colonized by bacteria and fungi, forming highly organized communities known as biofilms (Shirtliff et al. 2009). Within this context are the fungi *Candida*, which colonize different mucosal surfaces as a component of the normal

D. R. Monteiro (✉) · A. C. M. Caldeirão · B. F. Gulart · J. de Oliveira · M. B. dos Santos
University of Western São Paulo (UNOESTE), Presidente Prudente, São Paulo, Brazil

L. S. Arias · H. C. Araujo · J. P. Pessan
School of Dentistry, Araçatuba, Department of Pediatric Dentistry and Public Health, São Paulo State University (Unesp), Araçatuba/São Paulo, Brazil

G. Ramage
School of Medicine, Dentistry and Nursing, MVLS, University of Glasgow, Glasgow, UK

© Springer Nature Singapore Pte Ltd. 2020
A. K. Shukla (ed.), *Nanoparticles and their Biomedical Applications*,
https://doi.org/10.1007/978-981-15-0391-7_7

human microflora (Ganguly and Mitchell 2011). However, under conditions of immune dysfunction or other local predisposing factors, *Candida* species increase in number and become invasive, causing local or systemic infections.

The success of antifungal therapy does not depend exclusively on the drugs used, but also on factors inherent to the host, such as the pH of the medium, cellular barriers and degrading enzymes that often prevent the drugs from reaching their maximum performance, since they are inactivated before reaching the target cell (Bowman and Leong 2006). In addition, the emergence of fungal resistance to conventional drugs has caused difficulties in the clinical treatment of candidiasis. Thus, the development of alternative antifungal therapies aiming to overcome the above-mentioned problems is crucial.

In recent years, nanotechnology has advanced in several fields, especially in industry, agriculture, energy, environmental protection and health (Kahan et al. 2009; Bourzac 2012). In the biomedical area, the synthesis of nanoparticles (NP) by chemical route or "green" (biosynthesis) has generated nanomaterials with antimicrobial properties and with capacity to improve the quality of image exams and the delivery of medicines (Suri et al. 2007; Niemirowicz et al. 2012). Smaller NP size (1–100 nm) provides a higher surface area ratio by volume and higher reactivity. In addition, NP improve endocytosis and cellular delivery, optimizing the effect of the drug and often reducing its concentration of use (Bowman and Leong 2006; Ali and Ahmed 2018). Considering the above, this chapter addresses the clinical significance of *Candida* biofilms and their resistance mechanisms to conventional antifungals, as well as the main NP used to fight different *Candida* species.

7.2 Clinical Significance of *Candida* Species Biofilms

Candida species can colonize several parts of the human body, such as the oral cavity, upper and lower airways, gastrointestinal and genitourinary tracts as well as skin (Williams and Ramage 2015). These species produce hydrolytic enzymes that digest and destroy cell membranes, facilitating their invasion into host tissues (Williams et al. 2011). Likewise, biofilm formation capacity on biotic and abiotic surfaces are considered the main virulence factors of *Candida* species (Williams et al. 2011). In a simple way, the formation process of fungal biofilms can be characterized by four successive and different phases: (1) adhesion of yeasts to a surface; (2) the beginning of the production of extracellular matrix and transformation of yeasts into hyphae, forming microcolonies; (3) maturation phase, where the biofilm is composed by a dense network of interconnected yeasts and hyphae surrounded by a thick extracellular matrix; and (4) dispersion phase, in which some cells are released from the biofilm, starting phase 1 in different sites (Chandra et al. 2001).

Candida albicans is the most studied yeast, and frequently associated with local and systemic fungal infections, mainly due to its cellular dimorphism that favours the development of structurally complex, dense and resistant biofilms (Ramage et al. 2014). In most patients, this fungus is found as a commensal; however, in situations of host imbalance or immunocompromising, it becomes pathogenic and

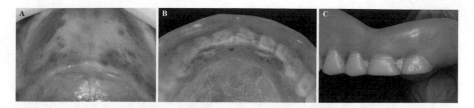

Fig. 7.1 (a) Palatal mucosa of complete denture wearer with erythematous lesions of denture stomatitis. Note biofilm accumulation on the inner surface of the denture (b), as well as between the artificial teeth (c)

contributes to the emergence of infectious diseases (Hirota et al. 2017). In addition, the role of other non-*albicans Candida* species in the pathogenesis of fungal diseases has increased in recent years, mainly due to more accurate diagnostic methods and the higher resistance of these species to some antifungals compared to *C. albicans* (Silva et al. 2012).

In the oral cavity, *Candida* biofilm formation on the surface of total or partial dentures is one of the etiological factors of denture stomatitis (Fig. 7.1). This infectious process mainly affects the mucosa that supports the denture, and patients may experience oral burning sensation, discomfort or bad breath (Gendreau and Loewy 2011), but in most cases they are not aware of the problem. The prevalence rates of stomatitis vary from 2.5% to 77.5% depending on factors such as the age of the patients, geographical region, diagnostic method used and type of denture (Gendreau and Loewy 2011). For individuals with denture stomatitis, *C. albicans* is the most frequently detected/isolated species, followed by *Candida glabrata, Candida tropicalis* and *Saccharomyces cerevisiae* (Marcos-Arias et al. 2009). Furthermore, mixed cultures of *C. albicans* and *C. glabrata* were associated with severe inflammation of the oral mucosa (Coco et al. 2008).

Gastrointestinal infections and aspiration pneumonia may be related to swallowing and aspiration of parts of the biofilms developed on dentures (Nikawa et al. 1998). *Candida* species also comprise biofilms of voice prosthesis, leading to losses in swallowing, respiration, phonetics and airflow in patients (Elving et al. 2001; Sayed et al. 2012).

Vulvovaginal candidiasis is another pathology associated with colonization by *Candida* species. It is characterized by the presence of signs and symptoms of inflammation, being pruritus and vulvar burning, dysuria and dyspareunia the most common clinical manifestations (Gonçalves et al. 2016). It is believed that the vagina of 10–15% of asymptomatic women is colonized by *Candida* species, and 70–75% of women will have a vulvovaginal candidiasis episode in their lives (Gonçalves et al. 2016). The most frequently detected species in vulvovaginal candidiasis is *C. albicans*, followed by *C. glabrata, C. tropicalis, Candida parapsilosis* and *Candida krusei*. The majority of these co-infections are related to the presence of *C. albicans* and *C. glabrata* (Gonçalves et al. 2016).

In addition, fungi of the genus *Candida* are among the most detected in blood cultures of hospitalized patients. The incidence of infections caused by them has increased in individuals with cancer, in intensive care units and postsurgical units (Rajendran et al. 2016a, b). Formation of *Candida* biofilms on medical devices (catheters, valves) can cause candiduria and candidemia, leading to systemic infections of different organs and tissues, and loss of biomaterials. On the epidemiology of invasive candidiasis, *C. albicans*, *C. glabrata*, *C. tropicalis* and *C. parapsilosis* have been collectively detected in 95% of the cases (Pfaller and Diekema 2007). Mortality rates associated with invasive candidiasis in adults and children can reach up to 50% and 30%, respectively, even with the current antifungal therapies (Costa et al. 2000; Viudes et al. 2002; Moran et al. 2009; Andes et al. 2012).

First reported in 2009, *Candida auris* has emerged as a drug-resistant pathogen with a strong potential for nosocomial transmission, which has caused hospital outbreaks and generated health concerns (Alfouzan et al. 2019). Surface disinfection is a major challenge as this species can tolerate clinically relevant concentrations of sodium hypochlorite and peracetic acid (Kean et al. 2018). The physiological characteristics of the strains of *C. auris* show differences in their pathogenicity, since non-aggregating strains present greater virulence than the aggregating ones (Borman et al. 2016). However, there is still much to be studied regarding its pathogenicity, treatment and genetic mechanisms of resistance.

7.3   Resistance of *Candida* Biofilms to Conventional Antimicrobials

The knowledge about *Candida* species and their mechanisms of resistance has made it possible to search for alternatives to control of fungal biofilms and their virulence factors. Studies highlight the role of biofilms in the resistance to conventional antimicrobials, many times exceeding the therapeutic dose allowed in a given type of treatment (Jabra-Rizk et al. 2004; Ramage et al. 2012). Adhered to each other and protected by their own extracellular matrix, the cells of a biofilm acquire certain advantages that make it difficult to fight them, among which stand out: physical shielding against the environment, protection of the cells from physical and chemical stress, nutritional and metabolic cooperation and regulation in gene expression coordinated by communication of the microbial community through the exchange of molecular signals, also known as *quorum sensing* (Ramage et al. 2012).

However, the advantages mentioned above do not act in isolation on the resistance of biofilms to antimicrobials. When penetration of fluconazole and flucytosine were evaluated in single or mixed biofilms of *C. albicans*, *C. glabrata*, *C. krusei*, *C. parapsilosis* and *C. tropicalis*, it was verified that, irrespective of penetration power (high or low) of the drug, this physical factor alone does not represent a significant impact on antimicrobial resistance (Al-Fattani and Douglas 2004). Thus, it is assumed that the variables that interfere in the architecture and composition of the biofilm are also related to the fungal resistance, such as temperature, pH, flow and nutrition of the environment in which the biofilm is inserted, osmolarity and host immune factors, among others (Ramage et al. 2012; Jones et al. 2015; Pumeesat et al. 2017).

Antimicrobial resistance may be considered as primary, when the microorganism is naturally drug resistant, or secondary (adaptive), when it develops resistance after exposure to the drug (Jabra-Rizk et al. 2004). These types of resistance can be verified for azole antifungals, since frequent use of this class of drugs in immunocompromised patients may lead to the development of resistance in *Candida* species (Zhang et al. 2014), while certain species of *Candida* non-*albicans* are intrinsically resistant to fluconazole (Arendrup and Patterson 2017).

Azoles act by inhibiting the enzyme sterol 14α-demethylase (14DM), a cytochrome P-450 enzyme, which is involved in the transformation of lanosterol into ergosterol. These antifungals bind to the active site of 14 DM, reducing ergosterol and affecting the integrity of the cell membrane (Morschhauser 2002; de Oliveira Santos et al. 2018). Resistance to azoles can occur by different factors, including (i) mutations in the 14DM enzyme, decreasing its affinity for fluconazole, (ii) increased expression or point mutations of the ERG11 gene, which encodes the 14DM enzyme and (iii) changes in ergosterol biosynthesis, by overexpression of efflux system genes, which encode for membrane-carrying proteins (*CDR1/CDR2*) or the main facilitator transporter (*MDR1*) (Morschhauser 2002). Furthermore, strains of *C. albicans* susceptible and resistant to azole showed differentiated gene expression for various functions (sterol metabolism, transcription factors) (Yan et al. 2008). Mutations or even the difference of a single amino acid may be responsible for the overexpression and inactivation of certain genes, such as the transcription factor Tac1p, which activates the expression of drug efflux pumps from the ABC transporter family, leading to resistance (Yan et al. 2008). Point mutations in the ERG11 gene in azole-resistant *C. albicans* can alter the affinity of CYP51A1 for an azole if the resultant amino acid substitutions lead to changes in the enzyme's tertiary structure (Xiang et al. 2013). Some of the amino acid mutations in the ERG11 gene reported in literature are S405F, Y132F, Y132H, Y257H, R467K, G448E, G464S, A114S, D116E, K128T, K143R, E266D and V437I (Ramage et al. 2012; Xiang et al. 2013).

In addition to the issue of microbial resistance, azoles can cause hepatotoxicity in patients (Haegler et al. 2017); this has led to the choice of other antifungals, such as echinocandins. This relatively new class of antifungal agent (micafungin, anidulafungin and caspofungin) targets the glucan synthase enzyme, which participates in the β-glucan biosynthesis of the fungal cell wall (Perlin 2015). Although resistance reports are less frequent, *C. glabrata* has shown high resistance to echinocandins alone or in combination with fluconazole; it is able to withstand exposure to the drug even before promoting mutations related to echinocandin resistance genes (FKS1 and FKS2) (Healey and Perlin 2018). However, when resistance occurs, it involves amino acid changes in "hot spot" regions of FKS-encoded subunits of glucan synthase, which decreases the sensitivity of enzyme to drug (Perlin 2015).

Another drug widely used in the treatment of fungal infections is amphotericin B. It belongs to the class of polyenes and causes direct damage to the plasma membrane through interaction with ergosterol, which generates extravasation of intracellular components such as magnesium, potassium and sugars, leading to cell death (Mesa-Arango et al. 2012). Amphotericin B may be used in combination therapy with fluconazole without antagonism, probably because its mechanism of action occurs on the cell surface and is not related to the activity of efflux mechanisms (Butts et al. 2019). However, some genetic mutations may result in fungal resistance to this drug. For example, enzymes encoded by the *ERG2* and *ERG3* genes influence the conversion of fecosterol, and consequently the amount of ergosterol required for the action of amphotericin B (de Oliveira Santos et al. 2018). Another mechanism of resistance involves increased catalase activity, causing reduction of oxidative damage to cells (Kanafani and Perfect 2008). Overexpression of *MDR1* and *FKS1* resistance genes in *C. albicans* biofilms exposed to amphotericin B has also been demonstrated (Watamoto et al. 2011).

Finally, allylamines and griseofulvin can also be used to fight fungal infections. While the first ones act by inhibiting enzymes involved in the synthesis of ergosterol, griseofulvin inhibits fungal cell mitosis (de Oliveira Santos et al. 2018). For these antimicrobials, in addition to genetic modifications, cells can develop resistance mechanisms associated with stress adaptation. In this sense, when there is inhibition of one or more components of the cell wall, the microorganism is able to compensate them by increasing the production of another component (e.g. such as chitin) in order to maintain its survival (Pfaller et al. 1989; Chamilos et al. 2007).

7.4   Nanoparticles Used to Manage *Candida* Species

Considering the increasing levels of resistance across the spectrum of antifungals, then there is a need for new and novel agents to augment these fungal active drugs. Some of them are presented below.

7.4.1   Gold Nanoparticles

In addition to the use in thermal therapies and cancer treatment, gold NP (Au-NP) have been tested as an alternative antimicrobial in the fight against planktonic cells and different *Candida* species biofilms.

The shape and size of Au-NP may influence their antifungal effect. Au nanocubes showed superior antifungal effect on *C. albicans*, *C. glabrata* and *C. tropicalis* compared to nanospheres and nanowires (Jebali et al. 2014). The presence of 12 edges in nanocubes probably favours the antifungal effect by the greater number of active faces in this type of nanoparticle (Jebali et al. 2014). For *C. glabrata*, *C. albicans*, *C. tropicalis* and *C. krusei*, Au nanodiscs with an average diameter of 25 nm resulted in lower values of minimum inhibitory concentration (MIC, 16–32 µg/mL) than those found for nanocrystals polyhedral with 30 nm (32–128 µg/mL) (Wani et al. 2013). These results may be associated with the larger surface area of smaller NP, which favours their interaction with plasma membrane binding sites (Wani et al. 2013). The smaller size of the NP can also favour their diffusion through the membrane into the microbial cell (Wani et al. 2013).

It is believed that Au-NP can strongly interact with proteins containing sulphur in their plasma membrane or phosphorus molecules in the DNA, leading to disorders in cellular functions and death of the microorganism (Tan et al. 2011). Using *C. albicans* as a model, it was found that Au-NP destroy the cell nucleus, the nucleic acids, and weaken mitochondrial homeostasis, generating apoptosis not linked to the production of reactive oxygen species (Seong and Lee 2018). In this case, the apoptotic response was related to the release of mitochondrial cytochrome C in the cytosol and the metacaspase activation (Seong and Lee 2018).

In biofilm tests, Au-NP colloidal suspensions with an average size of 10–20 nm promoted significant reductions (> 80%) in *C. albicans* biofilm metabolism when applied at concentrations equal to or greater than 20 ppm (Yu et al. 2016). These nanoparticles can also be used as carriers of antimicrobial peptides for the treatment of hospital-acquired infections associated with *Candida* biofilms. Therefore, when Au-NP were coated with the indolicidin peptide, the nanocomplex significantly reduced the metabolic activity of *C. albicans* biofilms at different development stages (24 and 48 h); it was even more effective than Au-NP and indolicidin alone (de Alteriis et al. 2018). The nanocomplex was also shown to decrease the expression of the genes *EFG1*, *HWP1*, *ALS1* and *ALS3*, which are associated to biofilm formation, development of hyphae and cell-cell adherence (de Alteriis et al. 2018).

Au-NP are also used in photodynamic therapy in order to enhance the antimicrobial effect of photosensitizing agents. The conjugation of Au-NP with methylene blue photosensitizer (MB) showed higher reductions in total biomass and metabolic activity of *C. albicans* biofilms compared to MB alone (Khan et al. 2012). In addition, the combined use of Au-NP with MB and toluidine blue was able to reduce significantly the burden of *C. albicans* hyphae in oral and skin infections of mice (Sherwani et al. 2015).

7.4.2   Selenium Nanoparticles

Selenium (Se) is considered an essential element for humans, and its incorporation (via diet) results in the production of numerous selenoproteins, which have several biological functions, including immunological defence (Guisbiers et al. 2017). In addition, the NP synthesis of Se (Se-NP) has been proposed as a promising alternative in medical applications. Se-NP synthesized by pulsed femtosecond laser ablation in deionized water (25 ppm) proved to be effective in adhering to *C. albicans* biofilms, penetrating and damaging the cellular structure of this microorganism, thus, inhibiting the formation of biofilms. Interestingly, the smaller-sized particles (separated by centrifugation) and those of predominantly crystalline structure (heated at 90 ° C for 2 h, converting amorphous particles to triangular) were more effective in reducing biofilm formation, indicating that size and crystallinity are important factors to consider (Guisbiers et al. 2017).

In addition to the Se-NP chemical synthesis, these can be obtained by biosynthesis methods, which involve the culture of Se-reducing bacteria, such as *Klebsiella pneumoniae*, *Bacillus* sp. MSh-1, *Stenotrophomonas maltophilia* and *Bacillus mycoides*, all of which promote intracellular biosynthesis of Se-NP, and are subsequently released from the cytoplasm using different physicochemical methods. The biosynthesis of Se-NP by *Bacillus* sp. MSh-1 highlighted the predominant production of spherical NP between 120 and 140 nm, whose MIC was determined at 70 µg/mL for *C. albicans* (Shakibaie et al. 2015). The antifungal effect of these particles (synthesized by *Bacillus* sp. MSH-1) was also evaluated in strains of *C. albicans* resistant and susceptible to fluconazole, whose MIC was, respectively, determined in 100 and 70 µg/mL; the antifungal effect of Se-NP was associated with reduced expression of two genes *(CDR1 ERG11)* related to azole resistance (Parsameher et al. 2017). On the other hand, a much higher MIC was obtained for Se-NP synthesized by *K. pneumoniae* (2000 µg/mL); the addition of Se-NP at 1, 2 and 4 times the MIC (1 h at 25 °C) was not only ineffective in inhibiting the growth of *C. albicans*, but also stimulated the growth of this fungus (Kazempour et al. 2013). These findings may be associated with physiological alterations of *C. albicans* promoted by exposure to Se, leading the authors to assume that the limited exposure of *C. albicans* to Se-NP during the treatment of some fungal infections may favour their recovery.

Although Se-NP synthesized by *S. maltophilia* and *B. mycoides* were not effective in inhibiting *C. albicans* clinical isolates (MIC>256 µg/mL), they were effective in reducing (60–70%) the formation of biofilms or eradicate pre-formed biofilms (45–60%) of this fungus using low concentrations (50 µg/mL), without significant improvement when higher concentrations were used (Cremonini et al. 2016). It is interesting that the MICs obtained by biosynthesis for both Se-NP were smaller than those observed for NP chemically obtained, which is possibly due to the role of biogenic Se-NP proteins associated with their antifungal action. Possibly, the interaction of the organic cover of the biogenic Se-NP with the outer layer of the *C. albicans* cell wall promotes the permeabilization of the cell wall, with subsequent disaggregation (Cremonini et al. 2016).

Se-NP were also effective in reducing *Candida* biofilms when co-administered with probiotic bacteria. *Lactobacillus plantarum* and *Lactobacillus johnsonii* grown in the presence of Se-NP promoted an increase in their antifungal potential in inhibition halo experiments on plates containing *C. albicans* (Kheradmand et al. 2014). Inhibitory effects were observed both for the treatment with the suspensions of these microorganisms (containing NP-Se) and for the culture medium of these bacteria after the removal of these by filtration, which promoted significant reductions in the biofilm biomass of *C. albicans*.

7.4.3 Silver Nanoparticles

The broad spectrum of activity against bacteria, fungi, protozoa and viruses is the main advantage of silver (Ag) use in the control of biofilms. The use of Ag-NP has been proposed to increase the antimicrobial effects and to reduce side effects of Ag (Monteiro et al. 2009; Pokrowiecki et al. 2017). It was demonstrated that Ag-NP with a mean diameter of 5 nm shows a fungicidal effect on planktonic cells of *C. albicans* and *C. glabrata* using very low concentrations (0.4–3.3 µg/mL). Such effects are more pronounced on the biomass of biofilms in early stages of formation

(2 h) compared to mature biofilms (48 h) (Monteiro et al. 2011). The exception was for the *C. glabrata* strain, for which Ag-NP showed reductions around 90% in both cases (2 or 48 h). A similar pattern was observed for cell viability, since Ag-NP showed to be more effective on *C. glabrata* compared to *C. albicans*, in both situations (2 or 48 h). Among the factors associated with the mechanism of action of Ag-NP on fungal cells, we highlight interactions with cell membrane proteins (leading to membrane damage and reduction in the level of intracellular ATP), as well as with the DNA of these microorganisms (consequently affecting cell multiplication) (Monteiro et al. 2009). In addition, Ag-NP may affect the expression of genes involved in the transformation of yeasts into hyphae in *C. albicans*, reducing the virulence of this species (Halbandge et al. 2019).

Regarding the size and shape of Ag-NP, previous studies have shown that both factors should be considered, since smaller particles with a triangular shape have a superior biocidal effect on planktonic cells when compared with larger particles or with spherical or rod-like shapes (Baker et al. 2005; Pal et al. 2007). However, the size of the NP (5, 10 or 60 nm) and the stabilizing agent (ammonia or polyvinylpyrrolidone) did not influence the antifungal activity of Ag-NP on mature biofilms of *C. albicans* and *C. glabrata* (Monteiro et al. 2012), which is probably due to the agglomeration of Ag-NP when in contact with biofilms (Fig. 7.2). Thus, it is possible that these factors (particle size and stabilizing agent) represent secondary aspects related to the fungicidal effect of Ag-NP on mature *Candida* biofilms.

In addition to the factors mentioned above, the antifungal effect of Ag-NP does not seem to be influenced by the stability of the treatment solutions or by the stage of *Candida* biofilm formation. Different temperatures (50 °C, 70 °C or 100 °C), pH (5.0 or 9.0) and contact time with the biofilm (5 h or 24 h) of 54 mg/L Ag-NP colloidal suspensions were tested on mature (48 h) biofilms of *C. albicans* and *C. glabrata* formed on acrylic resin. It was shown that these parameters did not affect the biofilm susceptibility and cellular viability (Monteiro et al. 2014). In addition, the effect of Ag-NP colloidal suspensions on biofilms formed on acrylic resin was not

Fig. 7.2 (a) Transmission electron microscopy image of spherical silver nanoparticles (diameter of 15–20 nm) synthesized via reduction of silver nitrate by sodium citrate; (b) scanning electron microscopy image showing clusters of silver nanoparticles (black arrows) in contact with a 48-h *Candida albicans* biofilm

affected by the biofilm formation stage (24 or 48 h) in terms of total biomass and cell viability (Monteiro et al. 2015).

It is important to point out that similar effects were observed for Ag-NP and conventional antifungal drugs. Single- and dual-species biofilms of *C. albicans* and *C. glabrata* formed on acrylic resin were treated with Ag-NP or nystatin, and no significant differences were observed between the biomass and cell viability in these therapies (Silva et al. 2013). In addition, the combined use of Ag-NP with nystatin or chlorhexidine promoted a synergistic antifungal effect on biofilms of *C. albicans* and *C. glabrata*, depending on the strain and concentration of the drugs (Monteiro et al. 2013). Such synergism is due to the different mechanisms of action of each drug on the analysed biofilms; it is a clinically relevant aspect to consider the possibility of using lower concentrations of the analysed drugs.

Although the effect of Ag-NP has been more frequently investigated in colloidal solutions, there is evidence that their incorporation into acrylic resins may be an effective strategy to reduce denture stomatitis, which is related to the formation of biofilms on these materials. The addition of Ag-NP (0.05%) to acrylic resins was effective in inhibiting the formation of *C. glabrata* biofilms on the resin surface. It directly influences the distribution and dispersion of the nanoparticles on the resin polymer matrix (de Souza-Neto et al. 2019).

7.4.4 Copper Nanoparticles

Among the metallic NP with antimicrobial activity on bacteria and fungi, copper (Cu) deserves to be mentioned, mainly due to its higher availability compared to other noble metals (Ag and Au). The literature shows that NP of Cu (Cu-NP) synthesized by chemical route in the presence of chitosan (CS) present antifungal activity on *C. albicans*, determined by the disc-diffusion method (Usman et al. 2013). Although inhibition halos of NP (~ 9 mm) are lower than those found for nystatin (22 mm), a major advantage of this synthesis method is that the CS acts as a stabilizer of NP, protecting them from aggregation and oxidation (Usman et al. 2013). Another relevant factor is the concentration of CS used in the synthesis, since this affects the size of NP and the rate of microbial growth (Usman et al. 2013).

When synthesized by chemical reduction of $CuSO_4.5H_2O$ with hydrazine monohydrate in micellar solution of sodium dodecyl sulphate, Cu-NP (50 nm) exhibited antifungal activity on *C. albicans* and *C. parapsilosis*, although with a higher MIC value (3.75 µg/mL) compared to fluconazole (1–2 µg/mL) (Kruk et al. 2015). Among the Cu compounds, colloidal suspensions of NP of Cu oxide (CuO-NP) with a mean size of 40 nm inhibited the growth of *C. albicans*, *C. glabrata* and *C. krusei* at concentrations ranging from 1 to 1000 µg/mL, with a MIC of 1000 µg/mL for the three species (Amiri et al. 2017). In addition, CuO-NP (30–50 nm, spherical shape) have the ability to reduce the production of *Candida* biofilms, especially when fungal cells are pretreated with NP, and completely inhibit (100%) the formation of germ tubes at a concentration of 300 mg/L compared to fluconazole (32 mg/L; ~ 80% inhibition), indicating the nanotherapy potential in reducing the pathogenicity of *Candida* (Mudiar and Kelkar-Mane 2018). It is believed that the production of reactive oxygen species, such as H_2O_2, is one of the mechanisms responsible for the antimicrobial activity of CuO-NP, in addition to damage in the cytoplasmic membrane. Furthermore, Cu ions (Cu^{2+}) can penetrate into the microbial cell and connect to DNA fragments, suppressing the nucleic acid chains (El-Batal et al. 2018).

Biosynthesis or "green" synthesis has also been used to obtain CuO-NP as a less toxic and greener alternative. Using *Acalypha indica* leaf extract, spherical CuO-NP with a size ranging from 26 to 30 nm were obtained, and these NP showed antifungal effect on planktonic cells of *C. albicans* at 25 μg/mL concentration (Sivaraj et al. 2014). Furthermore, Cu-NP synthesized by natural polysaccharides (pectin and alginate) and aqueous extract of *Pleurotus ostreatus* showed strong antimicrobial activity on microorganisms associated with burn skin infections, such as *C. albicans* (El-Batal et al. 2018). For this fungus, the MIC value for Cu-NP was 1.93 μg/mL, with an antimicrobial effect superior to that observed for nystatin (El-Batal et al. 2018).

Otherwise, Cu-NP can be produced simultaneously with polymers, such as polyaniline, forming a composite in which NP are uniformly dispersed in the polymer and present synergistic antifungal activity on *C. albicans* (Bogdanović et al. 2015).

7.4.5   Bismuth Nanoparticles

Bismuth (Bi) can be found in various forms, including Bi (bismuthinite) sulphide, Bi (bismuthite) carbonate and Bi oxide (Bi_2O_3; bismite). In the form of subsalicylate, Bi has been used to control of stomach pains, nausea and vomiting (Hernandez-Delgadillo et al. 2013). Bi_2O_3 is also found in glass and ceramic products, as a catalyst in the oxidation of hydrocarbons, in microelectronics and in optical technologies (Hernandez-Delgadillo et al. 2013; El-Batal et al. 2017). In addition, Bi_2O_3 in nanoparticulate form has been tested as an alternative antimicrobial against bacteria and fungi.

For *C. albicans*, NP of Bi_2O_3 (Bi_2O_3-NP; 77 nm, needle form) revealed a MIC of 1.5 mM (Hernandez-Delgadillo et al. 2013). At the 2 mM concentration, these NP reduced the growth of *C. albicans* by 85% compared to the untreated group, being more effective than chlorhexidine (44%), terbinafine (51%) and non-nanoparticulate Bi_2O_3 (Hernandez-Delgadillo et al. 2013). However, for *C. albicans* biofilms analysed by fluorescence microscopy, Bi_2O_3-NP were as effective as chlorhexidine and terbinafine in inhibiting biofilm formation (Hernandez-Delgadillo et al. 2013). In aqueous suspension, Bi_2O_3-NP present highly negative electrical surface potential, forming strongly basic aggregation points that lead to the death of *C. albicans* (Hernandez-Delgadillo et al. 2013).

With regard to material incorporation, lipophilic NP of Bi (Bi-NP, spherical shape, diameter of 29.3 nm) have been associated with mineral trioxide aggregate (MTA), a biomaterial used in dentistry to seal dental perforations. These NP at 100 mg/mL showed high antifungal activity on *C. albicans*, with inhibition halo of 23 mm (Hernandez-Delgadillo et al. 2017). When MTA was supplemented with

Bi-NP (100 mg/mL), the nanocomposite created was also able to inhibit fungal growth in single culture (24 mm halo) and mixed with *Enterococcus faecalis* and *Escherichia coli* (22 mm halo), and the physical properties of the material (roughness and microhardness) were not affected. In contrast, MTA alone was not able to inhibit microbial growth (Hernandez-Delgadillo et al. 2017).

7.4.6   Zinc Oxide Nanoparticles

Zinc oxide (ZnO), as well as several metal oxides, has been used as NP and studied for its antibiofilm potential. Its effectiveness has been proven on several microorganisms, including fungi *C. albicans* (Cierech et al. 2016; Dananjaya et al. 2018; Hosseini et al. 2018) and *C. tropicalis* (Jothiprakasam et al. 2017), and also in cases of strains resistant to antifungal, such as fluconazole. In addition, it is a non-toxic and biocompatible material (Dananjaya et al. 2018).

The treatment of *C. tropicalis* biofilms with NP of ZnO (ZnO-NP) and ethylene-diamine tetra-acetic acid (EDTA, chelating agent that influences filamentation in *C. albicans* biofilms) has shown interesting results. It was found that spherical ZnO-NP (5–50 nm) and EDTA suppressed 48-h biofilms of *C. tropicalis* strains that are susceptible and resistant to fluconazole at concentrations ranging from 5.2 to 10.8 µg/mL (Jothiprakasam et al. 2017). Spherical ZnO-NP (20–50 nm) also showed antifungal effects on strains of *C. albicans* isolated from urinary catheters with mean MIC values of 28 and 47 µg/mL, respectively, for fluconazole-susceptible and -resistant isolates (Hosseini et al. 2018). For biofilms developed on the urinary catheter surface, ZnO-NP were effective in reducing the total biomass and number of cultivable cells; the reductive effect on the biofilm biomass of the resistant strains is greater than on the susceptible ones. This shows that ZnO-NP are promising in the treatment of urinary tract infections (Hosseini et al. 2018).

The antimicrobial mechanism of action of ZnO-NP is multifactorial and includes direct interaction of the NP with microbial cells or the release of cations (Zn^{2+}) from its surface, cell membrane penetration, increased induction of oxidative stress and increased production of reactive oxygen species, which leads to damage to the cell membrane and to intracellular components (Cierech et al. 2016; Dananjaya et al. 2018). On the other hand, the antifungal effect of ZnO-NP can be maximized when used in combination with other compounds. That way, the association of these NP with CS has shown lower MIC values on *C. albicans* (75 µg/mL) compared to ZnO-NP alone (200 µg/mL), highlighting the possibility of synergistic effects for biomedical applications (Dananjaya et al. 2018).

ZnO-NP have been incorporated into dental materials, such as polymethylmethacrylate (PMMA) used in the preparation of denture bases, aiming at the prevention of biofilm formation and the development of denture stomatitis. When PMMA was coated (by spraying) with ZnO-NP at 2.5%, 5% and 7.5% concentrations, the generated nanocomposites significantly reduced the biofilm formation of *C. albicans* compared to the control group (PMMA without ZnO-NP), with a dose-dependent effect (Cierech et al. 2016). Furthermore, coating with 7.5% ZnO-NP increased

PMMA hardness and hydrophobicity without significant changes in its roughness (Cierech et al. 2018).

7.4.7   Boron Nitride Nanoparticles

Colloidal suspensions of NP of boron nitride (BN-NP) have been tested on oral pathogens for their use in oral hygiene products. In this sense, BN-NP with mean diameter of 121 nm were able to inhibit the growth (fungiostatic effect) of *Candida* sp. M25 in the planktonic state at a concentration of 0.00325 mg/mL (Kıvanç et al. 2018). Concentrations above the minimum inhibitory did not promote significant reductions in the formation of *Candida* biofilms. However, when applied to 24-h pre-formed fungal biofilms, BN-NP led to significant reductions in total biomass at concentrations ranging from 0.8 to 0.00156 mg/mL (Kıvanç et al. 2018). Probably, the best effect on pre-formed biofilms is related to the ability of these NP to bind themselves to the extracellular polysaccharide, reducing sugar consumption of microorganisms (Kıvanç et al. 2018).

7.4.8   Nanoparticles of *Melaleuca alternifolia*

Plants and their extracts are a viable option to obtain alternative drugs in the fight against *Candida* biofilms, since these natural compounds have an antifungal potential and may be less toxic to human cells. The nanostructure of these compounds can improve their antimicrobial efficiency and stability, while reducing possible side effects.

Melaleuca alternifolia oil (MAO), also known as tea tree, has been nanoparticulated and tested on polymicrobial biofilms and different *Candida* species. For polymicrobial oral biofilms formed in situ, 0.3% MAO NP (MAO-NP) were more effective in reducing cell viability than chlorhexidine (0.12%) and non-nanoparticulate MAO (0.3%) (de Souza et al. 2017). Moreover, at concentrations of 3.9–31.2%, MAO-NP (158 nm) significantly reduced total biomass of pre-formed biofilms (24 h) of *C. albicans*, *C. glabrata*, *C. parapsilosis*, *C. tropicalis* and *Candida membranifaciens* (Souza et al. 2017). For *C. albicans* and *C. glabrata*, reductions in biomass reached values of 67 and 72%, respectively. Proteins and exopolysaccharides from biofilms were also reduced after treatment with MAO-NP (Souza et al. 2017). The antifungal effect of this essential oil may be related to the reduction of surface hydrophobicity of fungal cells, decreasing their adhesion capacity (Sudjana et al. 2012), or to the inhibition the formation of germ tubes during biofilm formation (Hammer et al. 2002). Comparing equal concentrations of MAO in micro and nanoparticulate forms, the NP had a higher reduction potential for *C. albicans* biofilms (75% versus 60%) (Souza et al. 2017). This increase in the antibiofilm activity can be explained by the smaller size of the NP, which facilitates their penetration into the cells and in the biofilm matrix, leading to membrane rupture and cell death (Seil and Webster 2012).

7.4.9   Chitosan Nanoparticles

CS is a polysaccharide derived from chitin, naturally originating from the shell of crustaceans. The antifungal and bactericidal activities of this compound are widely reported, as well as its haemostatic properties (Ardila et al. 2017; Costa et al. 2017; Gondim et al. 2018; Hu et al. 2018). CS also shows higher biocompatibility and biodegradability than synthetic polymers (Rabea et al. 2003). Added to all these characteristics, the literature points out the low toxicity of CS as one of its main advantages (Kong et al. 2010; Elieh-Ali-Komi and Hamblin 2016; Regiel-Futyra et al. 2017). However, pH is a determining factor for its solubility, since it is insoluble at neutral pH and positively charged at acidic pH (Hejazi and Amiji 2003; Kalliola et al. 2017). This aspect, associated with its high viscosity, may limit the application of CS, which has stimulated alternatives aiming for the modification of its preparation and solubilization for clinical applications (Lin Teng Shee et al. 2006; Li et al. 2009).

In the last few years, different theories have been proposed about the exact mechanism of action of CS and its derivatives. It has been considered that positively charged CS molecules interact with negatively charged cell membranes, leading to extravasation of proteinaceous and intracellular constituents, causing cell death (Rabea et al. 2003; Kong et al. 2010; Ma et al. 2017). After release of the cell wall by enzymes of the pathogen itself, CS is able to penetrate the nucleus of the fungal cell, interfering in the synthesis of mRNA and proteins (Muzzarelli et al. 1986). Moreover, a recent study proposed that a potential antifungal effect of CS against *C. albicans* occurs through the inhibition of Spt-Ada-Gcn5-acetyltransferase (SAGA) complex gene expression, which leads to the alteration of the cell surface integrity, decreasing its protection against CS (Shih et al. 2019).

NP-based CS (CS-NP) nanosystems have been developed, showing clinically promising results. In vivo tests with CS-NP associated with miconazole nitrate revealed an efficacy in the treatment of vulvovaginal candidiasis similar to miconazole alone, but at a seven-fold lower concentration (Amaral et al. 2019). In vitro tests for the formation of *Candida* biofilms on acrylic resin for dentures showed that CS-NP inhibited the formation of *Candida* biofilms similarly to sodium hypochlorite, but at a lower concentration and causing minor changes in hardness and roughness of the material (Gondim et al. 2018). CS-NP (40–100 nm) have also been incorporated into a tissue conditioner used in total dentures, being able to completely inhibit the growth of *C. albicans* at 5% concentration after 24 and 48 h in contact with the microorganism (Mousavi et al. 2018).

It is important to point out that the CS molecular weight plays a key role in its clinical performance. When associated with simvastatin, low, medium and high molecular weight CS-NP performed differently in drug release in human nasal cell line, with slower release of simvastatin to higher molecular weight CS (Bruinsmann et al. 2019). In the same sense, Se-NP associated with CS with different molecular weights indicated higher antioxidant activity when associated with higher molecular weight (510 kDa) CS, highlighting the importance of adequate planning for the design of CS-based nanosystems (Chen et al. 2019).

7.4.10 Iron Oxide Magnetic Nanoparticles

Iron oxide magnetic NP (IONPs) have been tested in several biomedical applications, such as magnetic resonance (Sousa et al. 2017), hyperthermia treatment (Wang et al. 2017), cell detection and separation (Gu et al. 2006) and drug delivery (Niemirowicz et al. 2016; Slavin et al. 2017). In general, the antimicrobial action of these NP involves three mechanisms: (1) ability to destabilize the membrane's lipid bilayer, promoting cell lysis; (2) binding ability to cytosolic proteins such as DNA, triggering cell death; and (3) production of oxygen-reactive species that causes increase in oxidative stress and cellular instability (Tokajuk et al. 2017) (Fig. 7.3). Interestingly, IONPs can bypass fungal resistance through inactivation of catalase (Cat-1) (Niemirowicz and Bucki 2017).

The antibiofilm effect of IONPs depends on the species and concentration evaluated. In this context, biofilms of *C. albicans* and *C. parapsilosis* were more susceptible to IONPs (1000–4000 µg/mL) than to fluconazole (512–2048 µg/mL), whereas for biofilms of *C. krusei, C. tropicalis* and *Candida lusitaniae*, the antifungal showed significantly higher reducing effects (Salari et al. 2018). For *C. glabrata*, both compounds showed similar effects (Salari et al. 2018). On the other hand, the use of lower IONPs concentration (110 µg/mL) did not significantly reduce the density and metabolism of the *C. albicans* biofilm (Vieira et al. 2019).

IONPs-based drug delivery nanosystems have been extensively studied in recent years, mainly because of their nanometer scale size and magnetic property, which increase the penetration power of biological barriers, facilitate the delivery of drugs to specific targets and reduce the concentrations of the conjugated drug, minimizing possible side effects (Khan et al. 2015; Arias et al. 2018). In order to make these nanosystems more biocompatible for clinical applications, IONPs have been coated with different compounds, including synthetic and natural polymers (such as CS), organic surfactants, inorganic compounds and bioactive molecules (Arias et al. 2018) (Fig. 7.4).

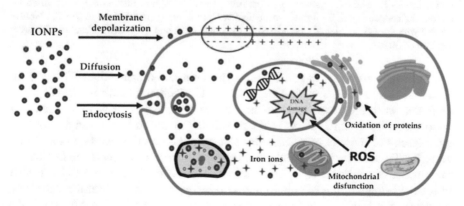

Fig. 7.3 Mechanisms of antimicrobial action of iron oxide nanoparticles. ROS: reactive oxygen species. (Source: Arias et al. 2018)

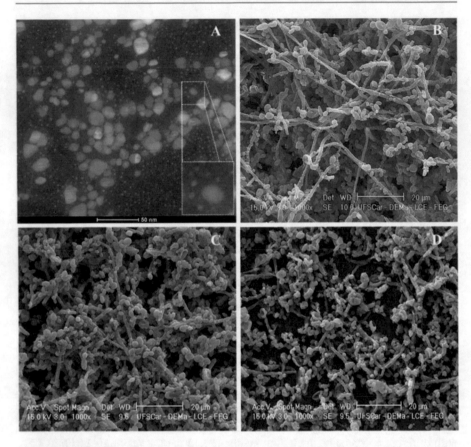

Fig. 7.4 (a) Transmission electron microscopy image illustrating a miconazole nanocarrier; image at the bottom right corner represents an increased view of iron oxide nanoparticle coated with chitosan, and miconazole particles adhered to chitosan; (b) scanning electron microscopy image of a 48-h dual-species biofilm of *Candida albicans* and *Candida glabrata* without treatment. Images (c) and (d) represent biofilms treated with 78 μg/mL miconazole or miconazole-containing nanocarrier at 78 μg/mL, respectively. Note a less compact structure for the biofilm treated with the nanocarrier (d), compared to the control (b) and miconazole alone (c). (Source: Arias et al. 2018)

Antifungal drug carrier nanosystems (nystatin and amphotericin B) based on IONPs showed a better antimicrobial effect on *C. albicans*, *C. glabrata* and *C. tropicalis* than each drug used alone (Niemirowicz et al. 2016). In this same sense, a nanocarrier based on IONPs and CS was able to provide an increase in the antifungal capacity of nystatin on *C. albicans* (Hussein-Al-Ali et al. 2014). In the dental area, IONPs coated with CS and chlorhexidine promoted significant reductions in the number of cultivable cells, total biomass and metabolic activity of *C. albicans* biofilms in single or mixed culture with *Streptococcus mutans* (Vieira et al. 2019). However, for most of the analysed parameters, the antibiofilm effect of this nanosystem did not differ significantly from that observed for chlorhexidine (Vieira et al. 2019).

7.4.11 Titanium Dioxide Nanoparticles

The antimicrobial effect of NP of titanium dioxide (TiO_2-NP) has been demonstrated in several studies evaluating their action on different strains of bacteria and fungi. These NP can be obtained by biosynthesis or from conventional chemical routes. Chemically synthesized TiO_2-NP were tested for their antifungal effect using different sizes (obtained by calcination at temperatures ranging from 400 to 800 °C) and concentrations. The resulting NP presented a superior antifungal effect compared to that of chloramphenicol against *C. albicans* (reduction in the number of colony forming units (CFUs)) in the presence of visible light; the size and concentration of TiO_2-NP were shown to be important factors that influence the antimicrobial effect (Priyanka et al. 2016).

Regarding the obtention of TiO_2-NP from biosynthesis methods, *Morinda citrifolia* leaf extract was effective in producing spherical nanoparticles (15–19 nm) with low amount of anthraquinone and phenolic compounds from leaf extract (Sundrarajan et al. 2017), and the obtained TiO_2-NP had a higher antifungal effect (zone of inhibition) against *C. albicans* than that observed for ketoconazole. In addition, TiO_2-NP biosynthesis can also be performed using $TiCl_3$ and biological yeast (ascomycete class), generating small NP (6.7 ± 2.2 nm) with a predominantly spherical shape (Peiris et al. 2018). It was demonstrated that TiO_2-NP had a potent antimicrobial effect on *C. albicans* (reduction of the number of CFU) both in presence and absence of ambient light, suggesting their potential as anti-*Candida* agents in future applications (Peiris et al. 2018).

Considering the above-mentioned advantages of TiO_2-NP on *C. albicans* in planktonic state, it was risen the possibility of evaluating the effect of these particles on the formation of biofilms of two strains of *C. albicans* (resistant and susceptible to antifungals) on the surface of catheters, PVC and glass. It was shown that the treatment with TiO_2-NP (70–100 nm) on the surface of these materials promoted a significant reduction in the adhesion of *C. albicans* whose biofilms were less aggregate compared to non TiO_2-NP-treated surfaces, especially for surfaces of catheter and glass (Haghighi et al. 2012).

In Addition to the TiO_2-NP effect directly on fungal cells, the addition of TiO_2-NP to dental materials has also proved to be an effective strategy to control *Candida* species. It has been recently demonstrated that the incorporation of TiO_2-NP and cellulose nanocrystals into a conventional glass ionomer cement was effective in reducing the number of *C. albicans* CFU, as well as increasing the mechanical properties of the material, without compromising the cytotoxic effect of the material when compared to the material without the addition of TiO_2-NP and cellulose (Sun et al. 2019). In addition, the incorporation of TiO_2-NP into acrylic resins used in the manufacture of "3D printing" (i.e. stereolithography) complete dentures was recently tested against *Candida scotti*; it has been observed that samples containing

0.4, 1 and 2,5% of TiO$_2$-NP inhibited the growth of this fungus (Totu et al. 2017). However, the use of higher concentrations (1 and 2.5%) promoted the agglomeration of NP, as opposed to the higher dispersion observed for 0.4% TiO$_2$-NP, which may reduce the mechanical properties of the resin.

7.4.12 Silica Nanoparticles

Silicon dioxide (SiO$_2$), also known as silica, is easily found the environment (e.g. soil and rocks), including crystalline forms such as quartz. Considering its oxidative potential and subsequent intracellular damage, SiO$_2$ NP (Si-NP) have been intensively investigated for their antifungal effect on *Candida* species, with promising results in several applications. They include the use in photodynamic therapy whose association with phthalocyanines (pigments) showed a significant antimicrobial effect on *C. albicans* (Baigorria et al. 2018). As for their use in dressings, an expressive antifungal effect was observed for Si-NP associated with metacyanic acid in different strains of *Candida*, including *C. albicans, C. parapsilosis, C. glabrata, C. krusei* and *Candida famata* (Krokowicz et al. 2015).

Si-NP are also used because of their ability to carry antimicrobial drugs, which was observed both for their administration alone and when incorporated into various materials. In this sense, the functionalization of Si-NP with amphotericin B and their incorporation into composite resins promoted long-lasting antifungal activity against five strains of *Candida* (*C. albicans, C. parapsilosis, C. tropicalis, C. glabrata* and *C. krusei*), especially for smaller NP (5 nm) compared to larger ones (80 nm) (Lino et al. 2013). In addition, the incorporation of Si-NP mesoporous (Si-M-NP) to acrylic resins was effective in reducing *C. albicans* adhesion in comparison to resins without Si-M-NP, with a long-lasting antimicrobial effect (2 weeks) when associated with amphotericin B (Lee et al. 2016). In addition, it has been recently shown that Si-M-NP functionalized with phenazine-1-carboxamide (PCA) were more effective in inhibiting the formation of mixed biofilms of *C. albicans* and *Staphylococcus aureus* than the compounds (Si-M-NP and PCA) alone (Kanugala et al. 2019). These effects were observed at very low concentrations (10 µg/mL), with no biofilm formation occurring in Si-M-NP + PCA-coated catheters. The effects were due to intracellular accumulation of reactive oxygen species, reduction in membrane permeability and changes in ionic homeostasis of *C. albicans* cells (Kanugala et al. 2019).

Besides the antifungal effects of Si-NP per se, their use as nitric oxide (NO) carriers has also been investigated, given the recognized antimicrobial role of this free radical. Biofilms of *C. albicans* formed in vitro were exposed to Si-NP releasing NO, leading to the death of 99% of the cells of these biofilms caused by the NO release, with cytotoxic effects to human fibroblasts lower than those observed for conventional antiseptics (Hetrick et al. 2009).

7.5   Conclusion

In view of the scientific evidence on the subject discussed in this chapter, it is possible to conclude that different types of NP have antifungal effects on planktonic cells and biofilms of *Candida* species. These effects depend on the concentrations tested, size and shape of NP, type of strain evaluated, and association or not with other materials. The mechanism of fungicidal action of NP is multifactorial, which grants a great advantage over conventional antifungals, in view of the lower probability of emergence of strains resistant to these particles. In addition, most studies have used *C. albicans* as a model to test the effectiveness of alternative nanotherapies, so that the effect on non-*albicans Candida* species and on polymicrobial infections should be further explored, especially when the relevant role of these species in fungal infections and its mechanisms of resistance to antifungals are considered. Finally, most of the available data on the antimicrobial efficiency of NP comes from in vitro studies, which points out the need to compare the effect of these NP with conventional antifungals in randomized clinical trials, aiming at safe clinical applications for the control or prevention of different candidiasis.

References

Al-Fattani MA, Douglas LJ (2004) Penetration of *Candida* biofilms by antifungal agents. Antimicrob Agents Chemother 48:3291–3297

Alfouzan W, Dhar R, Albarrag A, Al-Abdely H (2019) The emerging pathogen *Candida auris*: A focus on the Middle-Eastern countries. J Infect Public Health. pii: S1876-0341(19)30118-2

Ali A, Ahmed S (2018) A review on chitosan and its nanocomposites in drug delivery. Int J Biol Macromol 109:273–286

Amaral AC, Saavedra PHV, Oliveira Souza AC, de Melo MT, Tedesco AC, Morais PC et al (2019) Miconazole loaded chitosan-based nanoparticles for local treatment of vulvovaginal candidiasis fungal infections. Colloids Surf B Biointerfaces 174:409–415

Amiri M, Etemadifar Z, Daneshkazemi A, Nateghi M (2017) Antimicrobial effect of copper oxide nanoparticles on some oral bacteria and *Candida* species. J Dent Biomater 4:347–352

Andes DR, Safdar N, Baddley JW, Playford G, Reboli AC, Rex JH et al (2012) Impact of treatment strategy on outcomes in patients with candidemia and other forms of invasive candidiasis: a patient-level quantitative review of randomized trials. Clin Infect Dis 54:1110–1122

Ardila N, Daigle F, Heuzey MC, Ajji A (2017) Antibacterial activity of neat chitosan powder and flakes. Molecules 22

Arendrup MC, Patterson TF (2017) Multidrug-resistant *Candida*: epidemiology, molecular mechanisms, and Treatment. J Infect Dis 216:S445–SS51

Arias LS, Pessan JP, Vieira APM, Lima TMT, Delbem ACB, Monteiro DR (2018) Iron oxide nanoparticles for biomedical applications: a perspective on synthesis, drugs, antimicrobial activity, and toxicity. Antibiotics (Basel) 7:46

Baigorria E, Reynoso E, Alvarez MG, Milanesio ME, Durantini EN (2018) Silica nanoparticles embedded with water insoluble phthalocyanines for the photoinactivation of microorganisms. Photodiagn Photodyn Ther 23:261–269

Baker C, Pradhan A, Pakstis L, Pochan DJ, Shah SI (2005) Synthesis and antibacterial properties of silver nanoparticles. J Nanosci Nanotechnol 5:244–249

Bogdanović U, Vodnik V, Mitrić M, Dimitrijević S, Škapin SD, Žunič V et al (2015) Nanomaterial with high antimicrobial efficacy--copper/polyaniline nanocomposite. ACS Appl Mater Interfaces 7:1955–1966

Borman AM, Szekely A, Johnson EM (2016) Comparative pathogenicity of United Kingdom isolates of the emerging pathogen *Candida auris* and other key pathogenic *Candida* species. mSphere 1:pii: e00189–16

Bourzac K (2012) Nanotechnology: carrying drugs. Nature 491:S58–S60

Bowman K, Leong KW (2006) Chitosan nanoparticles for oral drug and gene delivery. Int J Nanomedicine 1:117–128

Bruinsmann FA, Pigana S, Aguirre T, Souto GD, Pereira GG, Bianchera A et al (2019) Chitosan-coated nanoparticles: effect of chitosan molecular weight on nasal transmucosal delivery. Pharmaceutics 11:86

Butts A, Reitler P, Nishimoto AT, DeJarnette C, Estredge LR, Peters TL et al (2019) A systematic screen reveals a diverse collection of medications induce antifungal resistance in *Candida* species. Antimicrob Agents Chemother 63(5):pii: e00054–19

Chamilos G, Lewis RE, Albert N, Kontoyiannis DP (2007) Paradoxical effect of Echinocandins across *Candida* species *in vitro*: evidence for echinocandin-specific and candida species-related differences. Antimicrob Agents Chemother 51:2257–2259

Chandra J, Kuhn DM, Mukherjee PK, Hoyer LL, McCormick T, Ghannoum MA (2001) Biofilm formation by the fungal pathogen *Candida albicans*: development, architecture, and drug resistance. J Bacteriol 183:5385–5394

Chen W, Yue L, Jiang Q, Xia W (2019) Effect of chitosan with different molecular weight on the stability, antioxidant and anticancer activities of well-dispersed selenium nanoparticles. IET Nanobiotechnol 13:30–35

Cierech M, Kolenda A, Grudniak AM, Wojnarowicz J, Woźniak B, Gołaś M et al (2016) Significance of polymethylmethacrylate (PMMA) modification by zinc oxide nanoparticles for fungal biofilm formation. Int J Pharm 510:323–335

Cierech M, Osica I, Kolenda A, Wojnarowicz J, Szmigiel D, Łojkowski W et al (2018) Mechanical and physicochemical properties of newly formed ZnO-PMMA nanocomposites for denture bases. Nanomaterials (Basel) 8:pii: E305

Coco BJ, Bagg J, Cross LJ, Jose A, Cross J, Ramage G (2008) Mixed *Candida albicans* and *Candida glabrata* populations associated with the pathogenesis of denture stomatitis. Oral Microbiol Immunol 23:377–383

Costa SF, Marinho I, Araújo EA, Manrique AE, Medeiros EA, Levin AS (2000) Nosocomial fungaemia: a 2-year prospective study. J Hosp Infect 45:69–72

Costa EM, Silva S, Vicente S, Neto C, Castro PM, Veiga M et al (2017) Chitosan nanoparticles as alternative anti-staphylococci agents: bactericidal, antibiofilm and antiadhesive effects. Mater Sci Eng C Mater Biol Appl 79:221–226

Cremonini E, Zonaro E, Donini M, Lampis S, Boaretti M, Dusi S et al (2016) Biogenic selenium nanoparticles: characterization, antimicrobial activity and effects on human dendritic cells and fibroblasts. Microb Biotechnol 9:758–771

Dananjaya SHS, Kumar RS, Yang M, Nikapitiya C, Lee J, De Zoysa M (2018) Synthesis, characterization of ZnO-chitosan nanocomposites and evaluation of its antifungal activity against pathogenic *Candida albicans*. Int J Biol Macromol 108:1281–1288

de Alteriis E, Maselli V, Falanga A, Galdiero S, Di Lella FM, Gesuele R et al (2018) Efficiency of gold nanoparticles coated with the antimicrobial peptide indolicidin against biofilm formation and development of *Candida* spp. clinical isolates. Infect Drug Resist 11:915–925

de Oliveira Santos GC, Vasconcelos CC, Lopes AJO, de Sousa Cartagenes MDS, Filho A, do Nascimento FRF et al (2018) *Candida* infections and therapeutic strategies: mechanisms of action for traditional and alternative agents. Front Microbiol 9:1351

de Souza ME, Clerici DJ, Verdi CM, Fleck G, Quatrin PM, Spat LE et al (2017) Antimicrobial activity of *Melaleuca alternifolia* nanoparticles in polymicrobial biofilm *in situ*. Microb Pathog 113:432–437 de Souza-Neto FN, Sala RL, Fernandes RA, Xavier TPO, Cruz SA, Paranhos CM et al (2019) Effect of synthetic coloidal nanoparticles in acrylic resin of dental use. Eur Polym J 112:531–538

El-Batal AI, El-Sayyad GS, El-Ghamry A, Agaypi KM, Elsayed MA, Gobara M (2017) Melanin-gamma rays assistants for bismuth oxide nanoparticles synthesis at room temperature for enhancing antimicrobial, and photocatalytic activity. J Photochem Photobiol B 173:120–139

El-Batal AI, Al-Hazmi NE, Mosallam FM, El-Sayyad GS (2018) Biogenic synthesis of copper nanoparticles by natural polysaccharides and *Pleurotus ostreatus* fermented fenugreek using gamma rays with antioxidant and antimicrobial potential towards some wound pathogens. Microb Pathog 118:159–169

Elieh-Ali-Komi D, Hamblin MR (2016) Chitin and chitosan: production and application of versatile biomedical nanomaterials. Int J Adv Res (Indore) 4:411–427

Elving GJ, van Der Mei HC, Busscher HJ, van Weissenbruch R, Albers FW (2001) Air-flow resistances of silicone rubber voice prostheses after formation of bacterial and fungal biofilms. J Biomed Mater Res 58:421–426

Ganguly S, Mitchell AP (2011) Mucosal biofilms of *Candida albicans*. Curr Opin Microbiol 14:380–385

Gendreau L, Loewy ZG (2011) Epidemiology and etiology of denture stomatitis. J Prosthodont 20:251–260

Gonçalves B, Ferreira C, Alves CT, Henriques M, Azeredo J, Silva S (2016) Vulvovaginal candidiasis: epidemiology, microbiology and risk factors. Crit Rev Microbiol 42:905–927

Gondim BLC, Castellano LRC, de Castro RD, Machado G, Carlo HL, Valenca AMG et al (2018) Effect of chitosan nanoparticles on the inhibition of Candida spp. biofilm on denture base surface. Arch Oral Biol 94:99–107

Gu H, Xu K, Xu C, Xu B (2006) Biofunctional magnetic nanoparticles for protein separation and pathogen detection. Chem Commun (Camb) 9:941–949

Guisbiers G, Lara HH, Mendoza-Cruz R, Naranjo G, Vincent BA, Peralta XG et al (2017) Inhibition of *Candida albicans* biofilm by pure selenium nanoparticles synthesized by pulsed laser ablation in liquids. Nanomedicine 13:1095–1103

Haegler P, Joerin L, Krähenbüll S, Bouitbir J (2017) Hepatocellular toxicity of imidazole and triazole antimycotic agents. Toxicol Sci 157:183–195

Haghighi F, Mohammadi SR, Mohammadi P, Eskandari M, Hosseinkhani S (2012) The evaluation of *Candida albicans* biofilms formation on silicone catheter, PVC and glass coated with titanium dioxide nanoparticles by XTT method and ATPase assay. Bratisl Lek Listy 113:707–711

Halbandge SD, Jadhav AK, Jangid PM, Shelar AV, Patil RH, Karuppayil SM (2019) Molecular targets of biofabricated silver nanoparticles in *Candida albicans*. J Antibiot (Tokyo), [epub ahead of print]. https://doi.org/10.1038/s41429-019-0185-9

Hammer KA, Carson CF, Riley TV (2002) *In vitro* activity of *Melaleuca alternifolia* (tea tree) oil against dermatophytes and other filamentous fungi. J Antimicrob Chemother 50:195–199

Healey KR, Perlin DS (2018) Fungal resistance to echinocandins and the MDR phenomenon in *Candida glabrata*. J Fungi (Basel) 4

Hejazi R, Amiji M (2003) Chitosan-based gastrointestinal delivery systems. J Control Release 89:151–165

Hernandez-Delgadillo R, Velasco-Arias D, Martinez-Sanmiguel JJ, Diaz D, Zumeta-Dube I, Arevalo-Niño K et al (2013) Bismuth oxide aqueous colloidal nanoparticles inhibit *Candida albicans* growth and biofilm formation. Int J Nanomedicine 8:1645–1652

Hernandez-Delgadillo R, Del Angel-Mosqueda C, Solís-Soto JM, Munguia-Moreno S, Pineda-Aguilar N, Sánchez-Nájera RI et al (2017) Antimicrobial and antibiofilm activities of MTA supplemented with bismuth lipophilic nanoparticles. Dent Mater J 36:503–510

Hetrick EM, Shin JH, Paul HS, Schoenfisch MH (2009) Anti-biofilm efficacy of nitric oxide-releasing silica nanoparticles. Biomaterials 30:2782–2789

Hirota K, Yumoto H, Sapaar B, Matsuo T, Ichikawa T, Miyake Y (2017) Pathogenic factors in *Candida* biofilm-related infectious diseases. J Appl Microbiol 122:321–330

Hosseini SS, Ghaemi E, Koohsar F (2018) Influence of ZnO nanoparticles on *Candida albicans* isolates biofilm formed on the urinary catheter. Iran J Microbiol 10:424–432

Hu Z, Zhang DY, Lu ST, Li PW, Li SD (2018) Chitosan-based composite materials for prospective hemostatic applications. Mar Drugs 16

Hussein-Al-Ali SH, El Zowalaty ME, Kura AU, Geilich B, Fakurazi S, Webster TJ et al (2014) Antimicrobial and controlled release studies of a novel nystatin conjugated iron oxide nanocomposite. Biomed Res Int 2014:651831

Jabra-Rizk MA, Falkler WA, Meiller TF (2004) Fungal biofilms and drug resistance. Emerg Infect Dis 10:14–19

Jebali A, Hajjar FH, Pourdanesh F, Hekmatimoghaddam S, Kazemi B, Masoudi A et al (2014) Silver and gold nanostructures: antifungal property of different shapes of these nanostructures on *Candida* species. Med Mycol 52:65–72

Jones EM, Cochrane CA, Percival SL (2015) The effect of pH on the extracellular matrix and biofilms. Adv Wound Care (New Rochelle) 4:431–439

Jothiprakasam V, Sambantham M, Chinnathambi S, Vijayaboopathi S (2017) *Candida tropicalis* biofilm inhibition by ZnO nanoparticles and EDTA. Arch Oral Biol 73:21–24

Kahan DM, Braman D, Slovic P, Gastil J, Cohen G (2009) Cultural cognition of the risks and benefits of nanotechnology. Nat Nanotechnol 4:87–90

Kalliola S, Repo E, Srivastava V, Heiskanen JP, Sirvio JA, Liimatainen H et al (2017) The pH sensitive properties of carboxymethyl chitosan nanoparticles cross-linked with calcium ions. Colloids Surf B Biointerfaces 153:229–236

Kanafani ZA, Perfect JR (2008) Antimicrobial resistance: resistance to antifungal agents: mechanisms and clinical impact. Clin Infect Dis 46:120–128

Kanugala S, Jinka S, Puvvada N, Banerjee R, Kumar CG (2019) Phenazine-1-carboxamide functionalized mesoporous silica nanoparticles as antimicrobial coatings on silicone urethral catheters. Sci Rep 9:6198

Kazempour ZB, Yazdi MH, Rafii F, Shahverdi AR (2013) Sub-inhibitory concentration of biogenic selenium nanoparticles lacks post antifungal effect for *Aspergillus niger* and *Candida albicans* and stimulates the growth of *Aspergillus niger*. Iran J Microbiol 5:81–85

Kean R, Sherry L, Townsend E, McKloud E, Short B, Akinbobola A et al (2018) Surface disinfection challenges for *Candida auris*: an *in-vitro* study. J Hosp Infect 98:433–436

Khan S, Alam F, Azam A, Khan AU (2012) Gold nanoparticles enhance methylene blue-induced photodynamic therapy: a novel therapeutic approach to inhibit *Candida albicans* biofilm. Int J Nanomedicine 7:3245–3257

Khan I, Khan M, Umar MN, Oh DH (2015) Nanobiotechnology and its applications in drug delivery system: a review. IET Nanobiotechnol 9:396–400

Kheradmand E, Rafii F, Yazdi MH, Sepahi AA, Shahverdi AR, Oveisi MR (2014) The antimicrobial effects of selenium nanoparticle-enriched probiotics and their fermented broth against *Candida albicans*. Daru 22:48

Kıvanç M, Barutca B, Koparal AT, Göncü Y, Bostancı SH, Ay N (2018) Effects of hexagonal boron nitride nanoparticles on antimicrobial and antibiofilm activities, cell viability. Mater Sci Eng C Mater Biol Appl 91:115–124

Kong M, Chen XG, Xing K, Park HJ (2010) Antimicrobial properties of chitosan and mode of action: a state of the art review. Int J Food Microbiol 144:51–63

Krokowicz L, Tomczak H, Bobkiewicz A, Mackiewicz J, Marciniak R, Drews M et al (2015) *In Vitro* studies of antibacterial and antifungal wound dressings comprising H_2TiO_3 and SiO_2 nanoparticles. Pol J Microbiol 64:137–142

Kruk T, Szczepanowicz K, Stefańska J, Socha RP, Warszyński P (2015) Synthesis and antimicrobial activity of monodisperse copper nanoparticles. Colloids Surf B Biointerfaces 128:17–22

Lee JH, El-Fiqi A, Jo JK, Kim DA, Kim SC, Jun SK et al (2016) Development of long-term antimicrobial poly(methyl methacrylate) by incorporating mesoporous silica nanocarriers. Dent Mater 32:1564–1574

Li H, Liu J, Ding S, Zhang C, Shen W, You Q (2009) Synthesis of novel pH-sensitive chitosan graft copolymers and micellar solubilization of paclitaxel. Int J Biol Macromol 44:249–256

Lin Teng Shee F, Arul J, Brunet S, Mateescu AM, Bazinet L (2006) Solubilization of chitosan by bipolar membrane electroacidification. J Agric Food Chem 54:6760–6764

Lino MM, Paulo CS, Vale AC, Vaz MF, Ferreira LS (2013) Antifungal activity of dental resins containing amphotericin B-conjugated nanoparticles. Dent Mater 29:e252–e262

Ma Z, Garrido-Maestu A, Jeong KC (2017) Application, mode of action, and in vivo activity of chitosan and its micro- and nanoparticles as antimicrobial agents: a review. Carbohydr Polym 176:257–265

Marcos-Arias C, Vicente JL, Sahand IH, Eguia A, De-Juan A, Madariaga L et al (2009) Isolation of *Candida dubliniensis* in denture stomatitis. Arch Oral Biol 54:127–131

Mesa-Arango AC, Scorzoni L, Zaragoza O (2012) It only takes one to do many jobs: amphotericin B as antifungal and immunomodulatory drug. Front Microbiol 3:286

Monteiro DR, Gorup LF, Takamiya AS, Ruvollo-Filho AC, de Camargo ER, Barbosa DB (2009) The growing importance of materials that prevent microbial adhesion: antimicrobial effect of medical devices containing silver. Int J Antimicrob Agents 34:103–110

Monteiro DR, Gorup LF, Silva S, Negri M, de Camargo ER, Oliveira R et al (2011) Silver colloidal nanoparticles: antifungal effect against adhered cells and biofilms of *Candida albicans* and *Candida glabrata*. Biofouling 27:711–719

Monteiro DR, Silva S, Negri M, Gorup LF, de Camargo ER, Oliveira R et al (2012) Silver nanoparticles: influence of stabilizing agent and diameter on antifungal activity against *Candida albicans* and *Candida glabrata* biofilms. Lett Appl Microbiol 54:383–391

Monteiro DR, Silva S, Negri M, Gorup LF, de Camargo ER, Oliveira R et al (2013) Antifungal activity of silver nanoparticles in combination with nystatin and chlorhexidine digluconate against *Candida albicans* and *Candida glabrata* biofilms. Mycoses 56:672–680

Monteiro DR, Takamiya AS, Feresin LP, Gorup LF, de Camargo ER, Delbem AC et al (2014) Silver colloidal nanoparticle stability: influence on *Candida* biofilms formed on denture acrylic. Med Mycol 52:627–635

Monteiro DR, Takamiya AS, Feresin LP, Gorup LF, de Camargo ER, Delbem AC et al (2015) Susceptibility of *Candida albicans* and *Candida glabrata* biofilms to silver nanoparticles in intermediate and mature development phases. J Prosthodont Res 59:42–48

Moran C, Grussemeyer CA, Spalding JR, Benjamin DK Jr, Reed SD (2009) *Candida albicans* and non-*albicans* bloodstream infections in adult and pediatric patients: comparison of mortality and costs. Pediatr Infect Dis J 28:433–435

Morschhauser J (2002) The genetic basis of fluconazole resistance development in *Candida albicans*. Biochim Biophys Acta 1587:240–248

Mousavi SA, Ghotaslou R, Kordi S, Khoramdel A, Aeenfar A, Kahjough ST et al (2018) Antibacterial and antifungal effects of chitosan nanoparticles on tissue conditioners of complete dentures. Int J Biol Macromol 118:881–885

Mudiar R, Kelkar-Mane V (2018) Targeting fungal menace through copper nanoparticles and Tamrajal. J Ayurveda Integr Med. pii: S0975-9476(17)30481-3

Muzzarelli RAA, Jeuniaux C, Gooday GW (1986) Chitin in nature and technology. Plenum Press, New York

Niemirowicz K, Bucki R (2017) Enhancing the fungicidal activity of antibiotics: are magnetic nanoparticles the key? Nanomedicine (Lond) 12:1747–1749

Niemirowicz K, Markiewicz KH, Wilczewska AZ, Car H (2012) Magnetic nanoparticles as new diagnostic tools in medicine. Adv Med Sci 57:196–207

Niemirowicz K, Durnas B, Tokajuk G, Gluszek K, Wilczewska AZ, Misztalewska I et al (2016) Magnetic nanoparticles as a drug delivery system that enhance fungicidal activity of polyene antibiotics. Nanomedicine 12:2395–2404

Nikawa H, Hamada T, Yamamoto T (1998) Denture plaque – past and recent concerns. J Dent 26:299–304

Pal S, Tak YK, Song JM (2007) Does the antibacterial activity of silver nanoparticles depend on the shape of the nanoparticle? A study of the gram-negative bacterium *Escherichia coli*. Appl Environ Microbiol 73:1712–1720

Parsameher N, Rezaei S, Khodavasiy S, Salari S, Hadizade S, Kord M et al (2017) Effect of biogenic selenium nanoparticles on ERG11 and CDR1 gene expression in both fluconazole-resistant and -susceptible *Candida albicans* isolates. Curr Med Mycol 3:16–20

Peiris M, Gunasekara T, Jayaweera PM, Fernando S (2018) TiO_2 nanoparticles from Baker's yeast: a potent antimicrobial. J Microbiol Biotechnol 28:1664–1670

Perlin DS (2015) Echinocandin resistance in *Candida*. Clin Infect Dis 61(Suppl 6):S612–S617

Pfaller MA, Diekema DJ (2007) Epidemiology of invasive candidiasis: a persistent public health problem. Clin Microbiol Rev 20:133–163

Pfaller M, Riley J, Koerner T (1989) Effects of cilofungin (LY121019) on carbohydrate and sterol composition of *Candida albicans*. Eur J Clin Microbiol Infect Dis 8:1067–1070

Pokrowiecki R, Zaręba T, Szaraniec B, Pałka K, Mielczarek A, Menaszek E et al (2017) *In vitro* studies of nanosilver-doped titanium implants for oral and maxillofacial surgery. Int J Nanomedicine 12:4285–4297

Priyanka KP, Sukirtha TH, Balakrishna KM, Varghese T (2016) Microbicidal activity of TiO_2 nanoparticles synthesised by sol-gel method. IET Nanobiotechnol 10:81–86

Pumeesat P, Muangkaew W, Ampawong S, Luplertlop N (2017) *Candida albicans* biofilm development under increased temperature. New Microbiol 40:279–283

Rabea EI, Badawy ME, Stevens CV, Smagghe G, Steurbaut W (2003) Chitosan as antimicrobial agent: applications and mode of action. Biomacromolecules 4:1457–1465

Rajendran R, Sherry L, Deshpande A, Johnson EM, Hanson MF, Williams C et al (2016a) A prospective surveillance study of Candidaemia: epidemiology, risk factors, antifungal treatment and outcome in hospitalized patients. Front Microbiol 7:915

Rajendran R, Sherry L, Nile CJ, Sherriff A, Johnson EM, Hanson MF et al (2016b) Biofilm formation is a risk factor for mortality in patients with *Candida albicans* bloodstream infection-Scotland, 2012–2013. Clin Microbiol Infect 22(1):87–93

Ramage G, Rajendran R, Sherry L, Williams C (2012) Fungal biofilm resistance. Int J Microbiol 2012:528521

Ramage G, Robertson SN, Williams C (2014) Strength in numbers: antifungal strategies against fungal biofilms. Int J Antimicrob Agents 43:114–120

Regiel-Futyra A, Kus-Liskiewicz M, Sebastian V, Irusta S, Arruebo M, Kyziol A et al (2017) Development of noncytotoxic silver-chitosan nanocomposites for efficient control of biofilm forming microbes. RSC Adv 7:52398–52413

Salari S, Sadat Seddighi N, Ghasemi Nejad Almani P (2018) Evaluation of biofilm formation ability in different *Candida* strains and anti-biofilm effects of Fe_3O_4-NPs compared with fluconazole: an *in vitro* study. J Mycol Med 28:23–28

Sayed SI, Datta S, Deore N, Kazi RA, Jagade MV (2012) Prevention of voice prosthesis biofilms: current scenario and future trends in prolonging prosthesis lifetime. J Indian Med Assoc 110:175–178, 180

Seil JT, Webster TJ (2012) Antimicrobial applications of nanotechnology: methods and literature. Int J Nanomedicine 7:2767–2781

Seong M, Lee DG (2018) Reactive oxygen species-independent apoptotic pathway by gold nanoparticles in *Candida albicans*. Microbiol Res 207:33–40

Shakibaie M, Salari Mohazab N, Ayatollahi Mousavi SA (2015) Antifungal activity of selenium nanoparticles synthesized by *Bacillus* species Msh-1 against *Aspergillus fumigatus* and *Candida albicans*. Jundishapur J Microbiol 8:e26381

Sherwani MA, Tufail S, Khan AA, Owais M (2015) Gold nanoparticle-photosensitizer conjugate based photodynamic inactivation of biofilm producing cells: potential for treatment of *C. albicans* infection in BALB/c mice. PLoS One 10:e0131684

Shih PY, Liao YT, Tseng YK, Deng FS, Lin CH (2019) A potential antifungal effect of chitosan against *Candida albicans* is mediated via the inhibition of SAGA complex component expression and the subsequent alteration of cell surface integrity. Front Microbiol 10:602

Shirtliff ME, Peters BM, Jabra-Rizk MA (2009) Cross-kingdom interactions: *Candida albicans* and bacteria. FEMS Microbiol Lett 299:1–8

Silva S, Negri M, Henriques M, Oliveira R, Williams DW, Azeredo J (2012) *Candida glabrata*, *Candida parapsilosis* and *Candida tropicalis*: biology, epidemiology, pathogenicity and antifungal resistance. FEMS Microbiol Rev 36:288–305

Silva S, Pires P, Monteiro DR, Negri M, Gorup LF, Camargo ER et al (2013) The effect of silver nanoparticles and nystatin on mixed biofilms of *Candida glabrata* and *Candida albicans* on acrylic. Med Mycol 51:178–184

Sivaraj R, Rahman PK, Rajiv P, Narendhran S, Venckatesh R (2014) Biosynthesis and characterization of *Acalypha indica* mediated copper oxide nanoparticles and evaluation of its antimicrobial and anticancer activity. Spectrochim Acta A Mol Biomol Spectrosc 129:255–258

Slavin YN, Asnis J, Hafeli UO, Bach H (2017) Metal nanoparticles: understanding the mechanisms behind antibacterial activity. J Nanobiotechnol 15:65

Sousa F, Sanavio B, Saccani A, Tang Y, Zucca I, Carney TM et al (2017) Superparamagnetic nanoparticles as high efficiency magnetic resonance imaging T_2 contrast agent. Bioconjug Chem 28:161–170

Souza ME, Lopes LQ, Bonez PC, Gündel A, Martinez DS, Sagrillo MR et al (2017) *Melaleuca alternifolia* nanoparticles against *Candida* species biofilms. Microb Pathog 104:125–132

Sudjana AN, Carson CF, Carson KC, Riley TV, Hammer KA (2012) *Candida albicans* adhesion to human epithelial cells and polystyrene and formation of biofilm is reduced by sub-inhibitory *Melaleuca alternifolia* (tea tree) essential oil. Med Mycol 50:863–870

Sun J, Xu Y, Zhu B, Gao G, Ren J, Wang H et al (2019) Synergistic effects of titanium dioxide and cellulose on the properties of glassionomer cement. Dent Mater J 38:41–51

Sundrarajan M, Bama K, Bhavani M, Jegatheeswaran S, Ambika S, Sangili A et al (2017) Obtaining titanium dioxide nanoparticles with spherical shape and antimicrobial properties using *M. citrifolia* leaves extract by hydrothermal method. J Photochem Photobiol B 171:117–124

Suri SS, Fenniri H, Singh B (2007) Nanotechnology-based drug delivery systems. J Occup Med Toxicol 2:16

Tan YN, Lee KH, Su X (2011) Study of single-stranded DNA binding protein-nucleic acids interactions using unmodified gold nanoparticles and its application for detection of single nucleotide polymorphisms. Anal Chem 83:4251–4257

Tokajuk G, Niemirowicz K, Deptula P, Piktel E, Ciesluk M, Wilczewska AZ et al (2017) Use of magnetic nanoparticles as a drug delivery system to improve chlorhexidine antimicrobial activity. Int J Nanomedicine 12:7833–7846

Totu EE, Nechifor AC, Nechifor G, Aboul-Enein HY, Cristache CM (2017) Poly(methyl methacrylate) with TiO_2 nanoparticles inclusion for stereolitographic complete denture manufacturing – the fututre in dental care for elderly edentulous patients? J Dent 59:68–77

Usman MS, El Zowalaty ME, Shameli K, Zainuddin N, Salama M, Ibrahim NA (2013) Synthesis, characterization, and antimicrobial properties of copper nanoparticles. Int J Nanomedicine 8:4467–4479

Vieira APM, Arias LS, de Souza Neto FN, Kubo AM, Lima BHR, de Camargo ER et al (2019) Antibiofilm effect of chlorhexidine-carrier nanosystem based on iron oxide magnetic nanoparticles and chitosan. Colloids Surf B Biointerfaces 174:224–231

Viudes A, Pemán J, Cantón E, Ubeda P, López-Ribot JL, Gobernado M (2002) Candidemia at a tertiary-care hospital: epidemiology, treatment, clinical outcome and risk factors for death. Eur J Clin Microbiol Infect Dis 21:767–774

Wang C, Hsu CH, Li Z, Hwang LP, Lin YC, Chou PT et al (2017) Effective heating of magnetic nanoparticle aggregates for in vivo nano-theranostic hyperthermia. Int J Nanomedicine 12:6273–6287

Wani IA, Ahmad T, Manzoor N (2013) Size and shape dependant antifungal activity of gold nanoparticles: a case study of *Candida*. Colloids Surf B Biointerfaces 101:162–170

Watamoto T, Samaranayake LP, Egusa H, Yatani H, Seneviratne CJ (2011) Transcriptional regulation of drug-resistance genes in *Candida albicans* biofilms in response to antifungals. J Med Microbiol 60:1241–1247

Williams C, Ramage G (2015) Fungal biofilms in human disease. Adv Exp Med Biol 831:11–27

Williams DW, Kuriyama T, Silva S, Malic S, Lewis MA (2011) *Candida* biofilms and oral candidosis: treatment and prevention. Periodontol 2000 55:250–265

Xiang MJ, Liu JY, Ni PH, Wang S, Shi C, Wei B et al (2013) Erg11 mutations associated with azole resistance in clinical isolates of *Candida albicans*. FEMS Yeast Res 13:386–393

Yan L, Zhang J, Li M, Cao Y, Xu Z, Cao Y et al (2008) DNA microarray analysis of flucon-
    azole resistance in a laboratory Candida albicans strain. Acta Biochim Biophys Sin Shanghai
    40:1048–1060
Yu Q, Li J, Zhang Y, Wang Y, Liu L, Li M (2016) Inhibition of gold nanoparticles (AuNPs) on
    pathogenic biofilm formation and invasion to host cells. Sci Rep 6:26667
Zhang L, She X, Merenstein D, Wang C, Hamilton P, Blackmon A et al (2014) Fluconazole resis-
    tance patterns in *Candida* species that colonize women with HIV infection. Curr Ther Res Clin
    Exp 76:84–89

Biomedical Applications of Lignin-Based Nanoparticles

8

Siavash Iravani

Abstract

Nanosystems have been developed and applied as promising vehicles for various important biomedical applications. Renewable resources are gaining increasing attention as a source for environmentally benign biomaterials, such as drug encapsulation/release compounds, scaffolds for tissue engineering, and drug delivery systems for cancers. Because of the remarkable absorption capacity, biodegradability, and non-toxicity, lignin nanoparticles can be applied as appropriate vehicles for drug molecules and inorganic particles. In this chapter, some important biomedical and therapeutic applications of lignin nanoparticles are highlighted, briefly.

Keywords

Lignin · Lignin-based NPs · Nanoparticles · Biomedical applications · Drug delivery · Cancer

8.1 Introduction

Plant biomass includes lignocellulose, carbohydrates, lignin, proteins, fats, vitamins, dyes, flavors, and aromatic essences of various chemical structures (Volf and Popa 2018; Sriram and Shahidehpour 2005; Iravani 2011; Mohammadinejad et al. 2016; Mohammadinejad et al. 2019; Varma 2012, 2014a, b, 2016, 2019). Lignocellulose, with an abundance of about 70% of the total plant biomass (Zhang 2008), is a complex material synthesized by plants during cell wall formation, which has three main components, including microfibrils of cellulose,

S. Iravani (✉)
Faculty of Pharmacy and Pharmaceutical Sciences, Isfahan University of Medical Sciences, Isfahan, Iran

© Springer Nature Singapore Pte Ltd. 2020
A. K. Shukla (ed.), *Nanoparticles and their Biomedical Applications*,
https://doi.org/10.1007/978-981-15-0391-7_8

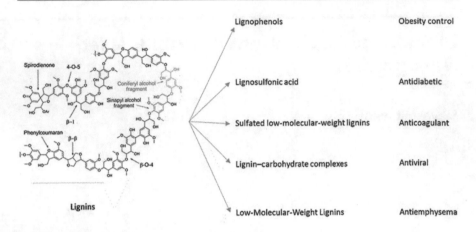

Fig. 8.1 Beneficial effects of lignins and their derivatives. (Reused from Vinardell and Mitjans (2017), An Open Access Article (CC BY 4.0))

hemicellulose, and lignin (Heredia et al. 1995). The concept of green technology biorefinery is to achieve complete utilization of most lignocellulosic biomass, such as lignin components by green production (Gregorová et al. 2005).

Lignin was more widely considered in pulp industry, but nowadays it is applied as building elements in the design of bio-based materials (lignin-based materials epoxy or polyurethane), UV stabilizer, antioxidant, and low- or high-value additive for polymers (Fig. 8.1) (Vinardell and Mitjans 2017). Important sources of lignin are wood, corn stalks, straw, bagasse, and etc. (Boerjan et al. 2003). Lignin is a complex and irregular biopolymer containing randomly cross-linked phenylpropanoid units (cumaryl, coniferyl, and sinapyl alcohol) and is detected in plant secondary cell walls. Based on these monolignol units, the lignin building blocks p-hydroxyphenyl, guaiacyl, and syringyl are generated. These building blocks are connected via several types of linkages, mainly ether bonds, such as aryl- or phenyl ether, and carbon-carbon bonds, such as biphenyls and diphenyl ethaneor pinoresinol (Holladay et al. 2007; Laurichesse and Avérous 2014). In this chapter, some important biomedical and therapeutic applications of lignins and lignin-based nanoparticles (NPs) are highlighted.

8.2 Biomedical Applications

Lignins can be applied in medicine and pharmaceutics to improve human health because of their antioxidant and antimicrobial characteristics. Additionally, lignins have biological activities, such as capability of reducing cholesterol by binding to bile acids in intestine. These activities of lignins offer their application in treatment of various diseases like obesity, diabetes, thrombosis, viral infections, and cancers (Vinardell and Mitjans 2017). The structure and functional modifications of lignin can increase its antidiabetic and antioxidant activities. For instance, alkali lignin, which was extracted from the deciduous plant *Acacia nilotica*, showed antioxidant and antidiabetic characteristics (Barapatre et al. 2016). This modified alkali lignin has α-amylase inhibitory activities and anti-hyperglycemic characteristics, which propose it as an appropriate candidate for healing diabetes (Barapatre et al. 2016). Lignosulfonic acid, a derivative of lignin, is a noncompetitive inhibitor of α-glucosidase, which has the capability of persuading a delay in glucose adsorption. This suggests that lignin and its derivatives, mainly lignosulfonic acid, can be applied for treatment of carbohydrate absorption and related diseases, such as diabetes (Hasegawa et al. 2015).

Kraft lignin is identified as a fat adsorbent that can control and prevent obesity. Lignophenols as a highly stable and antioxidant derivative of lignin have the capability of reducing cholesterol levels in HepG2 cells in a dose-dependent manner (Norikura et al. 2010). The experiment with rats that were fed a high-dose fat diet has shown that this lignin could suppress adipose tissues, reduce the triglyceride level in plasma, and attenuate the hepatic expression of SREBP-1c mRNA. However, the exact mechanism of LP effects on lipid metabolism and obesity control still needs to be further assessed (Sato et al. 2012). Additionally, investigations have shown the antiviral activity of lignin-carbohydrate-protein complexes (LC) against some viruses, such as herpes simplex virus types 1 and 2 (HSV-1 and -2), human cytomegalovirus (HCMV), and measles virus (Lee et al. 2011). Lignosulfonic acid, which is most applied in paper industry and formation of artificial vanilla flavor (Fargues et al. 1996), has anti-HIV activity when applied in low dosage (Gordts et al. 2015). The potential role of its therapeutic anti-viral properties has been approved in clinical trials and its safety in prevention of sexual transmission of HIV-1 has been established (Karim et al. 2010).

Sulfated low-molecular-weight lignins (LMWLs) have shown anticoagulant and immunomodulatory mediator properties. They are composed of oligomeric chains of different lengths and different inter-monomeric linkages, such as β-*O*-4 and β-5 (Henry et al. 2010). They interact with heparin-binding domain to inhibit a variety of serine coagulation proteases. Although certain ligands have the ability to bind to the heparin-binding site of serine coagulation proteases, none has been reported to have direct anticoagulant effects (Henry and Desai 2014). Sulfated β-*O*-lignins also possess anticoagulant properties. They concurrently stimulate anticoagulation and antiplatelet purposes (Mehta et al. 2016).

Different reports have shown the capacity of lignin-based NPs for the controlled drug release, which are important in human medicine (Fernández-Pérez et al. 2007). NPs that are synthesized from lignin are non-toxic, highly biodegradable, stable, and inexpensive, four major advantages that represent them as potent drug delivery systems in human diseases (Frangville et al. 2012; Lievonen et al. 2016). Recently, water-dispersed lignin NPs have been made to carry silver ions, which are significant in antimicrobial applications or in cancer treatments (Frangville et al. 2012). Additionally, lignins are appropriate for transferring both hydrophobic and hydrophilic drugs. For instance, the pH-sensitive polymers that are added to lignin NPs make it easier to load hydrophilic drugs (Richter et al. 2016). Poorly water-soluble drugs or water-soluble anticancer drugs can be loaded on lignin NPs, leading to an increase in their anticancer and growth inhibitory effects in different cancer cell lines (Figueiredo et al. 2017). Pure lignin NPs, iron (III)-complexed lignin NPs, and Fe_3O_4-infused lignin NPs have been produced with round shape, narrow size distribution, reduced polydispersity, and good stability at pH 7.4. It was reported that the produced lignin NPs had low cytotoxicity in all the tested cell lines and hemolytic rates below 12% after 12 h of incubation. Pure lignin NPs demonstrated the capacity to powerfully load poorly water-soluble drugs and other cytotoxic agents, including sorafenib and benzazulene, and improved their release profiles at pH 5.5 and 7.4 in a sustained manner. Additionally, the benzazulene-pure lignin NPs exhibited an accelerated anti-proliferation influence in various cells compared to the pure benzazulene and demonstrated most inhibitory concentration ranging from 0.64 to 12.4 µM after 24 h incubation. Lignin NPs are also used in cosmetic products due to its UV-absorbing property (Gutiérrez-Hernández et al. 2016). Lignin has the ability to hybrid with nano silver or nano chitosan to increase their antibacterial activity (Kim et al. 2013; Klapiszewski et al. 2015). Lignin was applied for transferring arthritis rheumatoid specific drug, methrotrexate, in a rat model as well. In this model, lignin could release the drug into blood vessels and inflamed tissues (Wahba et al. 2015). The capacity of lignin in the drug delivery system is being developed, which present lignin as a suitable candidate for more usage in medicine (Răschip et al. 2015).

The alkali lignin was applied for producing NPs with perfect spheres and good dispersibility through a simple self-assembly approach by adding water to a methanol solution of alkali lignin. Self-assembly of alkali lignin with the bioactive molecule resveratrol and Fe_3O_4NPs led to the preparation of a stable nano-drug carrier. In cytological and animal analyses, the magnetic resveratrol-loaded lignin NPs showed appropriate anticancer influences and accelerated *in vitro* resveratrol release and stability, drug accumulation, and suitable tumor reduction, and lower adverse effects than free drugs (Dai et al. 2017). Moreover, the lignin NPs that have been synthesized contained a hydrodynamic diameter ranging from ca. 80 to 230 nm, produced by self-assembly in a recyclable and non-toxic aqueous sodium p-toluenesulfonate solution at room temperature, with a lowest concentration of up to 48 g/L. The drug-encapsulated lignin NPs demonstrated improved characteristics, with sustained drug-releasing capability and biocompatibility. Additionally, the unloaded drugs and free p-toluenesulfonate could be simply recycled for multiple applications, thus gaining environmental sustainability (Chen et al. 2018). In another study, lignin-based targeted polymeric NPs platform, folic acid-polyethylene glycol-alkaline lignin conjugates, was produced through self-assembly for delivery of anticancer drug (hydroxyl camptothecin, HCPT). The lignin-based NPs (about 150 nm) showed outstanding biocompatibility, high drug loading efficiency, prolonged blood circulation time, and improved cellular uptake; these generated systems can be applied as promising candidate for anticancer drugs delivery (Liu et al. 2018).

The preparation of targeted lignin-based drug delivery NPs was accomplished for loading doxorubicin hydrochloride. The lignin hollow NPs have been applied as a platform for the production of targeted delivery material by incorporating

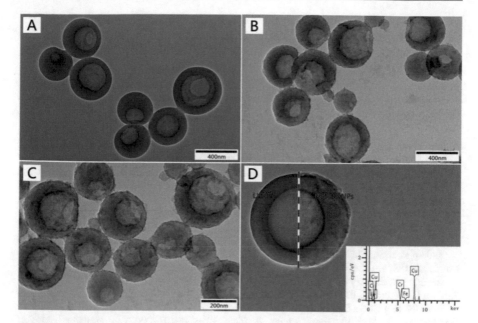

Fig. 8.2 Transmission electron microscope (TEM) images of (**a**) lignin hollow NPs, (**b**) magnetic-functionalized lignin hollow NPs, and (**c**) folic-magnetic-functionalized lignin hollow NPs. (Reused from Zhou et al. (2019), An Open Access Article (CC BY 4.0))

magnetic NPs and folic acid through layer-by-layer self-assembling. It was reported that the surface of lignin hollow NPs was covered uniformly by Fe_3O_4 NPs and grafted with folic acid. The folic-magnetic-functionalized lignin hollow NPs could respond to magnetic field and folic acid receptors. Additionally, the targeting performance of the folic-magnetic-functionalized lignin hollow NPs accelerated the cellular uptake of NPs in the case of HeLa cells (Fig. 8.2).

In one study, lignin NPs have been produced as oral drug delivery system for curcumin. The particle size of curcumin-loaded lignin NPs was about 104 nm, and the encapsulation efficiency of curcumin in the NPs was 92%. It was revealed that curcumin-loaded lignin NPs had appropriate stability in simulated gastric fluid and slow release under intestinal conditions as desirable. In vivo pharmacokinetics studies showed that the lignin NP system accelerated the bioavailability of curcumin by tenfold compared with the administration of unformulated curcumin (Alqahtani et al. 2019).

8.3   Conclusion

Lignin demonstrates precise antioxidant and antimicrobial activities. Currently, elaborative efforts in research and industries are focused on lignin applications as renewable macromolecular building blocks for the production of polymeric drug encapsulation and scaffold materials. Lignin and lignin-based NPs (as renewable green materials) can be applied as promising candidates for various biomedical applications, including drug delivery, cancer therapy, and diagnosis. Furthermore, lignin NPs show great potential to be applied as suitable oral drug delivery systems, especially for poorly soluble drugs with limited bioavailability.

References

Alqahtani MS, Alqahtani A, Al-Thabit A, Roni M, Syed R (2019) Novel lignin nanoparticles for oral drug delivery. J Mater Chem B. https://doi.org/10.1039/C9TB00594C

Barapatre A, Meena AS, Mekala S, Das A, Jha H (2016) In vitro evaluation of antioxidant and cytotoxic activities of lignin fractions extracted from Acacia nilotica. Int J Biol Macromol 86:443–453

Boerjan W, Ralph J, Baucher M (2003) Lignin biosynthesis. Annu Rev Plant Biol 54(1):519–546

Chen L, Zhou X, Shi Y, Gao B, Wu J, Kirk TB, Xu J, Xue W (2018) Green synthesis of lignin nanoparticle in aqueous hydrotropic solution toward broadening the window for its processing and application. Chem Eng J 346:217–225

Dai L, Liu OR, Hu L-Q, Zou Z-F, Si C-L (2017) Lignin nanoparticle as a novel green carrier for the efficient delivery of resveratrol. ACS Sustain Chem Eng 5:8241–8249

Fargues C, Mathias Á, Rodrigues A (1996) Kinetics of vanillin production from kraft lignin oxidation. Ind Eng Chem Res 35(1):28–36

Fernández-Pérez M, Villafranca-Sánchez M, Flores-Céspedes F (2007) Controlled-release formulations of cyromazine-lignin matrix coated with ethylcellulose. J Environ Sci Health 42(7):863–868

Figueiredo P, Lintinen K, Kiriazis A, Hynninen V, Liu Z, Bauleth-Ramos T, Rahikkala A, Correia A, Kohout T, Sarmento B (2017) In vitro evaluation of biodegradable lignin-based nanoparticles for drug delivery and enhanced antiproliferation effect in cancer cells. Biomaterials 121:97–108

Frangville C, Rutkevičius M, Richter AP, Velev OD, Stoyanov SD, Paunov VN (2012) Fabrication of environmentally biodegradable lignin nanoparticles. ChemPhysChem 13(18):4235–4243

Gordts SC, Férir G, D'huys T, Petrova MI, Lebeer S, Snoeck R, Andrei G, Schols D (2015) The low-cost compound lignosulfonic acid (LA) exhibits broad-spectrum anti-HIV and anti-HSV activity and has potential for microbicidal applications. PLoS One 10(7):e0131219

Gregorová A, Cibulková Z, Košíková B, Šimon P (2005) Stabilization effect of lignin in polypropylene and recycled polypropylene. Polym Degrad Stab 89(3):553–558

Gutiérrez-Hernández JM, Escalante A, Murillo-Vázquez RN, Delgado E, González FJ, Toríz G (2016) Use of Agave tequilana-lignin and zinc oxide nanoparticles for skin photoprotection. J Photochem Photobiol B Biol 163:156–161

Hasegawa Y, Kadota Y, Hasegawa C, Kawaminami S (2015) Lignosulfonic acid-induced inhibition of intestinal glucose absorption. J Nutr Sci Vitaminol 61(6):449–454

Henry BL, Desai UR (2014) Sulfated low molecular weight lignins, allosteric inhibitors of coagulation proteinases via the heparin binding site, significantly alter the active site of thrombin and factor Xa compared to heparin. Thromb Res 134(5):1123–1129

Henry BL, Aziz MA, Zhou Q, Desai UR (2010) Sulfated, low molecular weight lignins are potent inhibitors of plasmin, in addition to thrombin and factor Xa: novel opportunity for controlling complex pathologies. Thromb Haemost 103(3):507

Heredia A, Jiménez A, Guillén R (1995) Composition of plant cell walls. Z Lebensm Unters Forsch 200(1):24–31

Holladay JE, White JF, Bozell JJ, Johnson D (2007) Top value-added chemicals from biomass-volume II—results of screening for potential candidates from biorefinery lignin. Pacific Northwest National Lab.(PNNL), Richland

Iravani S (2011) Green synthesis of metal nanoparticles using plants. Green Chem 13:2638–2650

Karim QA, Karim SSA, Frohlich JA, Grobler AC, Baxter C, Mansoor LE, Kharsany AB, Sibeko S, Mlisana KP, Omar Z (2010) Effectiveness and safety of tenofovir gel, an antiretroviral microbicide, for the prevention of HIV infection in women. Science 329:1193748

Kim S, Fernandes MM, Matamá T, Loureiro A, Gomes AC, Cavaco-Paulo A (2013) Chitosan–lignosulfonates sono-chemically prepared nanoparticles: characterisation and potential applications. Colloids Surf B: Biointerfaces 103:1–8

Klapiszewski Ł, Rzemieniecki T, Krawczyk M, Malina D, Norman M, Zdarta J, Majchrzak I, Dobrowolska A, Czaczyk K, Jesionowski T (2015) Kraft lignin/silica–AgNPs as a functional material with antibacterial activity. Colloids Surf B: Biointerfaces 134:220–228

Laurichesse S, Avérous L (2014) Chemical modification of lignins: towards biobased polymers. Prog Polym Sci 39(7):1266–1290

Lee J-B, Yamagishi C, Hayashi K, Hayashi T (2011) Antiviral and immunostimulating effects of lignin-carbohydrate-protein complexes from Pimpinella anisum. Biosci Biotechnol Biochem 75(3):459–465

Lievonen M, Valle-Delgado JJ, Mattinen M-L, Hult E-L, Lintinen K, Kostiainen MA, Paananen A, Szilvay GR, Setälä H, Österberg M (2016) A simple process for lignin nanoparticle preparation. Green Chem 18(5):1416–1422

Liu K, Zheng D, Lei H, Liu J, Lei J, Wang L, Ma X (2018) Development of novel lignin-based targeted polymeric nanoparticle platform for efficient delivery of anticancer drugs. ACS Biomater Sci Eng 4(5):1730–1737

Mehta AY, Mohammed BM, Martin EJ, Brophy DF, Gailani D, Desai UR (2016) Allosterism-based simultaneous, dual anticoagulant and antiplatelet action: allosteric inhibitor targeting the glycoprotein Ibα-binding and heparin-binding site of thrombin. J Thromb Haemost 14(4):828–838

Mohammadinejad R, Karimi S, Iravani S, Varma RS (2016) Plant-derived nanostructures: types and applications. Green Chem 18:20–52

Mohammadinejad R, Shavandi A, Raie DS, Sangeetha J, Soleimani M, Hajibehzad SS, Thangadurai D, Hospet R, Popoola JO, Arzani A, Gómez-Lim MA, Iravani S, Varma RS (2019) Plant molecular farming: production of metallic nanoparticles and therapeutic proteins using green factories. Green Chem 21:1845–1865

Norikura T, Mukai Y, Fujita S, Mikame K, Funaoka M, Sato S (2010) Lignophenols decrease Oleate-induced Apolipoprotein-B secretion in HepG2 cells. Basic Clin Pharmacol Toxicol 107(4):813–817

Răschip IE, Panainte AD, Pamfil D, Profire L, Vasile C (2015) In vitro testing of xanthan/lignin hydrogels as carriers for controlled delivery of bisoprolol fumarare. Rev Med Chir Soc Med Nat Iasi 119:1189–1194

Richter AP, Bharti B, Armstrong HB, Brown JS, Plemmons D, Paunov VN, Stoyanov SD, Velev OD (2016) Synthesis and characterization of biodegradable lignin nanoparticles with tunable surface properties. Langmuir 32(25):6468–6477

Sato S, Mukai Y, Tokuoka Y, Mikame K, Funaoka M, Fujita S (2012) Effect of lignin-derived lignophenols on hepatic lipid metabolism in rats fed a high-fat diet. Environ Toxicol Pharmacol 34(2):228–234

Sriram N, Shahidehpour M (2005) Renewable biomass energy. In: Power Engineering Society General Meeting, IEEE, 2005. IEEE, pp 612–617

Varma RS (2012) Greener approach to nanomaterials and their sustainable applications. Curr Opin Chem Eng 1:123–128

Varma RS (2014a) Greener and sustainable chemistry. Appl Sci 4:493–497

Varma RS (2014b) Journey on greener pathways: from the use of alternate energy inputs and benign reaction media to sustainable applications of nano-catalysts in synthesis and environmental remediation. Green Chem 16:2027–2041

Varma RS (2016) Greener and sustainable trends in synthesis of organics and nanomaterials. ACS Sustain Chem Eng 4:5866–5878

Varma RS (2019) Biomass-derived renewable carbonaceous materials for sustainable chemical and environmental applications. ACS Sustain Chem Eng 7:6458–6470

Vinardell MP, Mitjans M (2017) Lignins and their derivatives with beneficial effects on human health. Int J Mol Sci 18(6):1219

Volf I, Popa VI (2018) Integrated processing of biomass resources for fine chemical obtaining: polyphenols. In: Biomass as renewable raw material to obtain bioproducts of high-tech value. Elsevier, Amsterdam, pp 113–160

Wahba SM, Darwish AS, Shehata IH, Elhalem SSA (2015) Sugarcane bagasse lignin, and silica gel and magneto-silica as drug vehicles for development of innocuous methotrexate drug against rheumatoid arthritis disease in albino rats. Mater Sci Eng C 48:599–610

Zhang Y-HP (2008) Reviving the carbohydrate economy via multi-product lignocellulose biorefineries. J Ind Microbiol Biotechnol 35(5):367–375

Zhou Y, Han Y, Li G, Yang S, Xiong F, Chu F (2019) Preparation of targeted lignin–based hollow nanoparticles for the delivery of doxorubicin. Nano 9:188. https://doi.org/10.3390/nano9020188

Green Nanoparticles for Biomedical and Bioengineering Applications

Luciano Paulino Silva, Gabriela Mendes da Rocha Vaz,
Júlia Moreira Pupe, Liana Soares Chafran,
Lucio Assis Araujo Neto, Thaís Ribeiro Santiago,
Thalita Fonseca Araujo, and Vera Lúcia Perussi Polez

Abstract

Green nanotechnology is a recent branch of nanotechnology in consonance with current concerns about sustainability issues using methods and materials that aim to generate eco-friendly nanosystems, with low environmental impact associated with significant economic and social gains. This concept offers opportunities for the use of nontoxic reagents and metabolites of living organisms in routes of green synthesis of nanosystems, including metallic nanoparticles, polymer nanoparticles, liposomes, and emulsions, because these materials enable a wide range of innovative applications, besides, in general, technologically desirable characteristics. Simplicity, scaling-up possibilities, and low cost of production, as well as the enhanced properties, thereby qualifying green nanoparticles as

L. P. Silva (✉)
Laboratório de Nanobiotecnologia (LNANO), Embrapa Recursos Genéticos e Biotecnologia, Brasília, DF, Brazil

Programa de Pós-graduação em Nanociência e Nanobiotecnologia, Universidade de Brasília, Instituto de Ciências Biológicas, Brasília, DF, Brazil

Programa de Pós-graduação em Ciências Biológicas, Universidade de Brasília, Instituto de Ciências Biológicas, Brasília, DF, Brazil

Programa de Pós-graduação em Ciências Farmacêuticas, Departamento de Farmácia, Universidade Federal do Paraná, Jardim Botânico, Curitiba, PR, Brazil

Instituto Nacional de Ciência e Tecnologia em Biologia Sintética, Brasília, DF, Brazil
e-mail: luciano.paulino@embrapa.br

G. M. da Rocha Vaz · J. M. Pupe
Laboratório de Nanobiotecnologia (LNANO), Embrapa Recursos Genéticos e Biotecnologia, Brasília, DF, Brazil

Programa de Pós-graduação em Ciências Biológicas, Universidade de Brasília, Instituto de Ciências Biológicas, Brasília, DF, Brazil promising candidates for unprecedented applications. Biomedical and bioengineering processes may directly benefit from this emerging nanotechnology field among other areas, such as agriculture and many industrial sectors. This chapter particularly focuses on subjects that were not covered by previous reviews about the area and presents a perspective that aligns high technology and sustainability in the development of nanotechnological products.

Keywords
Green chemistry · Green nanotechnology · Nanoparticles · Nanomaterials · Eco-friendly

9.1   Introduction

In the last few years, green nanotechnology emerges as an exciting and powerful branch of nanotechnology that deals with green chemistry principles, rational use of biological resources, and accomplishment of the Sustainable Development Goals (SDGs) as the pillars for innovation, economic development, and sustainable future. Hence, designing, creating, and developing new products and processes for industrial and other purposes with sustainability in mind offer positive environmental,

L. S. Chafran
Laboratório de Nanobiotecnologia (LNANO), Embrapa Recursos Genéticos e Biotecnologia, Brasília, DF, Brazil

Instituto Nacional de Ciência e Tecnologia em Biologia Sintética, Brasília, DF, Brazil

L. A. A. Neto
Laboratório de Nanobiotecnologia (LNANO), Embrapa Recursos Genéticos e Biotecnologia, Brasília, DF, Brazil

Programa de Pós-graduação em Ciências Farmacêuticas, Departamento de Farmácia, Universidade Federal do Paraná, Jardim Botânico, Curitiba, PR, Brazil

T. R. Santiago
Programa de Pós-graduação em Fitopatologia, Universidade de Brasília, Instituto de Ciências Biológicas, Brasília, DF, Brazil

T. F. Araujo
Laboratório de Nanobiotecnologia (LNANO), Embrapa Recursos Genéticos e Biotecnologia, Brasília, DF, Brazil

Programa de Pós-graduação em Nanociência e Nanobiotecnologia, Universidade de Brasília, Instituto de Ciências Biológicas, Brasília, DF, Brazil

V. L. P. Polez
Laboratório de Nanobiotecnologia (LNANO), Embrapa Recursos Genéticos e Biotecnologia, Brasília, DF, Brazil

Laboratório de Prospecção de Compostos Bioativos (LPCB), Embrapa Recursos Genéticos e Biotecnologia, Brasília, DF, Brazil health, and business benefits that go beyond mere convenience, opportunity, or beautification and thus bringing expressive progress and adding real value. Indeed, virtually all types of nanoparticles can be, in principle, produced by green synthesis routes. This way, they share in common the use of methods and reagents with low or absent adverse impacts to the health and environment aiming to produce nano-systems with broad technological applicability and overcoming some issues and challenges related to traditional chemical synthesis methods.

Currently, most of the nanosystems obtained through green synthesis routes are extremely eco-friendly (routes of synthesis use less toxic solvents and nontoxic reagents), biocompatible (they can be used directly to living organisms), and biode-gradable (can be degraded by biological routes to innocuous products) and have a low cost of production and typically high yield (compared to other technologies). Among the types of nanosystems that can be developed using green nanotechnology-based routes are the metallic nanoparticles, polymeric nanoparticles, emulsions, and liposomes based on the use of environment-friendly solvents, nontoxic reagents, and metabolites from bioresources. The successful development and scaling-up of green nanoparticle production can only be confirmed when actual applications of these advanced materials become apparent and overcome some current challenges, including the needs for the increase of available options for inputs and raw materials that can provide opportunities for academics and industries developing research, development, and innovation (RD&I) projects in this field.

Agricultural, food, and forestry products and their co-products, by-products, and wastes are now recognized as a rich source of primary and secondary metabolites that can be used in synthetic routes for the production of green nanoparticles. In fact, in several cases, the concept of biological synthesis of nanoparticles simply merges with green synthesis concept to the development of novel generations of sustainable nanomaterials. From these biological resources, it is possible to take up, in many cases at low cost, large amounts of molecules that can serve as raw materi-als for the development of new materials at the nanoscale with the potential to gen-erate products with high value-added and competitive advantages. This chapter describes, in particular, those research subjects in this field which were relatively neglected and not deeply discussed in previous reviews from the state of the art, and also to present some possible applications of green nanoparticles which are still beyond the reality of the market, but probably will represent the next wave of dis-ruption in nanotechnology. Anyway, this chapter will not exhaust the diversity of topics and research themes on this subject, but instead, it will present possibilities and perspectives to be explored aiming to expand the frontiers of research on green nanotechnology.

9.2   Green Chemistry for the Synthesis of Nanoparticles

The study of sustainable chemical compounds, whose formulation and use is of low or no risk to human health and the environment, is inserted within the context of green chemistry. This denomination created in the late 1990s is one of the main landmarks to the creation and publication of the 12 principles of green chemistry, which serve as a reference for processes and practices that reduce or eliminate the negative environmental impact caused by the use of hazardous chemical substances. However, the historical context which led to the creation of the laws and principles of green chemistry began in 1939, due to the discovery of the insecticidal properties of the compound dichlorodiphenyltrichloroethane, commonly known as DDT, carried out by the Swiss chemist Paul Hermann Müller, winner of the Nobel Prize in Physiology or Medicine, in 1948 (Kinkela 2016).

Dichlorodiphenyltrichloroethane, whose chemical structure can be observed from Fig. 9.1, is a hydrophobic organochlorine chemical compound with molecular formula $C_{14}H_9Cl_5$, crystalline, and highly toxic. It was widely used during World War II to control malaria and typhus and had its large-scale commercialization approved in the United States in 1945 for direct use in agriculture toward the control of pests (Kinkela 2016).

However, the use of DDT as well as other pesticides in agriculture was heavily criticized in the 1960s, especially by environmental activist Rachel Carson who published in 1962 her book titled *Silent Spring*, which presented real stories of the damage caused by the use of DDT in different communities. Carson denounced in her book how pesticides modified the daily lives of the population as well as wildlife, farm animals, birds, pets, and bees, among others. Despite being written as a "fable," *Silent Spring* served as an alert to the general public, becoming a bestseller and inspiring the modern environmental movement. After 8 years, in 1970, the President of the United States, Richard Nixon, established the US Environmental Protection Agency (EPA), whose purpose would be to adopt regulatory laws dedicated to protecting human health and the environment. In 1998, almost 30 years after the creation of the EPA, American chemists Paul Anastas and John C. Warner published the book *Green Chemistry: Theory and Practice*, in which the 12 principles of green chemistry were first presented, acting as a motivating guide to the use of substances and processes considered clean throughout the world. Since then, not only the production of insecticides but all sectors of industrial development, whose use of chemical substances is part of one or more stages of the production system, have become somewhat dependent on the parameters adopted in *Green Chemistry* (DeMarco 2017).

In the last decades, the study and development of functional nanoparticulate systems for application in sensors and electronic and catalytic systems, besides biomedical industry, have received great attention regarding the use of chemical

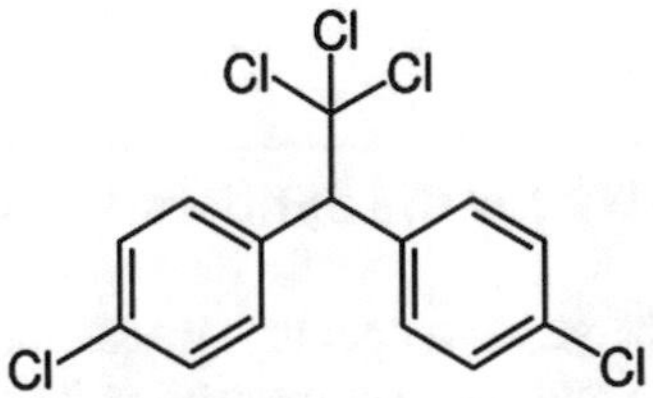

Fig. 9.1 Dichlorodiphenyltrichloroethane chemical structure ($C_{14}H_9Cl_5$), DDT

compounds obtaining monodisperse nanoparticles with controlled size and morphology. In physicochemical systems, the synthesis of nanoparticles often occurs through the use of toxic solvents such as formaldehyde, methanol, sodium borohydride, and hydrazine, among others. An example of this is the chemical synthesis of metallic nanoparticles such as silver nanoparticles, in bottom to top approach, which occurs from the addition of relatively toxic organic and inorganic reducing agents such as sodium citrate, Tollens' reagent, and block copolymers such as poly(ethyleneglycol) or poly(ethylene oxide)-poly(propylene oxide) (PEO–PPO), besides N, N-dimethylformamide (DMF) both in aqueous and nonaqueous medium (Iravani et al. 2014; Sakai et al. 2015). This leads to the generation of hazardous by-products to human health and the environment and moreover the possibility of increasing the reactivity of the particle, since experimentally the amount of reducing agent in the reaction is relatively greater than that required by stoichiometry, which can lead to a lack of predictability in its activity, especially when applied in biological systems (Duan et al. 2015). Among the most used physical and chemical methods for the synthesis of nanoparticles are laser desorption, lithographic techniques, wet chemical method, layer-by-layer growth, electrodeposition, chemical deposition (solution or vapor deposition), sol–gel process, Langmuir–Blodgett method, catalytic route, and hydrolysis coprecipitation method, among others (Dhand et al. 2015). In order to find cleaner synthesis routes, researchers around the world have developed safer, energy-efficient systems with reduced use of toxic solvents by adapting the principles of green chemistry to nanoscience, the correlation of which can be seen in Table 9.1, adapted from McKenzie and Hutchison (2004).

Green synthesis of nanoparticles using biological material (also termed biological synthesis or biosynthesis) such as algae, yeasts, plant extracts, bacteria, fungi, and viruses has been an alternative to the conventional physicochemical methods,

Table 9.1 Adaptation of the green chemistry principles to nanoscience and nanotechnology

	Green chemistry principles	Greener nanoscience methods
P1.	Prevention	Design for waste reduction
P2.	Atom economy	Design for materials efficiency
P3.	Less hazardous chemical synthesis	Design for process safety
P4.	Designing safer chemicals	Design for safer nanomaterials
P5.	Safer solvents and auxiliaries	Design for process safety and waste reduction
P6.	Design for energy efficiency	Design for energy-efficient nanomaterials
P7.	Use of renewable feedstocks	Design for process safety and for reduced environment impact
P8.	Reduce derivatives	Design for waste reduction
P9.	Catalysis	Design for materials efficiency and energy efficiency
P10.	Design for degradation	Design for reduced environmental impact
P11.	Real-time analysis for pollution prevention	Design for materials efficiency and energy efficiency
P12.	Inherently safer chemistry for accident prevention	Design for safer nanomaterials and process safety

exhibiting the main advantages as follows: the use of milder reaction conditions, such as low temperatures and environmental pressure; use of cleaner solvents like water; easy recovery of nanoparticles by centrifugation; reduced release of toxic waste such as by-products; scaling-up facility for large-scale synthesis; and being a low-cost and eco-friendly approach, since it is possible even to recover expensive metal salts, such as those from gold and silver, contained in the residual flows (Silva et al. 2015; Kaur et al. 2018). The main solvents used in green synthesis are water, supercritical fluids, and ionic liquids. Furthermore, in order to increase the degree of stabilization of the nanoparticles and colloidal systems, it is still possible to add capping and coating agents in the reaction medium, such as polymers, peptides, proteins, and dendrimers, among others. This allows greater control of the size, morphology, and degree of particle aggregation (Abdelghany et al. 2018).

The use of plant extracts as reducing agents has gained great attention in the academic field due to its easy access as a raw material besides the facility of synthesis, which commonly uses a one-step process with biocompatible products. Another point to be highlighted is the need for reduced amounts of metal ions in each synthesis and the use of water as the main reaction solvent (Oliveira et al. 2019). Green synthesis using plant extracts commonly involves the purification of bio-reducing agents, followed by their addition to the aqueous solution containing the precursor metal in a controlled manner. The reaction may occur spontaneously at room temperature or by gentle heating followed by stirring. However, most of the characteristics associated to the nanoparticles produced by green synthesis are directly related to the life cycle of the plant used as raw material and its behavior against biotic and abiotic factors, which determine the biosynthesis of the primary and secondary metabolites during its development (Silva et al. 2015). A possible mechanism of metallic nanoparticles biosynthesis can be seen from Fig. 9.2.

The use of water as the main solvent for the synthesis of nanoparticles has advantages and disadvantages. Although it is a nontoxic, non-flammable, and low-cost solvent (abundant source), water exhibits a high thermal capacity, making it necessary to apply a significant amount of energy in the reaction when compared to the same reaction using more volatile organic solvents (Zain et al. 2014). Silveira et al. (2018) synthesized silver nanoparticles (AgNPs) using aqueous extract of *Ilex paraguariensis* and evaluated its antimicrobial potential in vitro against Gram-negative and Gram-positive bacteria. The synthesis was performed by adding *I. paraguariensis* material to the boiling ultrapure water followed by filtration and the addition of 1 mM silver nitrate ($AgNO_3$). The syntheses were given in dark conditions with a reaction temperature of 50 °C for a period of 3.5 h. The authors observed that the AgNPs produced had a variable dry diameter, between 4 and 30 nm, with predominantly spherical morphology, high reproducibility, and great colloidal stability analyzed for 1 year. In addition, they were highly efficient in inhibiting the growth of *Escherichia coli* and *Staphylococcus aureus* when in concentrations close to 256 µM. The authors suggest that the antibacterial activity of the synthesized AgNPs occurred through a process known as molecular crowding. In this, the cells, when in contact with the AgNPs, lose homeostatic control, reducing diffusion rates, and

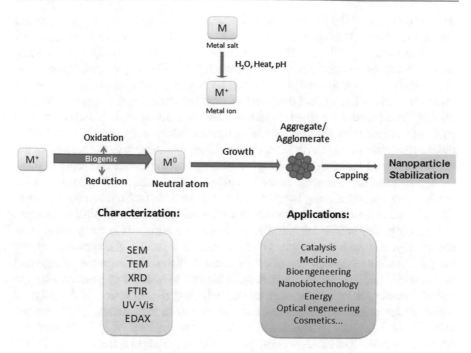

Fig. 9.2 Diagram representing a possible biosynthetic pathway for metallic nanoparticle synthesis

their biochemical activities, which can lead to changes in electrostatic interactions as well as inactivation of their vital activity.

Supercritical fluids in the green synthesis of nanoparticles involve the use of solvents with temperatures above their critical temperature, in which they exhibit behavior of gases and liquids simultaneously. In the supercritical temperature, the variation of temperature and pressure allows obtaining solvents with maximum solubility (liquid behavior) and easy compressibility (gas behavior). The variation of the reactional parameters makes it possible to synthesize nanoparticles of different size, morphology, composition, structure, and architecture in a controlled manner. Among the most commonly used supercritical solvents are supercritical water (*scWater*) and supercritical carbon dioxide (*scCO$_2$*). The supercritical water has as main advantage the possibility of complete solubilization of non-polar solvents, since, in the critical temperature, the water has significant differences in its dielectric constant.

Supercritical carbon dioxide as a solvent is interesting due to the easy recovery by depressurizing, besides its low reactivity when in supercritical temperatures (74 bar and 304 K). Demirdogen et al. (2018) synthesized poly(3-hydroxybutyrate-*co*-3-hydroxyhexanoate) [P(3HB-*co*-3HHx)] nanoparticles for sustained release of bortezomib using supercritical CO$_2$ as solvent combined with the electrospraying technique (Carbon Dioxide Assisted Nebulization-Electrodeposition, CAN-ED). The copolymer [P(3HB-*co*-3HHx)] belongs to the family of polyhydroxyalkanoate

(PHA), polymers widely known for their application in the biological environment due to their biodegradability, biocompatibility, and degradation of their chains by surface erosion. In addition, PHAs are interesting hydrophobic drug-encapsulating agents, such as bortezomib or (1R)-3-methyl-1-({{(2S)-3-phenyl-2-[(pyrazin-2-ylcarbonyl) amino]proponoyl}amino)butyl boronic acid, an antineoplastic protease inhibitor used for the clinical treatment of multiple myeloma and mantle cell lymphoma. The authors were able to obtain efficient and reproducible polymeric micro- and nanospheres with reduced solvent utilization when using supercritical carbon dioxide in syntheses by electrospraying, with less impairment to the activity of the active material, besides achieving higher loading capacity when compared to other techniques commonly adopted in the literature. The schematic representation of the electrospray apparatus using supercritical carbon dioxide ($scCO_2$) can be seen from Fig. 9.3. The technique consists the use of $scCO_2$ at high pressure to push the plunger of the syringe that acts as a reaction vessel in the conventional electrospinning technique. High voltage is then applied to the nozzle at the end of the syringe, ionizing the particles. As a consequence of the pressure difference between the syringe and the external medium, all the carbon dioxide used as solvent evaporates after the material passes through the syringe nozzle, allowing the collection of the polymer nanoparticles without the interference of the solvent in the reaction. The technique allowed obtaining dry nanoparticles, with a mean diameter of 150 nm, with the controllable release of bortezomib by pH variation, at temperatures close to 37 °C.

The development of methods and processes to nanoparticle production employing low toxicity and safe and easy recovery solvents has now become a major

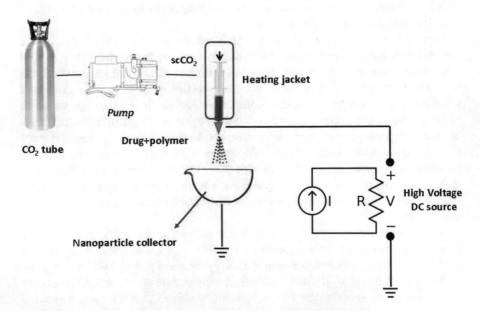

Fig. 9.3 Schematic representation of a possible electrospray apparatus using $scCO_2$ as the solvent

concern in the academia and industry. The toxicological potential of nanoparticles formed by physicochemical methods has been extensively researched in recent years mainly due to their effects not only on humans but also on animals and the environment. In this way, new synthetic routes have been developed to make the synthesis more efficient, clean, and sustainable through the use of natural materials and renewable substances of low energy consumption. Thus, it is possible to prevent the production of toxic waste and favor the economic, social, health, and environmental premises associated with the production and use of nanoparticles.

9.3   Bioresources for the Synthesis of Green Nanoparticles

Green synthesis of nanoparticles commonly exploits biological organisms, parts of them (organs, tissues, cells, biomolecules, and secondary metabolites), or biological waste products (Fig. 9.4) (Silva et al. 2015; Saif et al. 2016; Das et al. 2017; Ebrahiminezhad et al. 2018). The bioresources may be obtained from ecosystems, food or non-food sources, and agroindustrial and livestock wastes (Silva et al. 2015; Mythili et al. 2018; Das et al. 2017; Saratale et al. 2018; Gour and Jain 2019).

Several types of metallic (silver, gold, copper, palladium, platinum, ferric oxide, magnesium, manganese, and zinc, among others) nanoparticles have been successfully synthesized using biological sources (Ghosh et al. 2017; Yadi et al. 2018; Andra et al. 2019; Oliveira et al. 2019; Gour and Jain 2019). Most of the extracts from those sources are used as reducing and stabilizing agents for nanoparticle synthesis, having diverse activities such as antimicrobial, anticancer, antioxidant, antidiabetic, and others (Adelere and Lateef 2016; Ghosh et al. 2017; Rehana et al. 2017; Ahmad et al. 2018; Mythili et al. 2018; Hembram et al. 2018; Zuorro et al. 2019; Suwan et al. 2019; Ahn et al. 2019).

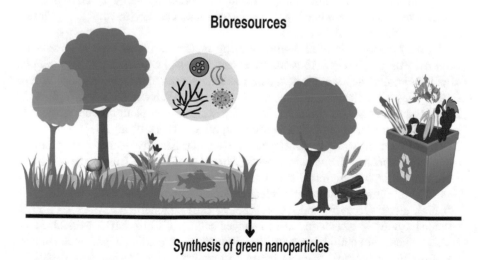

Fig. 9.4 Some potential bioresources for the synthesis of green nanoparticles

Nanoparticle synthesis using plants (gymnosperms to angiosperms) as a biological source represents a significant expression in scientific studies when compared with bacteria, algae, fungi, viruses, and animals (Silva et al. 2015; Rasheed et al. 2017; Mousavi et al. 2018; Yadi et al. 2018; Mousavi et al. 2018; Oliveira et al. 2019). The main phytochemical agents related to producing nanoparticles are alkaloids, polyphenols, phenolic acids, simple or polymeric carbohydrates, amino acids, and proteins (Silva et al. 2015; Andra et al. 2019). The phytochemical profile can be modified for biotic (pests and phytopathogens) or abiotic (pH, drought, temperature, salinity, and others) stress factors (Suzuki et al. 2014), such as different seasons (summer and winter), which influence the nanoparticle biosynthesis for modulating the final properties (Oliveira et al. 2019). Moreover, the antioxidant potential of the plant can increase the efficiency to produce nanoparticles by high reducing capacity (Harshiny et al. 2015; Muthukumar and Matheswaran 2015; Ebrahiminezhad et al. 2018). Other factors such as temperature, pH, metal, reaction time, and plant extract concentrations in the reaction medium could influence the biosynthesis (Bonatto and Silva 2014; Shah et al. 2015a, b; Khan et al. 2018).

Plant extracts used to produce nanoparticles may be obtained from their different parts such as fruits, flowers, stems, roots, seeds, latex, and especially leaves (Rajkuberam et al. 2015; Silveira et al. 2018; Santiago et al. 2019; Wang et al. 2019). Moreover, isolated compounds are also used for nanoparticle biosynthesis such as vitamins (ascorbic acid), amino acids (L-lysine), synthetic metabolites (tannic and gallic acid), simple carbohydrates (glucose), or biopolymers (starch and chitosan) (Park et al. 2011; Saif et al. 2016; Andra et al. 2019). Plants can contain a high quantity of antioxidants, which may be used as nutraceuticals, food additive drugs, as well as a natural agent to reduce and stabilize nanoparticles (Kirubaharan et al. 2012; Tahir et al. 2017; Yadi et al. 2018). From a technological viewpoint, the main advantages of the use of plants for the nanoparticle synthesis include (1) availability, (2) safety in handling, and (3) metabolite variability that may aid in quickly reducing metals and maintain highly stabilized nanoparticles (Iravani 2011; Chung et al. 2016; Yadi et al. 2018).

Recently, nanoparticles biosynthesis using medicinal plants such as *Artemisia vulgaris*, *Panax ginseng*, *Abutilon indicum*, *Acalypha indica*, *Erythrina indica*, *Melia dubia*, *Ocimum tenuiflorum*, *Solanum tricobatum*, and *Ziziphora tenuior* has shown potential anticancer, antimicrobial, and antioxidant capabilities (Singh et al. 2016; Rasheed et al. 2017). Furthermore, the utilization of plant product waste can be an important strategy for nanoparticle synthesis (Park et al. 2011; Saif et al. 2016). Biological waste products can be an eco-friendly strategy to produce nanoparticles such as *Trapa natans* peels (Ahmad et al. 2018). This species presents features such as antimicrobial and anti-inflammatory activities as well as cancer-protective abilities. Transmission electron microscopy revealed the size of gold (25 nm), silver (15 nm), and Au-Ag (26–90 nm) nanoparticles which exhibited potential cytotoxic effects in various cancer cells (HCT116, MDA-MB-231, and HeLa). These bimetallic nanoparticles induce ROS-mediated p53-independent apoptosis in cancer cells that can be used in cancer therapy (Ahmad et al. 2018). Suwan et al. (2019) used different cellulose derivatives and various preparation parameters to produce silver nanoparticles, which showed distinct inhibition potential against *Escherichia coli* and *Staphylococcus aureus*. Such results suggest that the type of cellulose derivatives and the reaction parameters of the nanoparticles synthesis, such as pH, temperature, and reaction period, play an important role in the yield and physicochemical properties of the obtained silver nanoparticles. Another group explored with success for nanoparticle synthesis is the microorganisms (Suwan et al. 2019).

Nanoparticle synthesis by microorganisms (bacteria, microalgae, yeast, and fungi) as well as macrofungi (mushrooms) and macroalgae can be intracellular (inside the organisms cells), extracellular (outside the organisms cells), or even on the surface of the cell (Silva et al. 2016; Elegbede et al. 2019; Puja and Kumar 2019). The extracellular synthesis of nanoparticles is faster and easier than intracellular synthesis because the former can secrete large amounts of enzymes and/or other compounds used for the synthesis process (Silva et al. 2016). Nanoparticle synthesis using fungi can occur by enzymatic reduction as well as a cell wall bound process using different biomolecules (Moghaddam et al. 2015; Silva et al. 2016; Elegbede et al. 2019). Elegbede et al. (2019) exploit xylanases of *Aspergillus niger* (NE) and *Trichoderma longibrachiatum* (TE) to produce silver-gold nanoparticles (Ag-AuNPs), mainly spherical shaped, ranged from 6.98 to 52.51 nm, which showed distinct inhibition potential against clinical bacteria (*Escherichia coli*, *Klebsiella granulomatis*, *Staphylococcus aureus*, and *Pseudomonas aeruginosa*) and fungi (*Aspergillus fumigatus* and *A. flavus*). Additionally, these nanoparticles also showed potent antioxidant activities as well as potential in the management of blood coagulation disorders (Elegbede et al. 2019).

There are a few studies using animals as a source to produce nanoparticles (Das et al. 2017). *Bombyx mori* silk fibroin was used with bioresource for AgNP synthesis (Shivananda et al. 2016). TEM images of nanoparticles showed the spherical shape with 35–40 nm. Silver nanoparticles showed enhanced antimicrobial activity against *Bacillus subtilis* and *Salmonella typhi* (Shivananda et al. 2016). Nadaroglu et al. (2017) have synthesized platinum nanoparticles from egg yolk, sources with high vitamin and protein quantity. Scanning electron microscopy analysis exhibited cubic nanoparticles with diameters ranging in size from 7 to 50 nm. Platinum nanoparticles have several applications such as pharmaceutical, chemical, energy, catalysis, among others (Nadaroglu et al. 2017). Other sources of biomolecules and compounds can be obtained from crustacean shell wastes from shrimp, crab, lobster, and krill that contain large amounts of chitin to produce chitosan that are widely employed as tissue-engineering scaffolds and in drug delivery applications such as micro/nanoparticles, micelles, hydrogels, and others (Ghosh et al. 2017). Kalaivani et al. (2018) used chitosan as a source to produce silver nanoparticles with mostly spherical shape and ranges from 10 to 60 nm. The nanoparticles showed antibacterial activity against *Bacillus* sp., *Staphylococcus* sp., *Pseudomonas* sp., *E. coli*, *Proteus* sp., *Serratia* sp., and *Klebsiella* sp., such as antifungal activity to control *Aspergillus niger*, *A. fumigatus*, *A. flavus*, and *Candida albicans* (Kalaivani et al. 2018). Thus, the diversity of biological organisms and compounds with potential for nanoparticle synthesis offers a wide application of nanotechnology in strategic areas such as biotechnology and its applications in the medical field and related areas.

9.4 Scaling-Up the Synthesis of Green Nanoparticles

The majority of scientific papers show that biological resources are valuable, abundant, and simple for large-scale synthesis of green nanoparticles, but they present it as a perspective specifically, they do not report rational experiments that confirm it (Mittal et al. 2013; Rajan et al. 2015; Gour and Jain 2019). This indicates that the green synthesis of nanoparticles in the laboratory scale is well established; however, converting it to an industrial scale is still a challenging issue. That's because nanoparticle properties, such as chemical, physical, and biological aspects, are difficult to reproduce and their application directly depends on these characteristics (Virkutyte and Varma 2013).

Therefore, to overcome large-scale obstacle, the scaling-up process of green nanoparticles synthesis, which means increasing the amount of production based on laboratory scale synthesis, needs further investigation, improvement, and validation. In addition, the scaling-up is a crucial step to make advances reported in scientific literature related to green nanotechnology processes that become tangible technologies and real commercial products (A matter of scale 2016).

Considering this, some few studies reported scaling-up methods that can be applied for nanoparticle manufacture and/or presented synthesis of green nanoparticles in semi-pilot or also large-scale amounts. In principle, different methods can be used for scaling-up synthesis of different types of green nanoparticles, and they are presented in Fig. 9.5. Paliwal et al. (2014) reviewed methods for the development of polymeric, lipidic, and metallic nanoparticles with therapeutics and diagnostics potential that can be applied for large-scale green synthesis of nanoparticles. Similarly, Vauthier and Bouchemal (2009) reported that emulsification–solvent diffusion, emulsification-reverse salting out, and nanoprecipitation methods have already been used as a scaling-up approach for polymeric nanoparticle synthesis.

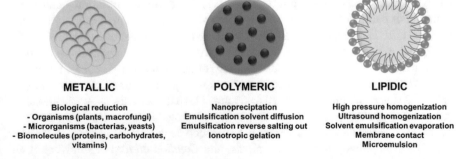

| METALLIC | POLYMERIC | LIPIDIC |

Biological reduction
- Organisms (plants, macrofungi)
- Microrganisms (bacterias, yeasts)
- Biomolecules (proteins, carbohydrates, vitamins)

Nanopreciptation
Emulsification solvent diffusion
Emulsification reverse salting out
Ionotropic gelation

High pressure homogenization
Ultrasound homogenization
Solvent emulsification evaporation
Membrane contact
Microemulsion

Fig. 9.5 Scheme showing different methods that can be used for pilot- and large-scale production of metallic, polymeric, and lipidic nanoparticles

Apart from that, Loh et al. (2010) presented a strategy to fabricate chitosan nanoparticles based on a spinning disc, an ionotropic gelation method, processing for possible commercialization of this nanomaterial as a drug delivery platform.

Regarding synthesis of lipid nanoparticles (LNP), Hu et al. (2016) developed a manufacturing strategy for the production of nanostructured lipid carriers loaded with coenzyme Q10 (CoQ10-NLC), using biocompatible reactants by high-pressure homogenization (HPH) technique. According to the procedure of large-scale production of CoQ10-NLC, five steps were considered: preparation of production, pre-emulsification, homogenization, cooling and inspection, and packing. Besides that, Esposito et al. (2017) compared HPH and ultrasound homogenization (UH) techniques for pilot-scale production of solid lipid nanoparticles and nanostructured lipid carriers loaded with progesterone. In conclusion, they presented that HPH was the best method because it enabled the production of smaller nanoparticles without agglomerates and an increase of 20-fold production volume of nanoparticles with respect to UH.

Moreover, in relation to metallic nanoparticles, Moon et al. (2010) showed that magnetic nanoparticles can be synthesized by bacteria capable of reducing Fe(III), and the scale-up process was performed in a 35 L reactor. Ould-Ely et al. (2011) developed a protocol for green synthesis of $BaTiO_3$ nanoparticles in a 2-liter reactor vessel at nearly room temperature (25 °C) based on a smaller-scale protocol. Additionally, Satapathy et al. (2015) exposed a strategy for AgNP production by an alga, *Chlorella vulgaris*, in a continuously stirring and non-aerated bioreactor, and Kisyelova et al. (2016) analyzed the possible influence of the use of a reactor on AgNP synthesis, in different operation conditions, considering pH, concentration of reactants ($AgNO_3$, starch, NaOH, and glucose), stirring, size, yield, and metal content weight.

Although scientific reports have addressed the process of scaling-up and large-scale production of some types of nanoparticles, much more have to be investigated, taking into account that the amount of different nanomaterials is constantly increasing. Therefore, approaches confronting this challenge must be exposed. Firstly, some scaling-up conditions/factors and their values should be defined in order to synthesize nanoparticles with similar properties to those of laboratory scale. Among various factors that can affect the yield and characteristics of green nanomaterials are temperature, pressure, pH, aeration, stirring, type, and concentration of reactants, type of reactor propeller, reaction time, type of biological resources, and environment conditions (Bonatto and Silva 2014; Patra and Baek 2014; Christopher et al. 2015).

Consequently, it is important to understand which variables are more significant and which ones can be influenced by others. This rational approach can be based on a deep review of scientific reports or mathematical and statistical models, as factorial or Plackett-Burman designs. For instance, El-Moslamy et al. (2016) presented a bioprocess for *Chlorella vulgaris* cultivation and biosynthesis of anti-phytopathogenic AgNPs. Those authors used the Plackett-Burman experimental design for screening seven parameters for the biosynthesis of AgNPs.

After choosing the factors that are more important to the process, the next step should be finding the ideal values for them and optimize the synthesis route. The optimization considers that the best values are those that can generate the desired characteristics in a process or final product (Araujo and Brereton 1996). This step also takes into account costs, productivity, time, human resources, and minimization of error sources. In spite of that, when it comes to optimization of semi-pilot or large-scale synthesis of green nanoparticles, it is not completely understood or even used.

The simplest nonstatistical method named one-factor-at-a-time approach could be used for the analysis of one variable each time, while the others should be kept fixed. However, this method is not capable of identifying the complex interactions between variables, and it needs more investments in time and money. Consequently, some studies began to use mathematical and statistical approaches, and computational tools to solve the scaling-up issue of green nanoparticle synthesis and, consequently, select the most relevant factors, and identify the best conditions (El-Moslamy 2018).

El-Moslamy et al. (2017) have applied the Taguchi design for optimization of large-scale production of AgNPs by *Trichoderma harzianum* fungi. Nevertheless, recently, El-Moslamy (2018) demonstrated a cost-effective bioprocess for magnesium oxide nanoparticles (MgONPs) synthesized by bacteria, and, for the optimization, statistical experimental designs such as Plackett-Burman design (PBD), Box-Behnken design (BBD), and Taguchi design were applied in the study. These statistical tools and others, like factorial and central composite design (CCD), should be more used in the study of semi-pilot and large-scale synthesis of green nanoparticles in order to simplify the analysis of variables that affect the process and also help to find the best value for each one.

Moreover, artificial neural network (ANN) is a computational tool inspired by biological neural systems and based on mathematical algorithms (Basheer and Hajmeer 2000). Since this technique can be applied to engineering and plant biology processes (Cruz et al. 2011; Gallego et al. 2011), it can be applied for nanotechnology and optimization of variables in laboratory-, semi-pilot-, and large-scale green synthesis of nanoparticles. ANN was applied for the productivity of iron oxide nanoparticles synthesized by *Coriandrum sativum* leaf extract applied for photocatalytic discoloration of red dye wastewater (Sathya et al. 2018). Furthermore, ANN was used for size prediction of AgNPs produced by *Curcuma longa* aqueous extract (Shabanzadeh et al., 2013a), starch, an aqueous solution of NaOH (Shabanzadeh et al. 2013b), and *Vitex negundo L.* aqueous extract (Shabanzadeh et al. 2015).

Considering the exposed, the number of scientific papers concerned about the scaling-up process and large-scale synthesis of green nanoparticles is insufficient. Researchers need to further investigate this topic in order to make green nanoproducts tangible innovations that can be applied in the real world, and a potentially feasible workflow is proposed in Fig. 9.6.

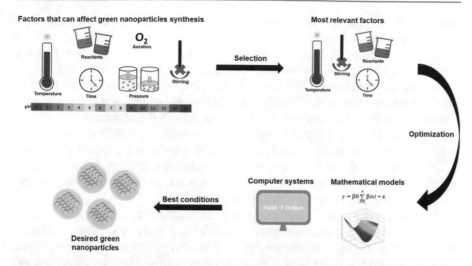

Fig. 9.6   Steps of a scaling-up strategy for obtaining the desired green nanoparticles

9.5   Uses of Green Nanoparticles for Cell Biology Research

Nowadays, there is a lot of research focusing on understanding greenly synthesized nanomaterials, such as carbon nanotubes, noble metal nanoparticles, metal oxide nanoparticles, and ferritic nanoparticles. The characteristics and properties of these nanomaterials depend directly on the methodology by which these are synthesized. Therefore, there are a lot of different applications based on how the nanomaterial behaves under different conditions. Nanoparticles have often been studied for therapeutic and diagnostic applications, and some big advances have been made in this field. However, the potential for application of these structures is much greater. In recent years, green nanoparticles have been used as research tools, providing new approaches for development and changes in established strategies.

Most published scientific reports demonstrate biological applications of nanoparticles synthesized through chemical routes, although it is hypothesized that there are no reasons that similar results cannot be achieved by using green synthesis routes. Even in 1999, Shenton et al. comment on how useful is the development of bio-derived routes to produce organized forms of inorganic matter and how the specificity of biomolecular interactions is a powerful tool toward target structures. In view of that, this research group demonstrated a different strategy, using antibody-antigen coupling for self-assembly of metallic nanoparticles into complex 3D networks.

The use of nanoparticles on the research field can be done exploring the nanoparticle-biomolecule interaction. Taking advantage of this, it is possible to control intracellular and extracellular processes, such as transcription regulation, enzymatic inhibition, delivery, and sensing. Nanoparticles are interesting due to their properties of being synthesized with different shapes, sizes, and compositions and also associable with different biomolecules through biofunctionalization

(De et al. 2008). These interactions can be useful for understanding complex cellular mechanisms and responses.

To achieve conjugation of biomolecules to nanoparticles, there are different strategies available: covalent binding taking advantage of the functional groups from both the biomolecule and the nanoparticles, non-covalent binding through affinity between receptor and ligand, electrostatic interactions, and through ligand-mediated binding (Ravindran et al. 2013). Faced with the enormous diversity of organic molecules found in nature and selected by evolution that already participate in biological process, these molecules can be used in order to functionalize nanoparticles. Biofunctionalization of nanoparticles was already performed through cysteine residues that are present in a protein surface, meaning that a green route can be used to achieve stable bioconjugation (Naka et al. 2003).

AuNP-oligonucleotide complexes have been already used for intracellular gene regulation controlling protein expression in cells (Rosi et al. 2006). The experimental results of this group demonstrate that by modifying nanoparticle surface, it was possible to achieve cooperative and incremental properties. These modifications led to enhancement of target binding and allowed the introduction of a variety of functional groups that can be useful in terms of understanding how structures work within a cell. Therefore, they demonstrated the possibility of using nanomaterials to control biological processes. However, it is important to emphasize that this study used a chemical route to achieve nanoparticle functionalization, but these could be achieved through a green route.

Nanoparticles are also greatly used for biosensing applications when properties such as malleability, high surface areas, a wide range of diameters, different shapes, and controlled composition are exploited. Commonly, a biosensor consists of two major components: a recognition element for the target binding and a signaling element when the event of binding happens (De et al. 2008). Various highly sensitive biosensing methods have been already developed exploring, among others, properties such as localized surface plasmon resonance (LSPR), fluorescence enhancement, electrochemical activity, or functionalizing these nanoparticles with biomolecules in order to increase specificity (Doria et al. 2012). For all these different properties, different types of sensing are available: colorimetric sensing (Vilela, González and Escarpa. 2012), fluorescence sensing (Ruedas-Rama et al. 2012), and electrochemical sensing (Luo et al. 2006), among others, which allow sensing intracellular biomolecules, enzyme activity, and pH for a better understanding of living cell processes (Ferreira 2009).

Various colorimetric methods of biosensing are used for detection of biomolecules, especially DNA (De et al. 2008). Typically, AuNP suspension presents a red color, while silver ones present a yellow color. A common procedure is the synthesis of nanoparticles biofunctionalized with DNA (Doria et al. 2012). When nanoparticles are in contact with the target DNA sequence, they aggregate and change color. This method is very sensitive and can detect very low levels of DNA without using PCR approaches and can also be used for protein detection. Therefore, it can be successfully used for research purposes (Fig. 9.7).

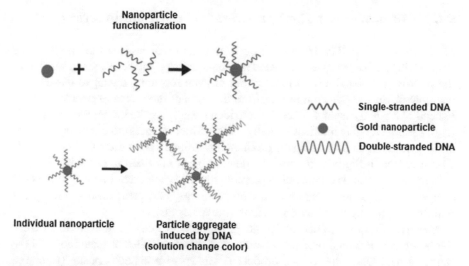

Fig. 9.7 Schematic illustration of detection of oligonucleotide through biofunctionalized nanoparticles

Raj and Sudarsanakumar (2017) developed a colorimetric sensor for cysteine using greenly synthesized AuNPs corroborating the rise of eco-friendly approaches. Nanoparticles alone formed a deep wine red-colored suspension that, when added to cysteine solution, changed to dark blue. In their assays, other amino acids such as tryptophan, histidine, and phenylalanine were also used as a control, with no color changes detected. These results indicate that cysteine could be selectively detected using greenly synthesized AuNPs and demonstrate a possible nanotool that can be adapted for each diagnostic context, depending on the requirements for biomolecule detection.

Fluorescent nanoparticles are other interesting and promising candidates to outstanding applications due to properties that confer advantages comparing to simple fluorescent molecules, such as high brightness and photostability. Although the first use of these particles was mainly as substitutes of traditional organic dyes, later, it was seen to have the potential for intracellular investigations (Ruedas-Rama et al. 2012), and they can be routinely produced through green synthesis routes.

Therefore, green nanoparticles have several applications for the biological research field. Their wide distribution of size, high surface-area-to-volume ratio, and a lot of possibilities of functionalization on the surface make sustainable nanoparticles so interesting and so applicable (Ferreira 2009). Such nanomaterials can be used to improve old techniques and increase the sensitivity of commonly used instruments. On the other hand, they can serve per se as a tool for studying cell-based processes.

9.6   Uses of Green Nanoparticles for Tissue Engineering

Tissue engineering (TE) is an exciting RD&I field that utilizes the principles of engineering and sciences associated to life (e.g., biology, medicine, veterinary) aiming to solve problems related to repairing or even replacing damaged tissues. For example, it has the potential to transform the lives of thousands of people who are in need of organ transplant, since the core idea is applying the individual's own cells (e.g., stem cells) plus matrices (usually made of biopolymers) to create new organs and to contribute in addressing the problem of waiting lists for transplants and offers a high compatibility with the patient (decreasing or eliminating the risks of rejection). Also, TE can be employed to promote tissue regeneration in the case of some injury. Thus, it is an emerging area with increasing and great results coming out mainly by applying novel multiple composite materials.

The use of more than one composite in formulations can be relevant to enhance the properties of a final product. When one of the composites or materials is in the range of nanoscale, the formed product is known as a nanocomposite (Camargo et al. 2009). Indeed, the current concept of several research groups in the field of TE involves joining the properties of materials to offer scaffolds with desirable characteristics. For example, greenly synthesized zinc oxide nanoparticles incorporated in polyurethane nanofibers produced by electrospinning loaded with virgin coconut oil were produced for TE purposes (Ghazali et al. 2018). Nanofibers produced by electrospinning present fibrillar structures that mimic the structure of the extracellular matrix, providing large surface areas and control of the mechanical properties, and are facile to functionalize (Agarwal et al. 2009). In Ghazali et al. (2018) study, the polymer polyurethane was employed because it offers great barrier properties and oxygen permeability; the zinc nanoparticles can offer antimicrobial activity, besides the ability to enhance growth factors release; and the virgin coconut oil offers antioxidant capability (helping in cell adhesion and proliferation, in case of fibroblasts). Therefore, the strategy to join materials with several characteristics of interest is extremely useful in the TE field.

Green nanoparticles, with potential use for TE processes, can also be produced using biopolymers, such as chitosan. This natural polymer is a polysaccharide derived from the partial deacetylation of chitin, a polymer produced by crustaceans that possess many advantages, including biodegradability, biocompatibility, is renewable, and environmentally friendly (Zargar et al. 2015). Sophisticated mechanisms have been proposed on the utilization of chitosan nanoparticles in TE. For example, it reported the production of gene-activated scaffold platforms with high potential on TE. Chitosan-DNA nanoparticles were applied as a gene delivery factor in mesenchymal stem cells and were introduced in three different collagen-based scaffolds, collagen alone, collagen + hydroxyapatite, and collagen + hyaluronic acid, creating a scaffold platform that can enhance bone and cartilage repair (Raftery et al. 2015).

Despite a large number of works that utilize nanoparticles soaked in scaffolds, nanoparticles can also be used as vehicles for delivering bioactive compounds useful for TE, as in the case of the work of Rajam et al. (2011) that produced chitosan nanoparticles for delivery of two growth factors (GFs). GFs are polypeptides that modulate cellular activities by signal transmission. In general, they can have either a stimulatory or inhibitory action over gene expression, cellular differentiation, migration, adhesion, and proliferation (Babensee et al. 2000). Therefore, GFs play an essential role in the TE field. In the study, epidermal growth factor (that stimulates the proliferation and differentiation of epithelial cells from the lung, cornea, tracheal tissue, gastrointestinal tract, and skin) and fibroblast growth factor (that can promote angiogenesis) were incorporated into chitosan nanoparticles either individually or in combination. Results showed that the nanosystem was successful in the incorporation of both GFs into chitosan nanoparticles. Besides, the system was biocompatible and nontoxic to cells, releasing the GFs in a sustained manner.

Other biopolymers can also be used alone or in combination with polymers like chitosan to produce green nanostructured scaffolds for TE applications, including cellulose that is probably one of the most (if not the most) abundant polymers produced by living organisms. Cellulose is a polysaccharide produced by plants that present several interesting characteristics such as biocompatibility, biodegradability in nature, and can be chemically or structurally modified, among others (Ioelovich 2014). Ko et al. (2018) reported the production of scaffolds for TE purposes, using cellulose nanocrystals extracted from *Lactuca sativa*, which were incorporated into scaffolds of chitosan and lactic acid that resulted in formulations with desirable characteristics such as porosity, biodegradability, drug release property, and high cell viability.

Besides, some green nanoparticles can be employed on scaffolds to introduce interesting properties, such as antimicrobial capability. In the process of TE, the cellular scaffolds are susceptible to contamination, and the addition of antimicrobial nanoparticles could avoid such inconveniences. There are many reports of entrapment of antimicrobial nanoparticles in hydrogels for several applications, including wound dressing purposes. Hydrogels are polymeric three-dimensional, hydrophilic networks with the ability to retain water or biological fluids (Caló and Khutoryanskiy 2015). In this manner, AgNPs are candidates with high potential to be applied in biodegradable scaffolds to form nanocomposites, since, in moderate concentrations, hydrogel-AgNPs are non-toxic (Dhar et al. 2012; Travan et al. 2009; Boonkaew et al. 2013) and exhibit antimicrobial activity against a wide range of pathogenic microorganisms (Haseeb et al. 2017), making them ideal candidates for addition in materials for TE purposes. Zulkifli et al. (2017) prepared scaffolds from hydroxyethyl cellulose-AgNPs for skin TE applications. In the study, AgNPs were synthesized by an environmentally friendly method, utilizing the hydroxyethyl cellulose as a reducing agent, and then, the scaffolds were formed by the freeze-drying method. The formed scaffolds presented good porosity, low cytotoxicity, and good biocompatibility for the growth and proliferation of human fibroblast cells. Thus, the scaffold of the study can be employed in tissue regeneration, especially in skin TE.

Thus, green nanoparticles can be employed in diverse bioengineering processes aiming to repair, reconstruct, or regenerate bones, cartilage, skin, and among other tissues and organs, in the form of nanofibers or porous scaffolds; or alone, carrying

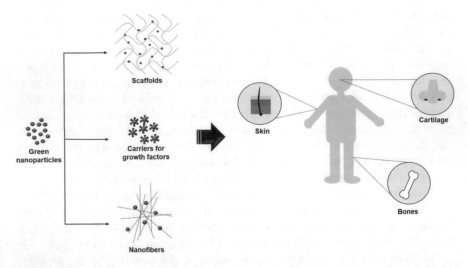

Fig. 9.8 Green nanoparticles can be employed in porous scaffolds, nanofibers, or alone as carriers for growth factors. The combination of those materials can enhance the properties of formulations for tissue engineering of several tissues and organs, such as the skin, bones, or cartilages, among others

GFs, for example (Fig. 9.8). The TE field is growing fast, and the addition of green nanoparticles with different and interesting intrinsic properties can revolutionize regenerative medicine as we currently know.

9.7 Uses of Green Nanoparticles for the Treatment of Plant Diseases

Despite the fact that the treatment of plant diseases does not explicitly relate to bio-medical and bioengineering applications of green nanoparticles, the relevance of them as models for the treatment of some human diseases (including those caused by microorganisms), and mainly their importance for food, feed, fuel, fiber, and other bio-based products that are essential to human health and well-being, justifies a subtopic of this chapter on this subject. If the current pace of consumption continues, by 2050, it will require 60% more food, 50% more energy, and 40% more water. To respond to global demand of nine billion people by 2050, efforts and investments are needed to promote this global transition to sustainable agriculture and land management systems. These measures imply an increase in the efficiency on the use of bioresources – mainly water, energy, and land – but also in the considerable reduction of food waste (Tilman et al. 2011).

One of the main problems that hinder the increase of the quality and quantity of crops is losses caused by phytopathogens, estimated at 20–40% per year all over the world (Myers et al. 2017). Chemical control is the main technology for the control of plant disease. Despite the advantages, pesticides have harmful effects on non-target organisms such as the resurgence of the pest population, development of resistance in phytopathogens, and high level of product waste in field application. As a consequence, it is necessary to develop new technologies with high efficiency, low cost, and few environmental damages.

The use of green nanotechnology can be an eco-friendly alternative to the development of a new generation of pesticides and other agrochemicals. Green nanotechnology apply natural agents as reducing, capping, and stabilizing constituents of plant extracts and microorganisms to produce nanoparticles or biological material, which are considered non-toxic, clean, and cheap compared with chemical syntheses (Hussain et al. 2016).

Nanoparticles are the basic unit of several nanotechnologies, presenting characteristics that differ from other materials, such as the high surface area/volume ratio, high reactivity, and easy delivery of organic and inorganic molecules. These characteristics allow the use of green nanoparticles in several applications in agriculture, gaining prominence in detection, transformation, and in the control of plant diseases (Elmer and White 2018).

In the detection, biosensors can be developed using nanoparticles to allow the early recognition of pathogens with reliability and reduced cost. In the transformation, nanocarriers are competent in carrying materials such as ssDNA (RNAi). However, the researchers are advancing mainly in the production of nanoparticles and nanomaterial applied in the management of pathogens. As a control agent, nanoparticles can be used against phytopathogens, presenting multisite action and contributing to the decrease in the development of resistance. The synthesized green nanoparticles have different compositions, shapes, sizes, and surface properties and can be applied directly to seeds, leaves, or soil with antimicrobial and anti-pest activities or inducing resistance in plants (Saratale et al. 2018). Among them, metallic nanoparticles (Ag, Al, Au, Ce, Cu, Fe, Mg, Mn, Ni, Ti, and Zn) are intensively studied due to their effects against a wide spectrum of phytopathogens through a multisite action, which include inhibiting the respiration, reducing enzyme activities and protein function, altering the properties of cell wall, and interfering cell replication, and frequently new mechanisms are reported (Worrall et al. 2018). Indeed, green nanoparticles are commonly employed to control bacteria, fungi, nematodes, and, more rarely, viruses that attack plants.

In an attempt to control bacteria, Masum et al. (2019) synthesize AgNPs using *Phyllanthus emblica* fruit extract that interfered in biofilm formation, swarming, and increase of the secretion of effector Hcp of *Acidovorax oryzae*. ZnO and TiO_2 nanoparticles reduced the severity of three bacterioses (*Pectobacterium betavasculorum, Xanthomonas campestris* pv. *beticola*, and *Pseudomonas syringae* pv. *aptata*) in beetroot. Moreover, these nanoparticles increased the chlorophyll and carotenoid contents, superoxide dismutase, catalase, ascorbate peroxidase, phenylalanine ammonia lyase, glutathione, proline, H_2O_2, and malondialdehyde in beetroot (Siddiqui et al. 2018).

When greenly synthesized copper-chitosan nanoparticles were sprayed in finger millet, they stimulated the growth and the suppression of blast disease. The decrease of symptoms was correlated with increase of chitinase, chitosanase, and β-1,3 glucanase activities (Sathiyabama and Manikandan 2018). In nematodes, Kalaiselvi et al. (2019) found indication that AgNPs interfere in reduction of gall formation, egg masses, number of eggs per egg mass, and J2 populations of *Meloidogyne incognita* within the plants. In virology, with the application of AuNPs in plants via mechanical abrasive process, it was observed to melt and dissolve the Barley yellow mosaic virus (Alkubaisaisi et al. 2015). Furthermore, it is common to observe some specific green nanoparticle reducing the severity of different groups of pathogens. Khan and Siddiqui (2018) report the reduction of incidence of diseases caused by *Ralstonia solanacearum*, *Phomopsis vexans*, and *Meloidogyne incognita* in eggplant after the application of ZnO nanoparticles.

Nanoparticles as carriers have additional advantages such as enhanced shelf-life, improved solubility of pesticides with low solubility, controlled and gradual release, low toxicity, and enhanced stability, boosting the site specific into the target pest of plants (Worrall et al. 2018). However, synthetic nanocarrier is as yet uncertain due to the high cost of manufacturing, making it financially infeasible for the application in agriculture. Aiming to resolve this problem, some recent nanomaterials used as carriers in agriculture contain biomolecules such as chitosan, biolipids, and protein coat of virus (Pérez-de-Luque and Rubiales 2009; Alemzadeh et al. 2018). They are grouped and considered as natural or green nanocarriers.

Chitosan has interesting properties favorable to be applied as a nanodelivery system, such as high biodegradability, low toxicity, biocompatibility, and antimicrobial activity. Moreover, chitosan increases the adherence in epidermis of leaves and stems, prolonging the contact time and consequently the activity time of the carrier biomolecule (Malerba and Cerana 2016). The synergistic antimicrobial activity of green AgNPs, associated with chitosan matrix, reduced the severity of tomato infested with *R. solanacearum* (Santiago et al. 2019). Choudhary et al. (2017) developed a Cu-chitosan with gradual release of copper over time to control *Curvularia* leaf spot and promote plant growth.

Liposomes are another promising nanocarrier for use in agriculture. This nanosystem is frequently studied as a carrier in the medical area; however, a few studies demonstrate its potential application in agriculture. Liposomes are biocompatible and biodegradable nanostructures capable of providing a matrix to encapsulate molecules without the use of organic solvents. Moreover, liposomes can also promote the controlled release of lipophilic components, due to the limited mobility in solid matrix (Mehnert and Mäder 2012). Recently, natural liposomes, composed of plant lipids, loaded with agriculture compounds were internalized into plant cells and translocated to leaves and roots (Karny et al. 2018). Another nanocarrier that is economically and environmentally viable in agriculture is plant viral nanoparticles. Plants' virus-based systems allow carrying and protecting the internal material in a wide range of temperature, pH, and salinity and in the presence of certain organic solvents. Chauriou and Steinmetz (2017) proposed the use of tobacco mild green mosaic virus (TMGMV) as delivery systems for nematicides. Other stable and robust virus-based material studied to date are Brome mosaic virus (BMV), Red clover necrotic mosaic virus (RCNMV), cowpea mosaic virus (CPMV), Hibiscus chlorotic rinspot virus (HCRSV), Tobacco mosaic virus (TMV), and Potato virus X (PVY) (Alemzadeh et al. 2018).

Green nanotechnology can contribute in different aspects: (1) increase the solubility of pesticides; (2) target specific delivery of compounds; (3) decrease the degradation of active molecules due to the variation in pH, UV, and rain-fastness; (4) develop new molecules; and (5) overcome pesticide resistance (Worrall et al. 2018). However, the studies and applications of this strategy have not yet reached its potential in phytopathology. This fact is observed through a reduced number of deposited patents, reduced number of articles published in high-impact journals of phytopathology, and limited number of field experiments. It demonstrates the need for advances and the absence of real-world studies, applying green nanotechnology-based strategies in the control of plant diseases.

However, it is clear that green nanotechnology represents a promising approach to revolutionize the management of plant pathogens. Some factors that limit the research and development of conventional nanopesticides are the fate, efficiency, and toxicity characteristics of some traditional nanoparticles in the long term in fields. Green nanoparticles have characteristics that completely contrast with chemical nanoparticles. These particles are inexpensive and environment-friendly and allow possible scale-up in production. Advances in the study of the synthesis of green nanoparticles have shown the potential in the production of nanoparticles with specific size, shape, and composition. Moreover, nanoparticles synthesized with living organisms or their parts thereof can be used efficiently to deliver antimicrobial compounds aiming to control phytopathogens. Despite the seemingly unlimited advantages of green nanoparticles over chemical nanoparticles, it is still necessary to have an extensive and integrated study between phytopathologists and nanotechnologists using new approaches, including green nanotechnology concerning the development of strategies meaningful to resolve the problem of particle size and shape consistency, legal uncertainties, absence of reproducibility, and analysis of interactions among nanoparticle-microorganism-environment to allow a deep knowledge about the complex bio-nano system and rational use of nanoparticles in a sustainable manner.

9.8 Uses of Green Nanoparticles for the Treatment of Human Diseases

Nanotechnology is emerging as a new RD&I field that deals with the production of nanomaterials and nanoparticles for applications in various areas due to their highly multifunctional, modular, and efficient properties and thus the field of biomedicine being well explored (Hoseinpour and Ghaemi 2018). Recently, nanoparticles, synthesized by green routes, have emerged as a non-toxic, environmentally friendly, clean, and less expensive nanomaterial solution for the treatment of human diseases (Nicolas et al. 2013). This technological approach can be considered as an alternative to the synthesis of biocompatible nanoparticles, which is the latest viable method of connecting materials science to biotechnology and biomedicine

(Narayanan and Sakthivel 2011). Since there is a vast array of experimental evidence and thousands of reports describing the use of several polymeric and lipidic nanoparticles as vehicles for delivery of bioactive molecules and drugs, it will not be possible to address them in this chapter. At the same time, the use of metallic nanoparticles to treat human diseases is still controversial and should be discussed more deeply. Indeed, how metallic nanoparticles are probably those less explored for the treatment of human diseases and, at the same time, are among the most studied and produced nanosystems by green routes, these were the major areas of focus of this subtopic. Certainly, in such respect, metallic nanoparticles synthesized from silver (Kumar et al. 2017), gold (Ahmed et al. 2016), iron (Naz et al. 2019), zinc (Azizi et al. 2014), and copper (Yallappa et al. 2013) are among the most used for the treatment of human diseases.

The most common targets for the action of nanoparticles synthesized by green routes are microorganisms, including viruses (Sharma et al. 2019), bacteria (Anwar et al. 2019), and fungi (Otari et al. 2014), but also some types of cancer (Benelli 2016), as illustrated in Fig. 9.9. A particular feature of these nanoparticles is their confirmed bactericidal action against microorganisms like *Escherichia coli* (Velusamy et al. 2016), *Streptococcus pneumoniae* (Pugazhendhi et al. 2019), *Staphylococcus aureus*, *Enterococcus faecalis*, *Pseudomonas aeruginosa*, and *Acinetobacter baumannii* (Abalkhil et al. 2017). Kanmani and Rhim (2014) reported that nanoparticles which exhibit toxic and antimicrobial activities have efficient properties due to their extremely large surface area, providing better contact with the microorganisms. According to Dakal et al. (2016), AgNPs are considered to be a nanomaterial with very high toxic properties for several bacterial strains, being their greatest efficacy due to the reduction of the particle size and consequent proportional increase in surface area. The study of this specific nanomaterial demonstrated that there are several distinct cytotoxic actions toward microorganisms, which are (1) adhesion in the cellular membrane and consequently causing its damage; (2) loss of protein stabilization; (3) penetration and accumulation within the

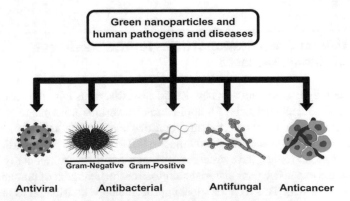

Fig. 9.9 Examples of different activities of nanoparticles synthesized by green routes against microorganisms and cancer

cell and nucleus; (4) mitochondrial and endoplasmic reticulum dysfunction; (5) alteration of internal cellular signaling pathways and; (6) intracellular reactive oxygen species (ROS) generation and oxidative stress (Baranwal et al. 2018).

Biological interactions are often quite versatile, such as the interaction between microorganisms and host cells, in which it involves multiple copies of receptors and ligands that bind in a coordinated fashion, resulting in the microbial agent taking possession of the cell under attack. This biological binding and virus entry into host cells represent a prime example of such versatile interactions between pathogens, surface components, and the cell membrane receptors (Fields et al. 2007). The intervention of these recognition events and thus, consequently, blocking viral entry into cells is one of the most promising strategies related to the development of new antiviral drugs (Melby and Westby 2009). The research done in recent years with the use of metallic nanoparticles, which may or may not have to be functionalized on its surface to optimize bio-interactions, is a process of increasing success (Tang et al. 2011). Some nanoparticles have antiviral activity on the following viruses that cause human diseases: human immunodeficiency virus (HIV) (Kasithevar et al. 2017), herpes simplex virus (HSV) (Orłowski et al. 2018), hepatitis B virus (HBV), human parainfluenza virus (Gaikwad et al. 2013), influenza virus (Mori et al. 2013), and Chikungunya virus (Sharma et al. 2019). Galdiero et al. (2011) report that metallic nanoparticles, especially those produced with gold or silver, present antiviral activities against a broad spectrum of viruses, reducing their potential for in vitro infection against cultured cells. In addition to direct interaction with viral surface glycoproteins, metallic nanoparticles can access the cell and present antiviral activity through interactions with the viral genome (DNA or RNA), as well as viral and cellular replicative factors.

Considering the effects of nanoparticles synthesized by green routes toward fungi control, thereby showing fungicide action, AgNPs are probably among the most effective nanomaterials, maybe excluding only those nanomaterials which carry classic fungicidal compounds (Singh et al. 2013). The investigations about the effects of AgNPs against this kingdom of living organisms involve mainly tests toward yeast species like *Candida albicans* and *Candida tropicalis* (Mallamann et al. 2015). According to Sharma et al. (2009), AgNPs exhibit antimicrobial activity; consequently, this property can be very useful mainly against microorganisms resistant to conventional antimicrobials. Furthermore, *C. albicans* and *C. tropicalis* showed high sensitivity to the AgNPs, comparable to the same sensitivity triggered by the activity of amphotericin B, a powerful antifungal compound. For the trials by Singh et al. (2013), in the presence of AgNPs, the antifungal activity of fluconazole increased significantly compared to the results with itraconazole, as observed for *C. albicans*. Moreover, the synergistic effect of the antifungal fluconazole and itraconazole associated with AgNPs against pathogenic fungi becomes a new and relevant finding in the field of drug development. In addition, nanoparticles stabilized by polymers and surfactants with incorporated antifungal drugs exhibited reasonably higher antifungal activity when compared to the drugs alone, as a result of their increased stability (Rank et al. 2017). In their work, they have been able to conclude that the nanoformulations tested can be used as effective therapeutic agents against human fungal pathogens.

In an exponential way, cancer cases in recent years have been increasing significantly, which, in large part, end with the death of the patient (Raghunandan et al. 2011). In several types of cancer, the manipulation of nanoparticles associated with drugs that are efficient may resemble the effect when compared to drugs that are administered alone, thus becoming a viable alternative to suitable chemotherapeutic agents (Abdel-Fattah et al. 2017). In contrast, there was a paradigm shift in cancer treatment strategies, in which the use of medicinal plants emerged as potential candidates for new treatments. Thus, with the concomitant advances in herbal research and nanotechnology, there is a great increase in the possible treatments of different types of cancer, benefiting patients economically and therapeutically (Chung et al., 2016). This interface for the manufacture of functional nanoparticles derived from plants has attracted many researchers and scientists in developing studies aiming the treatment of cancer (Gobbo et al. 2015).

Recently, the use of AgNPs combined with therapeutic drugs in the treatment of cancer increased the chemotherapeutic efficacy against multiple drug-resistant cancer cells, further enhancing their potential as combinatorial strategies (Igaz et al. 2016). Moreover, the anticancer effects may be due to genotoxicity of AgNPs that is supported by the induction of DNA double-strand breaks, along with the chromosomal instability that drives the onset of apoptosis (Jiang et al. 2013). Thus, it is implied that AgNPs have at least one known mechanism of action mutually associated with a large number of anticancer drugs that are also directed to DNA (Souza et al. 2016). In Table 9.2 some metallic nanoparticles that participate in this cytotoxic mechanism for cancer cells are shown, as well as the respective types of cancerous tumor that can be potentially treated.

Abdel-Fattah and Ali (2018) argue that AgNPs mainly demonstrated unique anticancer activity against different types of tumor cells. In addition, the diverse

Table 9.2 Types of cancers that have toxicity or treatment with metallic nanoparticles

Types of cancer	Types of nanoparticles	References
Breast	AgNPs	Kajani et al. (2014), Jang et al. (2016), Aceituno et al. (2016), and Jacob et al. (2017)
Cervical	AgNPs	Rónavári. et al. (2017), Nakkala et al. (2017), Singh et al. (2017), and Al-Sheddi et al. (2018)
Colon	AgNPs	Prabhu et al. (2013) and Nakkala et al. (2017)
Colon	AuNPs	González-Ballesteros et al. (2017) and Dey et al. (2018)
Gastric	AgNPs	He et al. (2017) and Mousavi et al. (2018)
Liver	AgNPs	Aceituno et al. (2016) and Nakkala et al. (2017)
Lung	AgNPs	Aceituno et al. (2016), Nakkala et al. (2017), and Cyril et al. (2019)
Neuroblastoma	AgNPs	Nakkala et al. (2017)
Prostate	AgNPs	He et al. (2016)

approaches of synthesis significantly affect the cytotoxic activity of AgNPs. Green nanomaterials are currently at a highly investigational stage for treatment, but that remains to be defined on the basis of clinical trials. New possibilities have been considered in relation to the use of these materials, due to their biocompatibility and effectiveness. In addition, many cancers that have no cure today could be treated by these nanomaterials in the future. Moreover, the full understanding of the major physiological barriers in vivo is the key to effectively deliver AgNPs to the tumor. In addition, current knowledge of the safety of nanomaterials is not sufficient, and the acute and chronic toxicity of clinical studies should be observed to identify the potential risks associated with the use of nanoparticles, including AgNPs. Figure 9.10 shows some of the proposed mechanisms of action of AgNPs against tumor cells.

The use of nanoparticles for the purpose of improving human health is currently increasing for the purpose of delivering drugs to cells (Ravichandran 2010). These nanoparticles are commonly designed to be attracted specifically to damaged cells, allowing direct control of these undesired cells, improving efficacy, reducing adverse side effects, and improving overall human health. In this sense, the use of green nanoparticles will potentially reduce some side effects of drugs in the body (Arruebo et al. 2007). Green chemistry, in recent years, has been introduced for the synthesis of nanoparticles for many fields. Traditional processes for the synthesis of nanoparticles sometimes produce large quantities of toxic and unnecessary and harmful substances (Shah et al. 2015a, b). However, green nanoparticles showed great promise in medicine, acting in the area of drug delivery and gene delivery, as

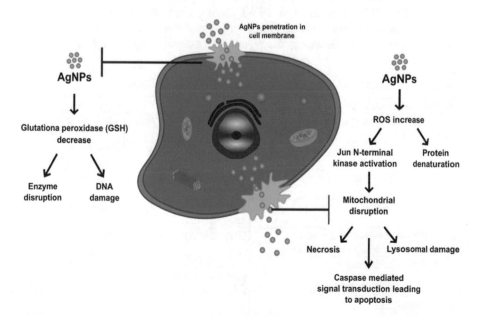

Fig. 9.10 Proposed mechanisms for activity demonstrated by AgNPs toward tumor cells

Fig. 9.11 Types of
nanoparticles for the
treatment of human
diseases like drug delivery
or gene delivery

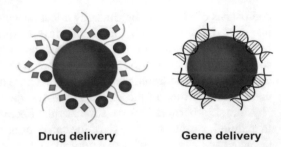

Drug delivery   **Gene delivery**

illustrated in Fig. 9.11. Table 9.3 compiles the types of nanomaterials synthesized
by green routes and their applications as therapeutic agents and delivery systems.

In addition, the combination of therapy and diagnosis, known as theranostics,
characterizes the most important, attractive, and challenging approach adopted by
researchers and health professionals, when related to the effective and personalized
therapy to some cancer type (Vedelago et al. 2018). Some nanoparticles, especially
metallic nanoparticles, are also plasmonic structures that are capable of dispersing
and absorbing particularly the light that is incident. After absorption into tumor
cells, diffuse light derived from metallic nanoparticles like greenly synthesized
AgNPs can be used for imaging purposes, while absorbed light can be used for
selective hyperthermia (Sharma et al. 2015).

In comparison to other activities and actions presented by nanoparticles synthesized by green routes, still within the field of biomedicine is its performance in
cardiovascular diseases (CVDs) that represent one of the leading causes of human
death worldwide, accounting for more than 17.9 million people that died from
CVDs in 2016 (Chamberlain et al. 2019). Lately, many studies have been grouped
in the evaluation of the effects of AgNPs on several types of cells found in the complex vascular system. However, the reported results were opposite to the original

Table 9.3 Applications of green nanocomposites in drug and gene delivery

Nanocomposite delivery	Use	References
Basal fibroblast growth factor – iron oxide nanoparticles	Cancer radiation therapy	Sun et al. (2016)
Magnesium nanoparticles	Hyperthermia therapy	Kumar and Mohammad (2011)
Iron oxide coating with PEG	Drug carrier for cancer treatment	Vangijzegem et al. (2018)
AgNPs in a polymer	Antiseptics and antimicrobial	Pozdnyakov et al. (2016)
Polyethylenimine-grafted chitosan oligosaccharide–hyaluronic acid–RNA	Gene delivery system	Zhao et al. (2016)
Core–shell micelle carriers	Delivery of cobalt pharmaceuticals	Nanaki et al. (2011)
Poly(ester amine)- and poly(amido amine)-based nanoparticles	siRNA delivery	Withey et al. (2009)
Chitosan–graphene oxide composite	Deliver drugs	Tu et al. (2015)

hypotheses. Gonzalez et al. (2016) reported that the data collected can provide substantial knowledge regarding the potential benefits of AgNPs for pathological and physiological stages related to the cardiovascular system, thus contributing to the development of new specific molecular therapies in vascular tone, permeability, and angiogenesis. In another study, Ge et al. (2014) developed the first silver-modified cardiovascular medical device that is an elemental silver-coated silicone valvular prosthesis developed to prevent valve-related bacterial infection and reduce the inflammatory response.

With regard to the tropical diseases that affect several populations in the world, AgNPs also demonstrate activity against certain parasites. In such respect, one of the most common infectious diseases found in tropical and subtropical regions, called malaria, has become a major global health concern. Rai et al. (2017) demonstrated that AgNPs have a huge activity against the protozoan that causes the disease (*Plasmodium falciparum*) and its related vector (female of the *Anopheles* mosquito). The intrinsic antiplasmodial effects exhibited by compounds and materials based on AgNPs represent a solid starting point for nanotechnology-derived therapy and worldwide control of malaria.

The performance of AgNPs in controlling human infections must be evaluated with caution, but organs naturally exposed to the microbial contamination could benefit, in theory, from the use of AgNP-based formulations. Weng et al. (2017) demonstrated that AgNPs show promising activity for the development of non-conventional therapies for eye-related infectious conditions. Indeed, the bactericidal effects related to AgNP-containing nanomaterials are essential aspects that should be considered for their exploitation as an improved class of antibacterial agent for ocular applications (Rizzello and Pompa 2014).

Nanoparticles synthesized by green routes are starting to be exploited for biomedical applications such as unconventional therapeutic strategies thanks to their attractive physicochemical properties related to the nanoscale and their biological functionality, such as their antimicrobial efficiencies and nontoxic nature. The investigations imply proving the beneficial effects of AgNPs on new biocompatible and nanostructured systems, developing modern and safety therapeutic strategies. Nevertheless, full investigations are needed on its possible short- and long-term toxicity, as well as on the toxic mechanisms responsible for some potential side effect.

9.9   Conclusion

Recently, green nanotechnology emerged from traditional nanotechnology by offering less expensive and more sustainable processes and products at nanoscale, which allow maximum benefits and minimum harmful impacts toward several applications, including those in biomedical and bioengineering fields. On the other hand, potential disadvantages include creating conditions for cultural disruption (e.g. overcome the resistance of some researchers to seek eco-friendly solutions) and possible threats to the nano-manufacturing processes (e.g. reproducibility, scaling-up, unavailability of some equivalent reagents, non-standardized raw materials, among others). All these challenging issues must be correctly addressed, and the specific needs of each application should be explored from the laboratorial scale to pilot scale of operations aiming at continuous production in paving the way toward achieving more expressive advances in green nanoparticles use in forthcoming years.

References

A matter of scale. [Editorial] (2016) Nat Nanotechnol 11:733

Abalkhil TA, Alharbi SA, Salmen SH et al (2017) Bactericidal activity of biosynthesized silver nanoparticles against human pathogenic bacteria. Biotechnol Biotechnol Equip 31:411–417

Abdel-Fattah WI, Ali GW (2018) On the anti-cancer activities of silver nanoparticles. J Appl Biotechnol Bioeng 5(1):43–46

Abdel-Fattah WI, Eid MM, El-Moez SA et al (2017) Synthesis of biogenic Ag@Pd core-shell nanoparticles having anti-cancer/anti-microbial functions. Life Sci 183:28–36

Abdelghany TM, Al-Rajhi AM, Al Abboud MA et al (2018) Recent advances in green synthesis of silver nanoparticles and their applications: about future directions. A review. Bionanoscience 8(1):5–16

Aceituno VC, Ahn S, Simu SY et al (2016) Anticancer activity of silver nanoparticles from *Panax ginseng* fresh leaves in human cancer cells. Biomed Pharmacother 84:158–165

Adelere IA, Lateef A (2016) A novel approach to the green synthesis of metallic nanoparticles: the use of agro-wastes, enzymes, and pigments. Nanotechnol Rev 5(6):567–587

Agarwal S, Wendorff JH, Greiner A (2009) Progress in the field of electrospinning for tissue engineering applications. Adv Mater 21:3343–3351

Ahmad N, Sharma AK, Sharma S et al (2018) Biosynthesized composites of Au-Ag nanoparticles using Trapa peel extract induced ROS-mediated p53 independent apoptosis in cancer cells. Drug Chem Toxicol 42(1):1–12

Ahmed S, Annu S, Ikram S et al (2016) Biosynthesis of gold nanoparticles: a green approach. J Photochem Photobiol 161:141–153

Ahn EY, Jin H, Park Y (2019) Green synthesis and biological activities of silver nanoparticles prepared by *Carpesium cernuum* extract. Arch Pharm Res

Alemzadeh E, Dehshahri A, Izadpanah K et al (2018) Plant virus nanoparticles: novel and robust nanocarriers for drug delivery and imaging. Colloid Surf B 167:20–27

Alkubaisi NAO, Aref NMMA, Hendi AA (2015) The polynucleotide confers pathogen or pest resistance. US Patent 9,198,434, 1 Dec 2015

Al-Sheddi ES, Farshori NN, Al-Oqail MM et al (2018) Anticancer potential of green synthesized silver nanoparticles using extract of *Nepeta deflersiana* against human cervical cancer cells (HeLA). Bioinorg Chem Appl 2018:1–13

Andra S, Balu SK, Jeevanandham J et al (2019) Phytosynthesized metal oxide nanoparticles for pharmaceutical applications. Naunyn Schm Arch Pharmacol 392(7):755–771

Anwar A, Masri A, Rao K et al (2019) Antimicrobial activities of green synthesized gums-stabilized nanoparticles loaded with flavonoids. Sci Rep 9(1):1–12

Araujo PW, Brereton RG (1996) Experimental design II. Optim Trends Anal Chem 15(2):63–70

Arruebo M, Fernández-Pacheco R, Ibarra MR et al (2007) Magnetic nanoparticles for drug delivery. Nano Today 2(3):22–32

Azizi S, Ahmad MB, Namvar F et al (2014) Green biosynthesis and characterization of zinc oxide nanoparticles using brown marine macroalga *Sargassum muticum* aqueous extract. Mater Lett 116:275–277

Babensee JE, McIntire LV, Mikos AG (2000) Growth factor delivery for tissue engineering. Pharm Res 17(5):497–504

Baranwal A, Srivastava A, Kumar P et al (2018) Prospects of nanostructure materials and their composites as antimicrobial agents. Front Microbiol 9:1–10

Basheer IA, Hajmeer M (2000) Artificial neural networks: fundamentals, computing, design, and application. J Microbiol Methods 43(1):3–31

Benelli G (2016) Green synthesized nanoparticles in the fight against mosquito-borne diseases and cancer—a brief review. Enzym Microb Technol 95:58–68

Bonatto CC, Silva LP (2014) Higher temperatures speed up the growth and control the size and optoelectrical properties of silver nanoparticles greenly synthesized by cashew nutshells. Ind Crop Prod 58:46–54

Boonkaew B, Suwanpreuska P, Cuttle L et al (2013) Hydrogels containing silver nanoparticles for burn wounds show antimicrobial activity without cytotoxicity. J Appl Polym Sci 131(9):1–10

Caló E, Khutoryanskiy VV (2015) Biomedical applications of hydrogels: a review of patents and commercial products. Eur Polym J 65:252–267

Camargo PC, Satyanarayana KG, Wypych F (2009) Nanocomposites: synthesis, structure, properties and new application opportunities. Mater Res 121:1–9

Chamberlain AM, Cohen SS, Weston SA et al (2019) Relation of cardiovascular events and deaths to low-density lipoprotein cholesterol level among statin-treated patients with atherosclerotic cardiovascular disease. Am J Cardiol 123:1739–1744

Chariou PL, Steinmetz NF (2017) Delivery of pesticides to plant parasitic nematodes using tobacco mild green mosaic virus as a nanocarrier. ACS Nano 11(5):4719–4730

Choudhary RC, Kumaraswamy RV, Kumari S et al (2017) Cu-chitosan nanoparticle boost defense responses and plant growth in maize (*Zea mays* L.). Sci Rep 7:1–11

Christopher JG, Saswati B, Ezilrani P (2015) Optimization of parameters for biosynthesis of silver nanoparticles using leaf extract of Aegle marmelos. Braz Arch Biol Technol 58(5):702–710

Chung IM, Park I, Seung-Hyun K et al (2016) Plant-mediated synthesis of silver nanoparticles: their characteristic properties and therapeutic applications. Nanoscale Res Lett 11(1):1–14

Cruz RM, Peixoto HM, Magalhães RM (2011) Artificial neural networks and efficient optimization techniques for applications in engineering. In: Suzuki K (ed) Artificial neural networks-methodological advances and biomedical applications. InTech, Rijeka, pp 45–68

Cyril N, George JB, Joseph L et al (2019) Assessment of antioxidant, antibacterial and anti-proliferative (lung cancer cell line A549) activities of green synthesized silver nanoparticles from *Derris trifoliata*. Toxicol Res 8(2):297–308

Dakal TC, Kumar A, Majumdar RS et al (2016) Mechanistic basis of antimicrobial actions of silver nanoparticles. Front Microbiol 7:1–17

Das RK, Pachapur VL, Lonappan L et al (2017) Biological synthesis of metallic nanoparticles: plants, animals and microbial aspects. Nanotechnol Environ Eng 2(18):1–21

De M, Ghosh PS, Rotello VM (2008) Applications of nanoparticles in biology. Adv Mater 20:4225–4241

DeMarco PM (2017) Rachel Carson's environmental ethic – a guide for global systems decision making. J Clean Prod 140(1):127–133

Demirdogen RE, Emen FM, Ocakoglu K et al (2018) Green nanotechnology for synthesis and characterization of poly(3-hydroxybutyrate-*co*-3-hydroxyhexanoate) nanoparticles for sustained bortezomib release using supercritical CO_2 assisted particle formation combined with electrodeposition. Int J Biol Macromol 107(A):436–445

Dey A, Yogamoorthy A, Sundarapandian SM (2018) Green synthesis of gold nanoparticles and evaluation of its cytotoxic property against colon cancer cell line. RJLSBPCS 4(6):1–17

Dhand C, Dwivedi N, Loh XJ et al (2015) Methods and strategies for the synthesis of diverse nanoparticles and their applications: a comprehensive overview. RSC Adv 5:105003–105037

Dhar S, Murawala P, Shiras A et al (2012) Gellan gum capped silver nanoparticle dispersions and hydrogels: cytotoxicity and in vitro diffusion studies. Nanoscale 4:563–567

Doria G, Conde J, Veigas B et al (2012) Noble metal nanoparticles for biosensing applications. Sensors 12:1657–1687

Duan H, Wang D, Li Y (2015) Green chemistry for nanoparticle synthesis. Chem Soc Rev 44:5778–5792

Ebrahiminezhad A, Zare-Hoseinabadi A, Sarmah AK et al (2018) Plant-mediated synthesis and applications of iron nanoparticles. Mol Biotechnol 60(2):154–168

Elegbede JA, Lateef A, Azeez MA et al (2019) Silver-gold alloy nanoparticles biofabricated by fungal xylanases exhibited potent biomedical and catalytic activities. Biotechnol Prog. https://doi.org/10.1002/btpr.2829

Elmer W, White JC (2018) The future of nanotechnology in plant pathology. Annu Rev Phytopathol 56:111–133

El-Moslamy SH (2018) Bioprocessing strategies for cost-effective large-scale biogenic synthesis of nano-MgO from endophytic Streptomyces coelicolor strain E72 as an anti-multidrug-resistant pathogens agent. Sci Rep 8(1):1–22

El-Moslamy SH, Kabeil SSA, Hafez EE (2016) Bioprocess development for *Chlorella vulgaris* cultivation and biosynthesis of anti-phytopathogens silver nanoparticles. J Nanomater Mol Nanotechnol 5:1

El-Moslamy SH, Elkady MF, Rezk AH et al (2017) Applying Taguchi design and large-scale strategy for mycosynthesis of nano-silver from endophytic Trichoderma harzianum SYA. F4 and its application against phytopathogens. Sci Rep 7:1–22

Esposito E, Sguizzato M, Drechsler M et al (2017) Progesterone lipid nanoparticles: scaling up and in vivo human study. Eur J Pharm Biopharm 119:437–446

Ferreira L (2009) Nanoparticle as tools to study and control stem cells. J Cell Biochem 108(4):746–752

Fields BN, Knipe DM, Howley PM (2007) Virus entry and uncoating. In: Knipe DM, Howley PM (eds) Fields virology, 5th edn. Lippincott Williams & Wilkins, Philadelphia, pp 99–118

Gaikwad S, Ingle A, Gade A et al (2013) Antiviral activity of mycosynthesized silver nanoparticles against herpes simplex virus and human parainfluenza virus type 3. Int J Nanomedicine 8:4303–4314

Galdiero S, Falanga A, Vitiello M et al (2011) Silver nanoparticles as potential antiviral agents. Molecules 16:8894–8918

Gallego PP, Gago J, Landín M (2011) Artificial neural networks technology to model and predict plant biology process. In: Suzuki K (ed) Artificial neural networks-methodological advances and biomedical applications. InTech, Rijeka, pp 197–216

Ge L, Li Q, Wang M et al (2014) Nanosilver particles in medical applications: synthesis, performance, and toxicity. Int J Nanomed 9:2399–2407

Ghazali NAB, Mani MP, Jaganathan SK (2018) Green-synthesized zinc oxide nanoparticles decorated nanofibrous polyurethane mesh loaded with virgin coconut oil for tissue engineering application. Curr Nanosci 14:280–289

Ghosh PR, Fawcett D, Sharma SB, Poinern GEJ (2017) Production of high-value nanoparticles via biogenic processes using aquacultural and horticultural food waste. Materials (Basel) 10(852):1–19

Gobbo OL, Sjaastad K, Radomski MW et al (2015) Magnetic nanoparticles in cancer theranostics. Theranostics 5(11):1249

Gonzalez C, Rosas-Hernandez H, Ramirez-Lee MA et al (2016) Role of silver nanoparticles (AgNPs) on the cardiovascular system. Arch Toxicol 90:493–511

González-Ballesteros N, Prado-López S, Rodríguez-González JB et al (2017) Green synthesis of gold nanoparticles using brown algae *Cystoseira baccata*: its activity in colon cancer cells. Colloid Surf B 153:190–198

Gour A, Jain NK (2019) Advances in green synthesis of nanoparticles. Artif Cell Nanomed B 47(1):844–851

Harshiny M, Matheswaran M, Arthanareeswaran G et al (2015) Enhancement of antibacterial properties of silver nanoparticles-ceftriaxone conjugate through *Mukia maderaspatana* leaf extract mediated synthesis. Ecotoxicol Environ Safe 121:135–341

Haseeb MT, Hussain MA, Abbas K et al (2017) Linseed hydrogel-mediated green synthesis of silver nanoparticles for antimicrobial and wound-dressing applications. Int J Nanomedicine 12:2845–2855

He Y, Du Z, Ma S et al (2016) Biosynthesis, antibacterial activity and anticancer effects against prostate cancer (pc3) cells of silver nanoparticles using *Dimocarpus longan Lour.* peel extract. Nanoscale Res Lett 11(1):300–310

He Y, Weia F, Mab Z et al (2017) Green synthesis of silver nanoparticles using seed extract of Alpinia katsumadai, and their antioxidant, cytotoxicity, and antibacterial activities. RSC Adv 7:39842–39851

Hembram KC, Kumar R, Kandha L et al (2018) Therapeutic prospective of plant-induced silver nanoparticles: application as antimicrobial and anticancer agent. Artif Cell Nanomed B 46(3):38–51

Hoseinpour V, Ghaemi N (2018) Green synthesis of manganese nanoparticles: applications and future perspective–a review. J Photochem Photobiol B 189:234–243

Hu C, Qian A, Wang Q et al (2016) Industrialization of lipid nanoparticles: from laboratory-scale to large-scale production line. Eur J Pharm Biopharm 109:206–213

Hussain I, Singh NB, Singh A et al (2016) Green synthesis of nanoparticles and its potential application. Biotechnol Lett 38:545–560

Igaz N, Kovács D, Rázga Z et al (2016) Modulating chromatin structure and DNA accessibility by deacetylase inhibition enhances the anti-cancer activity of silver nanoparticles. Colloid Surf B 146:670–677

Ioelovich M (2014) Peculiarities of cellulose nanoparticles. TAPPI J 13(5):45–52

Iravani S (2011) Green synthesis of metal nanoparticles using plants. Green Chem 10:2638–2650

Iravani S, Korbekandi H, Mirmohammadi SV et al (2014) Synthesis of silver nanoparticles: chemical, physical and biological methods. Res Pharm Sci 9(6):385–406

Jacob JBS, Prasad VLS, Sivasankar S et al (2017) Biosynthesis of silver nanoparticles using dried fruit extract of *Ficus carica* – screening for its anticancer activity and toxicity in animal models. Food Chem Toxicol 109:951–956

Jang SJ, Yang IJ, Tettey CO et al (2016) In vitro anticancer activity of green synthesized silver nanoparticles on MCF-7 human breast cancer cells. Mater Sci Eng C 68:430–435

Jiang X, Foldbjerg R, Miclaus T et al (2013) Multi-platform genotoxicity analysis of silver nanoparticles in the model cell line CHO-K1. Toxicol Lett 222(1):55–63

Kajani AA, Bordbar AK, Esfahani SHZ et al (2014) Green synthesis of anisotropic silver nanoparticles with potent anticancer activity using *Taxus baccata* extract. RSC Adv 4:61394–61403

Kalaiselvi D, Mohankumar A, Shanmugam G et al (2019) Green synthesis of silver nanoparticles using latex extract of *Euphorbia tirucalli*: a novel approach for the management of root knot nematode, Meloidogyne incognita. Crop Prot 117:108–114

Kalaivani R, Maruthupandy M, Muneeswaran T et al (2018) Synthesis of chitosan mediated silver nanoparticles (Ag NPs) for potential antimicrobial applications. Front Lab Med 2(1):30–35

Kanmani P, Rhim JW (2014) Physicochemical properties of gelatin/silver nanoparticle antimicrobial composite films. Food Chem 148:162–169

Karny A, Zinger A, Kajal A et al (2018) Therapeutic nanoparticles penetrate leaves and deliver nutrients to agricultural crops. Sci Rep 8(1):1–10

Kasithevar M, Saravanan M, Prakash P et al (2017) Green synthesis of silver nanoparticles using *Alysicarpus monilifer* leaf extract and its antibacterial activity against MRSA and CoNS isolates in HIV patients. J Interdiscip Nanomed 2:131–141

Kaur P, Thakur R, Malwal H et al (2018) Biosynthesis of biocompatible and recyclable silver/iron and gold/iron core-shell nanoparticles for water purification technology. Biocatal Agric Biotechnol 14:189–197

Khan M, Siddiqui ZA (2018) Zinc oxide nanoparticles for the management of Ralstonia solanacearum, Phomopsis vexans and Meloidogyne incognita incited disease complex of eggplant. Indian Phytopathol 71(3):355–364

Khan M, Shaik MR, Adil SF et al (2018) Plant extracts as green reductants for the synthesis of silver nanoparticles: lessons from chemical synthesis. Dalton Trans 35:11988–12010

Kinkela D (2016) Banned: a History of pesticides and the science of toxicology. By Frederick Rowe Davis. Environ Hist 21(2):401–403

Kirubaharan CJ, Kalpana D, Lee YS et al (2012) Biomediated silver nanoparticles for the highly selective copper (II) ion sensor applications. Ind Eng Chem Res 51(21):7441–7446

Kisyelova T, Novruzova A, Hajiyeva F et al (2016) Effect of the reactor configuration on the production of silver nanoparticles. Chem Eng Trans 47:121–126

Ko SW, Soriano JPE, Lee JY et al (2018) Nature derived scaffolds for tissue engineering applications: design and fabrication of a composite scaffold incorporating chitosan-g-D,L-lactic acid and cellulose nanocrystals from *Lactuca sativa* L. cv green leaf. Int J Biol Macromol 110:504–513

Kumar CSSR, Mohammad F (2011) Magnetic nanomaterials for hyperthermia-based therapy and controlled drug delivery. Adv Drug Deliv Rev 63(9):789–808

Kumar V, Singh DK, Mohan S et al (2017) Photoinduced green synthesis of silver nanoparticles using aqueous extract of Physalis angulata and its antibacterial and antioxidant activity. J Environ Chem Eng 5:744–756

Loh JW, Schneider J, Carter M (2010) Spinning disc processing technology: potential for large-scale manufacture of chitosan nanoparticles. J Pharm Sci 99(10):4326–4336

Luo X, Morrin A, Killard AJ et al (2006) Application of nanoparticles in electrochemical sensors and biosensors. Electroanalysis 18:319–326

Malerba M, Cerana R (2016) Chitosan effects on plant systems. Int J Mol Sci 17(7):1–15

Mallamann EJ, Cunha FA, Castro BNMF et al (2015) Antifungal activity of silver nanoparticles obtained by green synthesis. Rev Inst Med Trop Sao Paulo 57(2):165–167

Masum MMI, Siddiqa MM, Ali KA et al (2019) Biogenic synthesis of silver nanoparticles using Phyllanthus emblica fruit extract and its inhibitory action against the pathogen Acidovorax oryzae strain RS-2 of Rice bacterial Brown stripe. Front Microbiol 10:1–18

McKenzie LC, Hutchison JE (2004) Green nanoscience. Chem Today 22(9):30–33

Mehnert W, Mäder K (2012) Solid lipid nanoparticles: production, characterization and applications. Adv Drug Deliv Rev 47(2–3):165–196

Melby T, Westby M (2009) Inhibitors of viral entry. In: Kräusslich HG, Bartenschlager R (eds) Antiviral strategies. Springer, Berlin, pp 177–202

Mittal AK, Chisti Y, Banerjee UC (2013) Synthesis of metallic nanoparticles using plant extracts. Biotechnol Adv 31(2):346–356

Moghaddam BA, Namvar F, Moniri M et al (2015) Nanoparticles biosynthesized by fungi and yeast: a review of their preparation, properties, and medical applications. Molecules 20(9):16540–16565

Moon JW, Rawn CJ, Rondinone AJ et al (2010) Large-scale production of magnetic nanoparticles using bacterial fermentation. J Ind Microbiol Biotechnol 37(10):1023–1031

Mori Y, Ono T, Miyahira Y et al (2013) Antiviral activity of silver nanoparticle/chitosan composites against H1N1 influenza a virus. Nanoscale Res Lett 8:1–6

Mousavi B, Tafvizi F, Zaker Bostanabad S (2018) Green synthesis of silver nanoparticles using Artemisia turcomanica leaf extract and the study of anti-cancer effect and apoptosis induction on gastric cancer cell line (AGS). Artif Cells Nanomed Biotechnol 46:499–510

Muthukumar H, Matheswaran M (2015) *Amaranthus spinosus* leaf extract mediated FeO nanoparticles: physicochemical traits, photocatalytic and antioxidant activity. ACS Sustain Chem Eng 3(12):3149–3156

Myers SS, Smith MR, Guth S et al (2017) Climate change and global food systems: potential impacts on food security and undernutrition. Annu Rev Public Health 38:259–277

Mythili R, Selvankumar T, Kamala-Kannan S (2018) Utilization of market vegetable waste for silver nanoparticle synthesis and its antibacterial activity. Mater Lett 225(15):101–104

Nadaroğlu H, Alayl Güngör A, Ince S (2017) Synthesis of nanoparticles by green synthesis method. INJIRR 1(1):6–9

Naka K, Itoh H, Tampo Y et al (2003) Effect of gold nanoparticles as a support for the oligomerization of L-Cysteine in an aqueous solution. Langmuir 19(13):5546–5549

Nakkala JR, Mata R, Sadras SR (2017) Green synthesized nano silver: synthesis, physicochemical profiling, antibacterial, anticancer activities and biological in vivo toxicity. J Colloid Interface Sci 499:33–45

Nanaki SG, Pantopoulos K, Bikiaris DN (2011) Synthesis of biocompatible poly (ε-caprolactone)-block-poly (propylene adipate) copolymers appropriate for drug nanoencapsulation in the form of core-shell nanoparticles. Int J Nanomedicine 6:2981–2995

Narayanan KB, Sakthivel N (2011) Green synthesis of biogenic metal nanoparticles by terrestrial and aquatic phototrophic and heterotrophic eukaryotes and biocompatible agents. Adv Colloid Interf Sci 169(2):59–79

Naz S, Islam M, Tabassum S et al (2019) Green synthesis of hematite (α-Fe₂O₃) nanoparticles using Rhus punjabensis extract and their biomedical prospect in pathogenic diseases and cancer. J Mol Struct 1185:1–7

Nicolas J, Mura S, Brambilla D et al (2013) Design, functionalization strategies and biomedical applications of targeted biodegradable/biocompatible polymer-based nanocarriers for drug delivery. Chem Soc Rev 42:1147–1235

Oliveira GCS, Lopes CAP, Sousa MH et al (2019) Synthesis of silver nanoparticles using aqueous extracts of Pterodon emarginatus leaves collected in the summer and winter seasons. Int Nano Lett 9:1–9

Orłowski P, Kowalczyk A, Tomaszewska E et al (2018) Antiviral activity of tannic acid modified silver nanoparticles: potential to activate immune response in herpes genitalis. Viruses 10(524):1–15

Otari SV, Patil RM, Ghosh SJ, Pawar SH (2014) Green phytosynthesis of silver nanoparticles using aqueous extract of *Manilkara zapota* (L.) seeds and its inhibitory action against *Candida* species. Mater Lett 166:367–369

Ould-Ely T, Luger M, Kaplan-Reinig L et al (2011) Large-scale engineered synthesis of BaTiO₃ nanoparticles using low-temperature bioinspired principles. Nat Protoc 6(1):97–104

Paliwal R, Babu RJ, Palakurthi S (2014) Nanomedicine scale-up technologies: feasibilies and challenges. AAPS PharmSciTech 15(6):1527–1534

Park Y, Hong YN, Weyers A et al (2011) Polysaccharides and phytochemicals: a natural reservoir for the green synthesis of gold and silver nanoparticles. IET Nanobiotechnol 5(3):69–78

Patra JK, Baek KH (2014) Green nanobiotechnology: factors affecting synthesis and characterization techniques. J Nanomater 2014:219–230

Pérez-de-Luque A, Rubiales D (2009) Nanotechnology for parasitic plant control. Pest Manag Sci 65(5):540–545

Pozdnyakov A, Emel'yanov A, Kuznetsova N et al (2016) Nontoxic hydrophilic polymeric nanocomposites containing silver nanoparticles with strong antimicrobial activity. Int J Nanomedicine 11:1295–1304

Prabhua D, Arulvasua C, Babua G et al (2013) Biologically synthesized green silver nanoparticles from leaf extract of *Vitex negundo* L. induce growth-inhibitory effect on human colon cancer cell line HCT15. Process Biochem 48(2):317–324

Pugazhendhi A, Prabhu R, Muruganantham K et al (2019) Anticancer, antimicrobial and photocatalytic activities of green synthesized magnesium oxide nanoparticles (MgONPs) using aqueous extract of *Sargassum wightii*. J Photochem Photobiol B 190:86–97

Puja P, Kumar P (2019) A perspective on biogenic synthesis of platinum nanoparticles and their biomedical applications. Spectrochim Acta A Mol Biomol Spectrosc 211:94–99

Raftery RM, Tierney EG, Curtin CM et al (2015) Development of a gene-activated scaffold platform for tissue engineering applications using chitosan-pDNA nanoparticles on collagen-based scaffolds. J Control Release 210:84–94

Raghunandan D, Ravishankar B, Sharanbasava G et al (2011) Anti-cancer studies of noble metal nanoparticles synthesized using different plant extracts. Cancer Nanotechnol 2(1–6):57–65

Rai M, Ingle AP, Paralikar P et al (2017) Recent advances in use of silver nanoparticles as antimalarial agents. Int J Pharm 526:254–270

Raj DR, Sudardanakumar C (2017) Colorimetric and fiber optic sensing of cysteine using green synthesized gold nanoparticles. Plasmonics 13:327–334

Rajam M, Pulavendran S, Rose C, Mandal AB (2011) Chitosan nanoparticles as a dual growth factor delivery system for tissue engineering applications. Int J Pharm 410(1–2):145–152

Rajan R, Chandran K, Harper SL et al (2015) Plant extract synthesized silver nanoparticles: an ongoing source of novel biocompatible materials. Ind Crop Prod 70:356–373

Rajkuberan C, Sudha K, Sathishkumar G et al (2015) Antibacterial and cytotoxic potential of silver nanoparticles synthesized using latex of *Calotropis gigantea L.* Spectrochim Acta A 136:924–930

Rank LA, Walsh NM, Liu R et al (2017) A cationic polymer that shows high antifungal activity against diverse human pathogens. Antimicrob Agents Chemother 61(10):e00204–e00217

Rasheed T, Bilal M, Iqbal HMN et al (2017) Green biosynthesis of silver nanoparticles using leaves extract of *Artemisia vulgaris* and their potential biomedical applications. Colloid Surf B 158:408–415

Ravichandran S (2010) Green chemistry–a potential tool for chemical synthesis. Int J ChemTech Res 2(4):2188–2191

Ravindran A, Chandran P, Khan SS (2013) Biofunctionalized silver nanoparticles: advances and prospects. Colloid Surf B 105:342–352

Rehana D, Mahendiran D, Kumar RS et al (2017) In vitro antioxidant and antidiabetic activities of zinc oxide nanoparticles synthesized using different plant extracts. Bioprocess Biosyst Eng 40(6):943–957

Rizzello L, Pompa PP (2014) Nanosilver-based antibacterial drugs and devices: mechanisms, methodological drawbacks, and guidelines. Chem Soc Rev 43:1501–1518

Rónavári A, Kovács D, Igaz N et al (2017) Biological activity of green-synthesized silver nanoparticles depends on the applied natural extracts: a comprehensive study. Int J Nanomedicine 12:871–883

Rosi NL, Giljohann DA, Thaxton CS et al (2006) Oligonucleotide-modified gold nanoparticles for intracellular gene regulation. Science 312:1027–1030

Ruedas-Rama MJ, Walters JD, Orte A et al (2012) Fluorescent nanoparticles for intracellular sensing: a review. Anal Chim Acta 751:1–23

Saif A, Tahir A, Chen Y (2016) Green synthesis of iron nanoparticles and their environmental applications and implications. Nanomaterials 6(209):1–26

Sakai T, Ishihara A, Alexandridis P (2015) Block copolymer-mediated synthesis of silver nanoparticles from silver ions in aqueous media. Colloids Surf A Physicochem Eng Asp 487:84–91

Santiago TR, Bonatto CC, Rossato M et al (2019) Green synthesis of silver nanoparticles using tomato leaves extract and their entrapment in chitosan nanoparticles to control bacterial wilt. J Sci Food Agric 99:4248

Saratale RG, Saratale GD, Shin HS et al (2018) New insights on the green synthesis of metallic nanoparticles using plant and waste biomaterials: current knowledge, their agricultural and environmental applications. Environ Sci Pollut R 25(11):10164–10183

Satapathy S, Shukla SP, Sandeep KP et al (2015) Evaluation of the performance of an algal bioreactor for silver nanoparticle production. J Appl Phycol 27(1):285–291

Sathiyabama M, Manikandan A (2018) Application of copper-chitosan nanoparticles stimulate growth and induce resistance in finger millet (*Eleusine coracana* Gaertn.) plants against blast disease. J Agric Food Chem 66(8):1784–1790

Sathya K, Saravanathamizhan R, Baskar G (2018) ANN modeling for scale-up of green synthesis of iron oxide nanoparticle and its application for decolorization of dye effluent. Desalin Water Treat 121:158–165

Shabanzadeh P, Senu N, Shameli K et al (2013a) Application of artificial neural network (ANN) for prediction diameter of silver nanoparticles biosynthesized in *Curcuma longa* extract. Dig J Nanomater Biostruct 8(3):1133–1144

Shabanzadeh P, Shameli K, Ismail F et al (2013b) Application of artificial neural network (ANN) for the prediction of size of silver nanoparticles prepared by green method. Dig J Nanomater Biostruct 8(2):541–549

Shabanzadeh P, Yusof R, Shameli K (2015) Modeling of biosynthesized silver nanoparticles in *Vitex negundo L.* extract by artificial neural network. RSC Adv 5(106):87277–87285

Shah M, Fawcett D, Sharma S et al (2015a) Green synthesis of metallic nanoparticles via biological entities. Materials (Basel) 8(11):7278–7308

Shah M, Fawcett D, Sharma S et al (2015b) Green synthesis of metallic nanoparticles via biological entities. Materials 8(11):7278–7308

Sharma VK, Yngard RA, Lin Y (2009) Silver nanoparticles: green synthesis and their antimicrobial activities. Adv Colloid Interf Sci 145:83–96

Sharma H, Mishra PK, Talegaonkar S et al (2015) Metal nanoparticles: a theranostic nanotool against cancer. Drug Discov Today 20:1143–1151

Sharma V, Kaushik S, Pandit P et al (2019) Green synthesis of silver nanoparticles from medicinal plants and evaluation of their antiviral potential against chikungunya virus. Appl Microbiol Biotechnol 103:881–891

Shenton W, Davis SA, Mann S (1999) Direct self-assembly of nanoparticles into macroscopic materials using antibody-antigen recognition. Adv Mater 11(6):449–452

Shivananda CS, Rao BL, Pasha A et al (2016) Synthesis of silver nanoparticles using Bombyxmori silk fibroin and their antibacterial activity. IOP Conf Ser Mater Sci Eng 149. https://doi.org/10.1088/1757-899X/149/1/012175

Siddiqui ZA, Khan MR, AbdAllah EF et al (2018) Titanium dioxide and zinc oxide nanoparticles affect some bacterial diseases, and growth and physiological changes of beetroot. Int J Veg Sci:1–22

Silva LP, Reis IG, Bonatto CC (2015) Green synthesis of metal nanoparticles by plants: current trends and challenges. In: Basiuk V, Basiuk E (eds) Green processes for nanotechnology. Springer, Cham, pp 259–275

Silva LP, Bonatto CC, Polez VLP (2016) Green synthesis of metal nanoparticles by fungi: current trends and challenges. In: Prasad R (ed) Advances and applications through fungal nanobiotechnology, Fungal biology. Springer, Cham, pp 71–89

Silveira AP, Bonatto CC, Lopes CAP et al (2018) Physicochemical characteristics and antibacterial effects of silver nanoparticles produced using the aqueous extract of *Ilex paraguariensis*. Mater Chem Phys 216:476–484

Singh M, Kumar M, Kalaivani R et al (2013) Metallic silver nanoparticle: a therapeutic agent in combination with antifungal drug against human fungal pathogen. Bioprocess Biosyst Eng 36(4):407–415

Singh P, Kim YJ, Wang C et al (2016) The development of a green approach for the biosynthesis of silver and gold nanoparticles by using Panax ginseng root extract, and their biological applications. Artif Cells Nanomed Biotechnol 44(4):1150–1157

Singh H, Du J, Yi TH (2017) Green and rapid synthesis of silver nanoparticles using *Borago officinalis* leaf extract: anticancer and antibacterial activities. Artif Cells Nanomed Biotechnol 45(7):1310–1316

Souza TAJ, Franchi LP, Rosa LR et al (2016) Cytotoxicity and genotoxicity of silver nanoparticles of different sizes in CHO-K1 and CHO-XRS5 cell lines. Mutat Res Genet Toxicol Environ Mutagen 795:70–83

Sun Z, Worden M, Thliveris JA et al (2016) Biodistribution of negatively charged iron oxide nanoparticles (IONPs) in mice and enhanced brain delivery using lysophosphatidic acid (LPA). Nanomedicine: NBM 12(7):1775–1784

Suwan T, Khongkhunthian S, Okonogi S (2019) Silver nanoparticles fabricated by reducing property of cellulose derivatives. Drug Discov Ther 13(2):70–79

Suzuki N, Rivero RM, Shulaev V et al (2014) Abiotic and biotic stress combinations. New Phytol 203(1):32–43

Tahir R, Bilal M, Iqbal H et al (2017) Green biosynthesis of silver nanoparticles using leaves extract of *Artemisia vulgaris* and their potential biomedical applications. Colloid Surf B 158:408–415

Tang B, Wang J, Xu S et al (2011) Application of anisotropic silver nanoparticles: multifunctionalization of wool fabric. J Colloid Interface Sci 356(2):513–518

Tilman D, Balzer C, Hill J et al (2011) Global food demand and the sustainable intensification of agriculture. Proc Natl Acad Sci 108(50):20260–20264

Travan A, Pellilo C, Donati I et al (2009) Non-cytotoxic silver nanoparticle-polysaccharide nanocomposites with antimicrobial activity. Biomacromolecules 10(6):1429–1435

Tu H, Lu Y, Wu Y et al (2015) Fabrication of rectorite-contained nanoparticles for drug delivery with a green and one-step synthesis method. Int J Pharm 493(1–2):426–433

Vangijzegem T, Stanicki D, Laurent S (2018) Magnetic iron oxide nanoparticles for drug delivery: applications and characteristics. Expert Opin Drug Deliv 16:69–78

Vauthier C, Bouchemal K (2009) Methods for the preparation and manufacture of polymeric nanoparticles. Pharm Res 26(5):1025–1058

Vedelago J, Gomez CG, Valente M et al (2018) Green synthesis of silver nanoparticles aimed at improving theranostics. Radiat Phys Chem 146:55–67

Velusamy P, Kumar GV, Jeyanthi V et al (2016) Bio-inspired green nanoparticles: synthesis, mechanism, and antibacterial application. Toxicol Res 32(2):95–102

Vilela D, González MC, Escarpa A (2012) Sensing colorimetric approaches based on gold and silver nanoparticles aggregation: chemical creativity behind the assay. A review. Anal Chim Acta 751:24–43

Virkutyte J, Varma RS (2013) Green synthesis of nanomaterials: environmental aspects. In: Shamim N, Sharma VK (eds) Sustainable nanotechnology and the environment: advances and achievements. American Chemical Society, Washington, pp 11–39

Wang L, Li C, Huang Q et al (2019) Biofunctionalization of selenium nanoparticles with a polysaccharide from *Rosa roxburghii* fruit and their protective effect against H_2O_2-induced apoptosis in INS-1 cells. Food Funct 10(2):539–553

Weng Y, Liu J, Jin S et al (2017) Nanotechnology-based strategies for treatment of ocular disease. Acta Pharm Sin B 7(3):281–291

Withey ABJ, Chen G, Nguyen TL et al (2009) Macromolecular cobalt carbonyl complexes encapsulated in a click-cross-linked micelle structure as a nanoparticle to deliver cobalt pharmaceuticals. Biomacromolecules 10(12):3215–3226

Worrall E, Hamid A, Mody K et al (2018) Nanotechnology for plant disease management. Agronomy 285(8):1–24

Yadi M, Mostafavi E, Saleh B et al (2018) Current developments in green synthesis of metallic nanoparticles using plant extracts: a review. Artif Cells Nanomed Biotechnol 46(3):336–343

Yallappa S, Manjanna J, Sindhe MA et al (2013) Microwave assisted rapid synthesis and biological evaluation of stable copper nanoparticles using *T. arjuna* bark extract. Spectrochim Acta A Mol Biomol Spectrosc 110:108–115

Zain NM, Stapley AGF, Shama G (2014) Green synthesis of silver and copper nanoparticles using ascorbic acid and chitosan for antimicrobial applications. Carbohydr Polym 112(4):195–202

Zargar V, Asghari M, Dashti A (2015) A review on chitin and chitosan polymers: structure, chemistry, solubility, derivatives, and applications. ChemBioEng Rev 3:204–226

Zhao M, Cheng JL, Yan JJ et al (2016) Hyaluronic acid reagent functional chitosan-PEI conjugate with AQP2-siRNA suppressed endometriotic lesion formation. Int J Nanomedicine 2016(11):1323–1336

Zulkifli FH, Hussain FSJ, Zeyohannes SS et al (2017) A facile synthesis method of hydroxyethyl cellulose-silver nanoparticle scaffolds for skin tissue engineering applications. Mater Sci Eng C 79:151–160

Zuorro A, Iannone A, Natali S et al (2019) Green synthesis of silver nanoparticles using bilberry and red currant waste extracts. Processes 193(7):1–12

Nanoparticles: A Boon to Target Mitochondrial Diseases

10

Swarupa Ghosh and Saptarshi Chatterjee

Abstract

Mitochondrial medicine is a rapidly growing area in biomedical research. Armed with the much needed tools for probing, accessing, and manipulating subcellular organelles, nanoscience has leaped into the realm of mitochondrial research. It has become increasingly evident that mitochondrial dysfunction causes a variety of human disorders, including neurodegenerative and neuromuscular diseases, obesity and diabetes, ischemia–reperfusion injury, cancer and inherited mitochondrial diseases.

Mitochondria are a major source of superoxide anion and other free radicals. This in situ-generated reactive oxygen species alters the function of many metabolic enzymes in the mitochondrial matrix, as well as those comprising the electron transport chain. Antioxidant supplements and drugs are generally believed to scavenge toxic free radicals from mitochondrial environment. Because of the complex nature of the mitochondrion, different strategies may be required for mitochondrial uptake of different pharmacotherapeutic agents.

A variety of small-molecule drugs have been investigated as potential therapeutic agents for mitochondrial diseases, but with obvious limitations. This chapter deals with effective nanoparticulated drug delivery system for targeting biologically active compounds to brain and/or liver mitochondria in the pathogenesis of mitochondrial diseases. The aim is to evaluate the efficacy of vesiculated drug formulations (liposomes, nanoparticles) against oxidative-damage-evoked mitochondrial damage and their possible protection mechanism in preclinical setting.

S. Ghosh · S. Chatterjee (✉)
Department of Microbiology, School of Life Science and Biotechnology, Adamas University, Kolkata, West Bengal, India

© Springer Nature Singapore Pte Ltd. 2020
A. K. Shukla (ed.), *Nanoparticles and their Biomedical Applications*,
https://doi.org/10.1007/978-981-15-0391-7_10

Keywords

Mitochondrial dysfunction · Nanoparticles · Reactive oxygen species · Oxidative stress · Antioxidant

10.1   Mitochondria: The Powerhouses of the Cell

Mitochondria are considered the powerhouses of the cell, supplying energy in the form of adenosine triphosphate (ATP). They are double membrane-bound organelles present in eukaryotes (Henze and Martin 2003). The structure of a mitochondrion is shown in Fig. 10.1a and its functions are schematically shown in Fig. 10.1b.

Structurally, there are five distinct compartments within the mitochondrion. They are as follows:

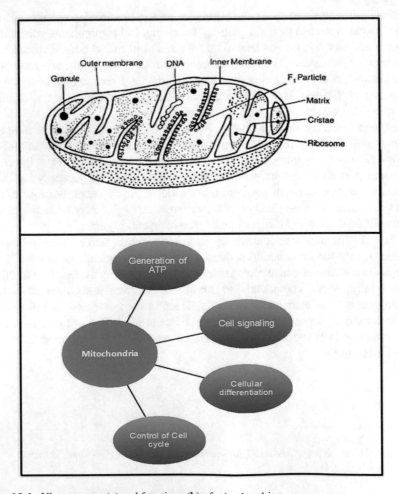

Fig. 10.1 Ultrastructure (**a**) and functions (**b**) of mitochondrion

1. Outer mitochondrial membrane (OMM)
2. Intermembrane space (the space between the outer and inner membranes)
3. Inner mitochondrial membrane (IMM)
4. Cristae space (formed by infoldings of the inner membrane)
5. Matrix (space within the inner membrane)

The inner membrane is rich in usual phospholipids called cardiolipin that makes the inner membrane impermeable (McMillin and Dowhan 2002). A membrane potential exists across the inner membrane by the action of enzymes of electron transport system. The inner membrane folds to form numerous cristae that increase the surface area. The invaginations of the inner membrane are responsible for the chemiosmotic function (Mannella 2006). The space enclosed in the inner membrane is called matrix. It contains a highly concentrated mixture of several enzymes, mitochondrial ribosomes, tRNAs, and several copies of the mitochondrial DNA genome. Mitochondria are unique in having their own genetic material and their ability to transcribe into RNA and further translate into protein. A human mitochondrial DNA revealed the presence of 37 genes (22 tRNAs, 2 rRNAs, 13 peptides) having a total size of approximately 16 kb (Anderson et al. 1981).

10.2 Mitochondrial DNA Diseases

Mitochondria substantially differ from other animal cell organelles. Any defect in the mitochondrial DNA (mtDNA) results in the reduction of energy production. The energy demand of the tissue and the magnitude of damage determine the clinical symptoms. The causal link between mtDNA defects and human diseases was described for the first time in 1988, and since then, the number of diseases linked with mtDNA has increased. Table 10.1 depicts a list of mitochondrial disorders.

10.3 Mitochondria: Source and Target for ROS

The inner mitochondrial membrane is a major intracellular source of reactive oxygen species (ROS) (Zhao et al. 2004) that nonspecifically hinders cellular function by damaging protein, lipid, or even DNA. Mitochondrial dysfunction has been related with both necrosis and apoptosis (Kroemer et al. 1997). The physiological and pathological conditions are greatly affected by the rate of mitochondrial ROS. ROS can be triggered by the inhibition of 3-nitropropionic acid (3NP), leading to death of neurons and occurrence of Huntington's disease (Beal et al. 1993). Calcium enters mitochondria by a uniporter in the inner mitochondrial membrane, resulting in increased levels of Ca^{2+} responsible for the generation of ROS that is associated with ischemia-reperfusion. The ROS within mitochondria that can lead to oxidative damage, thereby releasing cytochrome c into cytosol resulting in apoptosis, is schematically shown is Fig. 10.2. It also plays a major role in the modulation of cell signaling pathway.

Table 10.1 Examples of mitochondrial diseases (Niyazov et al. 2016)

Type	Disease	Features	References
Primary mitochondrial disease	Kearns-Sayre syndrome	Progressive external ophthalmoplegia, pigmentary retinitis, heart block	Parikh (2016)
	Alpers-Huttenlocher syndrome	Hypotonia, seizures, liver failure, mtDNA deletion/depletion (secondary)	Seneto et.al. (2013)
	Ataxia neuropathy syndrome	Sensory ataxia neuropathy, dysarthria, ophthalmoplegia (SANDO)	Bargiela et al. (2015)
	Mitochondrial encephalomyopathy with lactic acidosis and stroke-like episodes (MEALS)	Encephalomyopathy, lactic acidosis, stroke-like episodes at age < 40 years	Joo et al. (2013)
Secondary mitochondrial dysfunction	Spinal muscular atrophy	Loss of motor neurons, hypotonia, muscle weakness	Malkki (2016)
	Friedreich's ataxia	Ataxia, progressive muscle weakness, hypertrophic *FXN* cardiomyopathy, diabetes mellitus	Delatycki et al. (2000)
	Wilson's disease	Liver disease due to copper deposition; neurological features: *ATP7B* tremors, ataxia, etc.; psychiatric features: neurosis, depression, etc.; Kayser-Fleischer rings	Wu et al. (2015)

10.4   Cell, Reactive Oxygen Species, and Oxidative Stress

Generally, the reducing environment prevails in the intercellular region of most life forms that is maintained by the action of several enzymes. Oxygen is the most essential molecule for the aerobic organisms. ROS is generated during oxidative phosphorylation. ROS includes not only the oxygen radicals like $O_2^{\cdot-}$ (superoxide radical), •OH (hydroxyl radical) but also H_2O_2, singlet oxygen, etc. Among these, superoxide is considered "primary," while the others are "secondary." The ROS-generated oxidative stress is considered the major factor that leads to normal senescence and pathologies that are of public health concerns (Starkov 2008). Mitochondria possess multiple sites that are capable of generating ROS that is controlled by a sophisticated defense mechanism. ROS is also involved in physiological signaling cascade (Valco 2007). A well-illustrated comparison of ROS generation in various mitochondrial sites is given in Table 10.2 (Starkov 2008). Moreover, the likelihood that ROS production by mitochondria is a redox signal integrating mitochondrial function with that of the rest of the cell (Balaban et al. 2005).

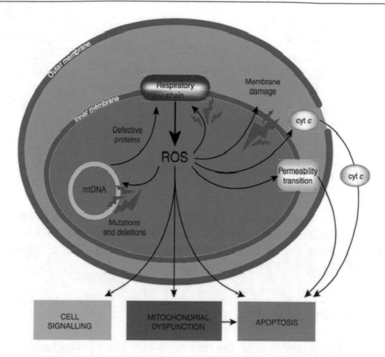

Fig. 10.2 Overview of mitochondrial ROS production. (Reproduced with permission from Murphy 2009)

10.4.1 Superoxide Radical

The monovalent reduction of O_2 gives $O_2{}^{•-}$ that is considered radical as well as anion with the radical sign (•) and a charge of -1. This free radical is of great significance in chemical and biological systems. Although maintaining $O_2{}^{•-}$ in a stable state for a long duration is difficult, $O_2{}^{•-}$ has attracted considerable attention (Hayyan et al. 2016).

10.4.2 Hydrogen Peroxide

The enzymes located in microsomes, peroxysome, and mitochondria are cable of producing H_2O_2 through enzymatic reaction. In plant or animal cells, superoxide

Table 10.2 ROS-generating capacities of a few mitochondrial sites (Starkov 2008)

Sl no	ROS-producing sites of mitochondria	Generation of H_2O_2 (nmol/min/mg)
1	Reverse electron transport	1–3
2	Forward electron transport	0.06–0.4
3	Complex I of respiratory chain	0.3–0.6
4	Mono amine oxidases (A&B)	0.7–1.5

Table 10.3 Enzymatic and non-enzymatic scavengers of antioxidant defense

Non-enzymatic scavenger	
Chemical name	**Common name**
Ascorbic acid	Vitamin C
Alpha-tocopherol	Vitamin E
Enzymatic scavenger	
Name	**Acronym**
Catalase	CAT
Glutathione peroxidase	GTPx
Superoxide dismutase	SOD
Glutathione transferase	GST

dismutase is able to produce H_2O_2 by dismutation of O_2^-, thus contributing to the lowering of oxidative reaction. Cell killing and DNA damage by H_2O_2 are mediated by intracellular iron (Mello Filho et al. 1984) and are attenuated by hypotonicity (Martins and Meneghini 1994).

10.4.3 Hydroxyl Radical

Hydroxyl radical is produced via the Fenton reaction in submitochondrial particles under oxidative stress (Thomas et al. 2009). The precursor and catalyst for Fenton reaction lie in the mitochondrial matrix through which sustained •OH (hydroxyl radical) is produced during oxidative stress. Tissue iron accumulation is of significant concern as Fe acts as catalyst.

10.4.4 Cellular Defense Against ROS

Since the generation of ROS creates myriads of physiological problems in the body contributing to several diseases, the body already possesses a defense mechanism to fight ROS. Enzymatic and non-enzymatic antioxidants play a major role in this context (Birben et al. 2012) Table 10.3 depicts a list of scavengers of antioxidant defense.

10.5 Therapeutic Application of Nanoparticles in Diseases Involving Mitochondrial Dysfunction

As already discussed, there are several diseases associated with mitochondrial dysfunction and several therapeutic options are available. Table 10.3 summarizes the application of nanoparticles in the therapeutics of diseases with mitochondrial dysfunction.

Sl. No.	Pathophysiological condition	Type of nanoparticles	Size of nanoparticles	Surfactants used	Administration	Remarks	References
1	Brain targeting and traumatic brain injury	Trefoil factor 3 loaded nanosized liposome	126 nm	–	Male Sprague-Dawley rats	Anti-depressant-like activity, better drug transport to brain	Qin et al. (2014)
2		Horseradish peroxidase in poly(n-butyl-2-cyanoacrylate) nanoparticles, enhanced green fluorescent protein in poly(n-butyl-2-cyanoacrylate) nanoparticles	150 nm	Polysorbate 80	iv to male Sprague-Dawley rats	Nanoparticles were widely distributed near injured sites	Lin et al. (2012)
3	Cerebral ischemia	Mannosylated liposomal citicoline	60–90 nm	–	iv to male Wistar rats	Protection against global moderate cerebral ischemia reperfusion-induced mitochondrial damage	Ghosh et al. (2010)
4		SOD-loaded poly(lactic co-glycolic acid) [PLGA] nanoparticles	81 nm	PVA	iv to male Sprague-Dawley rats	Animals had higher survival rates, regained most vital neurological functions	Reddy and Labhasetwar (2009)
5		Catalase-loaded PLGA nanoparticles	280 nm	PVA	In vitro to primary human cultured neurons	Reduced H_2O_2-induced protein oxidation, DNA damage, mitochondrial membrane transition pore opening and loss of cell membrane integrity, restored neuronal morphology, neurite network, and microtubule-associated protein-2 levels	Singhal et al. (2013)

(continued)

Sl. No.	Pathophysiological condition	Type of nanoparticles	Size of nanoparticles	Surfactants used	Administration	Remarks	References
6		Triphenylphosphonium-coated nano-quercetin	42 nm	DMAB	Oral delivery to male Wistar rats	Mitochondrial structural and functional integrity was retained, mitochondrial ROS-mediated apoptotic cell death prevented	Ghosh et al. (2017)
7	Alzheimer's disease	Mementine in PLGA nanoparticles	<200 nm	PEG	Intra-cranial, transgenic APPswe/PS1dE9 mice		Sánchez-López et al. (2018)
8		CoQ10-loaded trimethylated chitosan (TMC)–PLGA	150 nm	–	iv in mice	Reduction in senile plaques, improved memory impairment	Wang et al. (2010)
9		PLGA-functionalized quercetin		PVA	iv in APP/PS1 mice	Inhibition of neurotoxicity of Zn^{2+} –$A\beta 42$ system, amelioration of $A\beta$-induced spatial learning and memory impairment	Sun et al. (2016)
10		Epigallocatechin-3-gallate-functionalized Se nanoparticles (EGCG@Se)	–	–	In vitro in PC12 and NIH/3 T3 cells	Effective inhibition of $A\beta$ fibrillation and disaggregation of preformed $A\beta$ fibrils into nontoxic aggregates in PC12 cells	Zhang et al. (2014)
11		Polyaspertamide co-polymer-based micelles for rivastigmine	34.1 nm	Polysorbate 80	In vitro Neura2A cells	Efficient drug uptake by neuroblastoma cells	Scialabba et al. (2012)

12	Parkinsons disease	Catalase-loaded self-assembled catalase/PEI-PEG complexes	60–100 nm	–	iv to MPTP-intoxicated C57Bl/6 mice	Cell-mediated delivery of antioxidant reduced oxidative stress in animal model of PD	Batrakova et al. (2007)
13	Liver cancer	Honokiol-loaded polymeric [epigallocatechin-3-gallate (EGCG) functionalized chitin] nanoparticles	80 nm	–	Intratumoral injection to HepG2 tumor-bearing mice	Better tumor selectivity and growth reduction	Tang et al. (2018)
14		Brucine immuno-nanoparticles (alpha-fetoprotein)			iv to human hepatoma SMMC-7721 cells transplanted to BALB/c nu/nu male nude mice	Significantly reduced α-fetoprotein secretion of tumor cells	Qin et al. (2018)
15		Liver-targeted nanoparticles with glycyrrhetinic acid-modified hyaluronic acid (GA-HA) for co-delivery of doxorubicin (DOX) and Bcl-2 siRNA	185 nm	–	BALB/c nu/nu male nude mice iv to H22-bearing mice	Cellular apoptosis, higher anti-tumor effect	Tian et al. (2019)
16		Ginkgolide B-loaded PLGA nanocapsules	39 nm, 132 nm	DMAB, PEG	Oral delivery to DEN-induced male Wistar rats	Mitochondrial protection against DEN-induced HCC, prevention of disease progression	Ghosh et al. (2013)
17		Curcumin-loaded PLGA nanoparticles	14 nm	DMAB	Oral delivery to rats	Prevented oxidative damage of hepatic cells and eliminated hepatocellular cancer cells in rat	Ghosh et al. (2012)

10.6 Conclusion

Generation of reactive oxygen species is a common feature in different chronic and acute diseases including neurological disorders and cancer. Amelioration of oxidative stress by targeting mitochondria thus appears to be a promising approach to counteract ROS-induced pathophysiology. Recent studies have shown the use of nanoscale delivery modalities to treat such dysfunctions. However, the usage of mitochondria-targeting nanoparticles in treating dysfunctional mitochondria is limited, yet there lies enormous potential to explore.

References

Anderson S, Bankier AT, Barrell BG, de-Bruijn MHL, Coulson AR et al (1981) Sequence and organization of the human mitochondrial genome. Nature 290(5806):427–465

Balaban RS, Nemoto S, Finkel T (2005) Mitochondria, oxidants, and aging. Cell 120:483–495

Bargiela D, Shanmugarajah P, Lo C, Blakely EL, Taylor RW, Horvath R, Wharton S, Chinnery PF, Hadjivassiliou M (2015) Mitochondrial pathology in progressive cerebellar ataxia. Cerebellum Ataxias 2:16. https://doi.org/10.1186/s40673-015-0035-x

Batrakova EV, Li S, Reynolds AD, Mosley RL, Bronich TK, Kabanov AV et al (2007) A macrophage-nanozyme delivery system for Parkinson's disease. Bioconjug Chem 18:1498–1506

Beal MF, Brouillet E, Jenkins BG, Ferrante RJ, Kowall NW, Miller JM, Storey E, Srivastava R, Rosen BR, Hyman BT (1993) Neurochemical and histologic characterization of striatal excitotoxic lesions produced by the mitochondrial toxin 3- nitropropionic acid. J Neurosci 13:4181–4192

Birben E, Sahiner UM, Sackesen C, Erzurum S, Kalayci O (2012) Oxidative stress and antioxidant defense. World Allergy Organ J 5(1):9–19

Delatycki MB, Williamson R, Forrest SM (2000) Friedreich ataxia: an overview. J Med Genet 37:1–8

Ghosh S, Das N, Mandal AK, Dungdung SR, Sarkar S (2010) Mannosylated liposomal cytidine 5′ diphosphocholine prevent age related global moderate cerebral ischemia reperfusion induced mitochondrial cytochrome c release in aged rat brain. Neuroscience 171(4):1287–1299

Ghosh D, Choudhury ST, Ghosh S, Mandal AK (2012) Nanocapsulated curcumin: oral chemopreventive formulation against diethylnitrosamine induced hepatocellular carcinoma in rat. Chem Biol Interact 195(3):206–214

Ghosh S, Dungdung SR, Choudhury ST, Chakraborty S, Das N (2013) Mitochondria protection with ginkgolide B-loaded polymeric nanocapsules prevents diethylnitrosamine-induced hepatocarcinoma in rats. Nanomedicine 9(3):441–456

Ghosh S, Sarkar S, Choudhury ST, Ghosh T, Das N (2017) Triphenylphosphonium coated nano-quercetin for oral delivery: neuroprotective effects in attenuating age related global moderate cerebral ischemia reperfusion injury in rats. Nanomedicine 13(8):2439–2450

Hayyan M, Hashim MA, AlNashef IM (2016) Superoxide ion: generation and chemical implications. Chem Rev 116:3029–3085. https://doi.org/10.1021/acs.chemrev.5b00407

Henze K, Martin W (2003) Evolutionary biology: essence of mitochondria. Nature 426(6963):127–128

Joo J-C, Do Seol M, Yoon JW, Lee YS, Kim D-K, Choi YH, SeongAhn H, Cho WH (2013) A case of myopathy, encephalopathy, lactic acidosis and stroke-like episodes (MEALS) syndrome with intracardiac thrombus. Korean Circ J 43(3):204–206. https://doi.org/10.4070/kcj.2013.43.3.204

Kroemer G, Zamzami N, Susin SA (1997) Mitochondrial control of apoptosis. Immunol Today 18(1):44–51

Lin Y, Pan Y, Shi Y, Huang X, Jia N, Jiang JY (2012) Delivery of large molecules via poly(butyl cyanoacrylate) nanoparticles into the injured rat brain. Nanotechnology 23:165101

Malkki H (2016) Mitochondrial dysfunction could precipitate motor neuron loss in spinal muscular atrophy. Nat Rev Neurol 12:556

Mannella CA (2006) Structure and dynamics of the mitochondrial inner membrane cristae. Biochimica et BiophysicaActa 1763(5–6):542–548

Martins EA, Meneghini R (1994) Cellular DNA damage by hydrogen peroxide is attenuated by hypotonicity. Biochem J 299(Pt 1):137–140. https://doi.org/10.1042/bj2990137

McMillin JB, Dowhan W (2002) Cardiolipin and apoptosis. Biochim Et Biophys Acta 1585(2–3):97–107

Mello Filho AC, Hoffmann ME, Meneghini R (1984) Cell killing and DNA damage by hydrogen peroxide are mediated by intracellular iron. Biochem J 218:273–275

Murphy MP (2009) How mitochondria produce reactive oxygen species. Biochem J 417:1–13

Niyazov DM, Kahler SG, Frye RE (2016) Primary mitochondrial disease and secondary mitochondrial dysfunction: importance of distinction for diagnosis and treatment. Mol Syndromol 7(3):122–137. https://doi.org/10.1159/000446586

Parikh S (2016) Kearns–Sayre syndrome, mitochondrial case studies. Ann Neurol:43–47. https://doi.org/10.1016/B978-0-12-800877-5.00005-X

Qin J et al (2014) Enhanced antidepressant-like effects of the macromolecule trefoil factor 3 by loading into negatively charged liposomes. Int J Nanomedicine 9:5247

Qin J, Yang L, Sheng X, Sa Z, Huang T, Li Q, Gao K, Chen Q, Ma J, Shen H (2018) Antitumor effects of brucine immuno-nanoparticles on hepatocellular carcinoma in vivo. Oncol Lett 15(5):6137–6146

Reddy MK, Labhasetwar V (2009) Nanoparticle-mediated delivery of superoxide dismutase to the brain: an effective strategy to reduce ischemia-reperfusion injury. FASEB J 23:1384–1395

Sánchez-López E, Ettcheto M, Egea MA, Espina M, Cano A, Calpena AC, Camins A, Carmona N, Silva AM, Souto EB, García ML (2018) Memantine loaded PLGA PEGylated nanoparticles for Alzheimer's disease: in vitro and in vivo characterization. J Nanobiotechnol 16(1):32. https://doi.org/10.1186/s12951-018-0356-z

Saneto RP, Cohen BH, Copeland WC, Naviaux RK (2013) Alpers-Huttenlocher syndrome: a review. Pediatr Neurol 48(3):167–178. https://doi.org/10.1016/j.pediatrneurol.2012.09.014

Scialabba C, Rocco F, Licciardi M, Pitarresi G, Ceruti M, Giammona G (2012) Amphiphilic polyaspartamide copolymer-based micelles for rivastigmine delivery to neuronal cells. Drug Deliv 19(6):307–316. https://doi.org/10.3109/10717544.2012.714813

Singhal A, Morris VB, Labhasetwar V, Ghorpade A (2013) Nanoparticle-mediated catalase delivery protects human neurons from oxidative stress. Cell Death Dis 4:e903

Starkov AA (2008) The role of mitochondria in reactive oxygen species metabolism and signaling. Ann N Y Acad Sci 1147:37–52. https://doi.org/10.1196/annals.1427.015

Sun D, Li N, Zhang W, Zhao Z, Mou Z, Huang D, Liu J, Wang W (2016) Design of PLGA-functionalized quercetin nanoparticles for potential use in Alzheimer's disease. Colloids Surf B Biointerfaces 148:116–129. https://doi.org/10.1016/j.colsurfb.2016.08.052

Tang P et al (2018) Honokiol nanoparticles based on epigallocatechin gallate functionalized chitin to enhance therapeutic effects against liver cancer. Int J Pharm 545(1–2):74–83

Thomas C, Mackey MM, Diaz AA, Cox DP (2009) Hydroxyl radical is produced via the Fenton reaction in submitochondrial particles under oxidative stress: implications for diseases associated with iron accumulation. Redox Rep 14(3):102–108. https://doi.org/10.1179/135100009X392566

Tian G, Pan R, Zhang B, Qu M, Lian B, Jiang H, Gao Z, Wu J (2019) Liver-targeted combination therapy basing on Glycyrrhizic acid-modified DSPE-PEGPEI nanoparticles for co-delivery of doxorubicin and Bcl-2 siRNA. Front Pharmacol 10:4. https://doi.org/10.3389/fphar.2019.00004

Valko M (2007) Free radicals and antioxidants in normal physiological functions and human disease. Int J Biochem Cell Biol 39:44–84

Wang ZH, Wang ZY, Sun CS, Wang CY, Jiang TY, Wang SL (2010) Trimethylated chitosan-conjugated PLGA nanoparticles for the delivery of drugs to the brain. Biomaterials 31(5):908–915. https://doi.org/10.1016/j.biomaterials.2009.09.104

Wu F, Wang J, Pu C, Quio L, Jiang C (2015) Wilson's disease: a comprehensive review of the molecular mechanisms. Int J Mol Sci 16(3):6419–6431

Zhang J, Zhou X, Yu Q, Yang L, Sun D, Zhou Y, Liu J (2014) Epigallocatechin-3-gallate (EGCG)-stabilized selenium nanoparticles coated with Tet-1 peptide to reduce amyloid-β aggregation and cytotoxicity. ACS Appl Mater Interfaces 6(11):8475–8487. https://doi.org/10.1021/am501341u

Zhao K, Zhao GM, Wu D, Soong Y, Birk AV, Schiller PW, Szeto HH (2004) Cell permeable peptide antioxidants targeted to inner mitochondria membrane inhibit mitochondrial swelling, oxidative cell death and reperfusion injury. J Biol Chem 279(33):34682–34690

Index

© Springer Nature Singapore Pte Ltd. 2020
A. K. Shukla (ed.), *Nanoparticles and their Biomedical Applications*,
https://doi.org/10.1007/978-981-15-0391-7

Printed in the United States
by Baker & Taylor Publisher Services